ANIMAL COLOUR

AND THEIR

NEUROHUMOURS

ANIMAL COLOUR CHANGES
AND THEIR
NEUROHUMOURS

A Survey of Investigations
1910 – 1943

BY

GEORGE HOWARD PARKER
Harvard University

CAMBRIDGE
AT THE UNIVERSITY PRESS
1948

CAMBRIDGE UNIVERSITY PRESS
Cambridge, New York, Melbourne, Madrid, Cape Town,
Singapore, São Paulo, Delhi, Tokyo, Mexico City

Cambridge University Press
The Edinburgh Building, Cambridge CB2 8RU, UK

Published in the United States of America by Cambridge University Press, New York

www.cambridge.org
Information on this title: www.cambridge.org/9781107613256

© Cambridge University Press 1948

First published 1948
First paperback edition 2011

A catalogue record for this publication is available from the British Library

ISBN 978-1-107-61325-6 Paperback

Contents

Preface

To any worker in a field of Science the scope of his subject as it opens before him soon compels a feeling of unlimited possibilities, at times seemingly far beyond the range of human endeavour. In consequence even the master-mind of Newton was brought to picture himself as a child wandering on the edge of a flood of unexplored reality. Von Baer, in his review of natural knowledge, declared that the stream of Science arises in springs that flow eternally, that it spreads and fertilizes the land beyond measure, and that its volume passes comprehension. Conceptions of this kind seem extravagant; yet they are profoundly true. They call forth in every real investigator a sense of reasonable humility and point to the immense distance of his final goal. No present worker will attain that goal, and yet his daily effort has in itself much that is of value. Lessing went so far as to declare that the search for truth is more precious than its possession, and Stevenson popularized this aphorism when he wrote, "to travel hopefully is better than to arrive". Yet along the road of scientific inquiry there are frequent turning points where it may not be amiss to stop and look back over the way thus far travelled. Such a review concerning chromatophores and their activities is attempted in this volume. If the outcome of this effort is a greater array of questions than of answers in the domain of colour-cells the apparent loss may be turned to gain, since this outcome may lead to renewed investigations.

The present volume is the result of suggestions from two colleagues, Professor Hans Winterstein of Istambul and Professor James Gray of the English Cambridge. The intention of these suggestions has been seriously interfered with by the war, but through the encouragement of Professor Gray I have been led to complete the manuscript begun some years ago, and to submit it as a volume to the Cambridge University Press. Such a step I am sure I could not have taken without the ever-present help of Dr Gray. To him I wish to express here my most sincere thanks. I am also under great obligations to the Harvard Biological Laboratories, whose officers have aided me most generously in providing ample facilities both as to laboratory rooms and clerical help which otherwise would have been a drain on my resources. I am especially under obligation to Mrs Natalie Garrity for her careful and conscientious preparation and typing of the manuscript. During the writing of the text I have drawn freely on the help and advice of my biological colleagues at Harvard, to all of whom I wish here to express my thanks.

G. H. PARKER

HARVARD BIOLOGICAL LABORATORIES
APRIL, 1944

POSTSCRIPT

The publication of the present volume has been carried out by the University Press of Cambridge, England. I am greatly indebted to the Press for the very searching and accurate revision of the original manuscript and for the resulting marked improvement in the composition.

I am also under obligation to Dr Lewis H. Kleinholz for invaluable aid in the preparation of the manuscript and for careful and critical reading of the proof, especially of the bibliography. The volume is not without faults, but its faults are mine and its virtues belong to others.

Chapter I

INTRODUCTION

A. PRELIMINARY SURVEY OF ANIMAL COLOUR CHANGES

ALTHOUGH colour changes in animals were known to classical antiquity and received attention from such an authority as Aristotle, they did not claim the serious consideration of naturalists till a little over a century ago. They were then taken up in such work as that of Cuvier (1817) on the cephalopods and of Stark (1830) on fishes, but it was the monographic treatment of the colour changes in the African chameleon by Brücke (1852) that put the subject on a thoroughgoing scientific basis. Brücke raised and discussed some of the most important questions in the physiology of animal chromatic responses and laid the foundations for much of the work that has been carried out in this field during the last three-quarters of a century.

The results of the work on animal colour changes have been ably summarized in two extended reviews: one by the European physiologist, van Rynberk (1906), and the other by the zoologist, Fuchs (1914). In these surveys, the contributions to this subject from ancient times to the dates of publication of the two compilations have been brought together. The present account covers in the main papers that have been published between 1910 and 1943. The initial date, 1910, overlaps the time of appearance of Fuchs' compilation by some four years. Hence the review of the subject herein contained continues without serious break the accounts given by van Rynberk and by Fuchs. The recent growth of the subject has been very considerable and may be inferred from the fact that in the bibliographical lists included in Fuchs' monograph, which are remarkably full and complete, there are in all some five hundred items. In the present survey, covering as it does only the period from 1910 to 1943, an interval of thirty-three years, there are in the bibliographical lists more than 1200 items, over twice the number given by Fuchs for the whole period to 1914.

The bibliographical titles brought together in the present survey are assembled at the end of the volume under three heads: first, surveys covering the general field of colour changes and chromatophores; next, important historical references from the time prior to 1910; and lastly, the large body of references from the period proper to this review, 1910 to 1943. An attempt has been made to give as complete a list for this period as possible. It is to be regretted that this list on scrutiny will undoubtedly show deficiencies. These are in part to be explained by the disappearance of many of the scientific bibliographical agencies as a result of the disturbed conditions of the times and in part to the limitations of the author for which he offers his apologies. It is nevertheless hoped that the lists, even with their imperfections, may be of service to scholars in this field of inquiry.

B. CHROMATOPHORES AS ACTIVATED EFFECTORS

The innumerable activities by which animals respond to the environmental changes about them are carried out by their effectors, of which there are some seven classes variously distributed throughout the animal kingdom. Glands and muscles are of universal occurrence. Cilia, the delicate means to motion, are more restricted, for, though they are found from protozoans to vertebrates, they are strangely absent from nematodes and arthropods. Nettling organs, highly efficient offensive and defensive weapons, are characteristically limited to the coelenterates. Electric organs, modified parts of the muscular system, occur only in fishes; and luminous organs, though found in many creatures of the sea and the air, are absent from fresh-water forms. Chromatophores, by which the marvellous colour changes of certain creatures are accomplished, appear mainly among the higher animals. A few sporadic instances are to be met with in what may be called the lower half of the animal kingdom, but well differentiated and extensive chromatophoral systems are found only among the cephalopods, the crustaceans, and the cold-blooded vertebrates.

The word chromatophore, though often used to indicate any animal pigment-cell, is more appropriately employed for those bodies in which colouring matter may be dispersed or concentrated in order that the possessor may change its tint. This restricted usage of the term will be the one employed in the present survey. It conforms well with the name as originally proposed, for it was first used in the Italian form *cromoforo* by Sangiovanni in 1819 for the very active chromatic bodies responsible for the colour changes in the cephalopods.

Chromatophores are of several kinds and are usually classified in accordance with the character of the colouring matter within them. The most usual colour-cells contain dark brown or black pigment and are called melanophores. Such a designation must not be taken to indicate a class whose representatives are necessarily uniform either morphologically or physiologically. Although it is often intimated that the pigment in these cells is melanin, it is by no means certain that this is always true. Furthermore, it is well known that melanophores may react very differently to the same agent. Thus to the extract from the eye-stalks of a crustacean the melanophores in the shrimp *Crangon* concentrate their pigment and those in the crab *Uca* disperse it. In lower vertebrates, such as frogs, an injection of adrenaline calls forth a concentration of the pigment in the integumentary melanophores and a dispersion of it in those of the retina. In the skin of a number of animals, some of the melanophores are characteristically large and others small, and in consequence they have been designated as macromelanophores and micromelanophores respectively (Gordon, 1927). Thus melanophores are chromatophores that in several important respects may differ from one another.

Chromatophores that contain red or yellow pigment soluble in alcohol, ether, or other like reagents, are designated collectively as lipophores. They are assumed to possess carotenoid colouring matter. When their pigment is reddish in colour, the cells are called erythrophores and when yellow, xanthophores. Ballowitz (1913 e) has pointed out a class of chromatophores in which the pigment, though red or reddish brown in tint, is not soluble in alcohol. For these he has proposed the name of allophores.

Chromatophores that contain guanine or guanine-like substances have been termed guanophores. When the guanine is in the form of fine granules and subject to change of position within the cell, the colourelement has been called a leucophore. When the guanine has the form of relatively large, plate-like crystals, the containing cells may be called iridophores. These are the colour cells to which Pouchet gave the name of iridocytes because of their iridescence, but in conformity with modern terminology they may well be designated iridophores.

All the classes of chromatophores thus far noted are represented by single cells containing a single chromatic substance characteristic for a given class. Such colour elements have been called monochromatic chromatophores. These unicellular elements may be contrasted with coloured bodies composed of groups of unicellular chromatophores so closely united as to constitute a single colour-unit. All the cells in such a unit may contain the same kind of colouring matter, the unit in consequence being of one tint, but more usually the cells contain some one colour and others another thus producing a parti-coloured effect. Such combinations have been called polychromatic chromatophores or, if more detailed descriptive terms are required, dichromatic, trichromatic, tetrachromatic chromatophores, etc., depending upon the number of colours present (Brown and Wulff, 1940). Such compound chromatophores are to be found in the crustaceans, in the amphibians, and in some fishes. Thus in the shrimp *Crangon*, in addition to monochromatic chromatophores, there are dichromatic (black and red), trichromatic (black, red, and yellow) and tetra-chromatic (black, white, red, and yellow) elements (Brown and Wulff, 1941 *a*, 1941 *c*). For these compound chromatophores, the name chromatosome has been proposed by Ballowitz (1914 *d*, 1931), a term which may well be retained, notwithstanding the fact that Sumner (1933 *b*) has suggested it for the mass of pigment within a single chromatophore.

By an appropriate combination of terms, chromatosomes may be named and their compositions indicated. Thus a melaniridosome is a combination of a melanophore and an iridophore, a not unusual occurrence in many teleost fishes. A melanoxantholeucosome would be a combination of melanophores, xanthophores and leucophores, as is to be met with in some crustaceans.

Although chromatosomes may consist of two or more different kinds of colour-cells, they can scarcely be called organs, for they are not composed of different classes of tissues. In the cephalopods, however, a chromatophore is made up of a colour-cell and a circlet of muscle-fibres and thus involves at least two sets of histological elements. These chromatophores are therefore truly diminutive organs. In this respect they are the most complex of all chromatophores and are the only ones that have reached such a structural dignity. All other chromatophores are either groups of colour-cells, chromatosomes, or single colour-cells, chromatophores in the restricted sense.

Chromatophores and chromatosomes are the essential elements in the colour changes of animals. Such changes range over not only the common spectral colours and their mixtures, but also over white, grey, and black. In fact, in some animals the change in tint is limited to white through grey to black and back again. Such ranges might well be said to be colourless, but the term colour has been used to include blacks, greys and whites as well as the spectral tints, and will be so employed in this review.

Two other terms call for a word of explanation. In the past it has been common for writers on animal chromatophores to designate the dot-like condition of the colour-cell as its contracted state and the fully branched one as its expanded state. The designations contracted and expanded as applied to chromatophores are unfortunate terms, for they imply an activity of the colour-cell that may not take place. These terms are usually employed by writers after a word of apology and are generally understood not to imply necessarily any change in the form of the cell. But opposition to these terms has been raised recently and the subject has been thrown open more or less to discussion (Sumner, 1933 b, 1934 a; Mast, 1933, 1934; Parker, 1934 c). Sumner, on the assumption that it was the mass of pigment in the chromatophore that contracted and expanded, proposed to designate this mass as the chromatosome and to describe it as contracting or expanding. This view was rejected by Mast, who suggested for the two processes the terms aggregation and distribution of pigment granules. Parker proposed that no mention at all should be made of the condition of the chromatophore, but that the pigment should be described as concentrated or dispersed, in doing which he unwittingly revived the identical terms that had been proposed for these states by Hewer in 1923.

In the older literature it was customary to describe chromatophores as open to either direct or indirect stimulation. By indirect stimulation was commonly meant activation through nerves, and by direct stimulation that which resulted from any non-nervous means such, for instance, as the application of chemicals or drugs in the blood to the chromatophores or the impingement of light or of heat from the exterior on these cells. Of recent years, direct stimulation has been limited to such obviously external agents as heat and light, which pass essentially unchanged through the immediate outer covering of the animal to the subjacent chromatophores. In this later usage indirect agents include not only nerves but any other means of chromatophoral stimulation that may arise within the body of the animal such as internal secretions.

Indirect stimuli in this sense are consequently often said to be either nervous or humoral; nervous when the response is due to a nerve terminal in immediate contact with the chromatophore and humoral when the reaction depends upon some substance carried by the blood or lymph to the colour-cell. Nervous stimulation, which in the earlier days was the only form of activation suspected, has been shown by the work of the last few decades to be really subordinate to humoral stimulation. The movement in favour of humoral stimulation began some half a century ago in the discovery by Corona and Moroni (1898) that when adrenaline was injected into the body of a frog the animal blanched. Redfield (1918) substantiated this discovery in his investigation of the colour changes in the lizard *Phrynosoma* and he was led to conclude still further that adrenaline was a normally produced activator in this lizard's own body. Through the researches of Adler (1914), P. E. Smith (1916 a) and Allen (1916), who developed the technique of hypophysectomy in anuran larvae, it was learned that tadpoles, without pituitary glands, were always pale. Krogh (1922) pointed out that the same was true of adult frogs. Meanwhile Hogben and Winton (1922 a, 1922 b, 1922 c), who had been actively engaged in experiments on the relation of nerves and pituitary secretion to colour changes in frogs, reached the important conclusion that nerves played a wholly insignificant part in the colour changes of these animals, if in fact they played any part at all. These investigators showed that such changes were in truth dependent upon humoral substances in the blood. This view gained general acceptance and was extended by Koller (1925) and particularly by Perkins (1928) to the crustaceans, where colour change was shown by Perkins to depend upon material that could be extracted from the eye-stalks of these animals. Notwithstanding this rapid accumulation of evidence in favour of a humoral interpretation of chromatophore activation, the nervous interpretation was believed still to hold for cephalopods, fishes, and lacertilians. The last two of these groups, however, were soon shown to involve with their nervous responses indubitable evidence of humoral effects. Thus a diversity of conditions seemed to prevail in that in certain animals the colour-cells were activated nervously, in others humorally, and in still others both ways. From the standpoint of innervation, chromatophores may be conveniently designated aneuronic when they are without nerves, mononeuronic when they possess a single class of nerves, and dineuronic when supplied with two kinds of nerves (Parker, 1943 b). So far as activation is concerned, aneuronic chromatophores may with equal propriety be called humoral.

But the distinction between nervous and humoral activation for chromatophores seems to be disappearing. It is becoming evident that the terminals of chromatic nerve-fibres excite their end-organs, the chromatophores, in the same way that many other effector nerve-fibres appear to excite their responding organs, namely, through minute amounts of substance which are passed from the terminal to the effector. Hence, between the nerve terminal and the chromatophores there appears to be the same relation as, for instance, between the adrenal gland and the chromatophore in that a substance produced by one activates the other. The one point of difference between these two instances is that in the case of the nerve terminal the source is very near to the colour-cell and in that of the adrenal gland it is far from this cell. This difference of nearness or remoteness is not, however, in any real sense important; in both instances a specific, secreted substance, a humour, is liberated and on reaching the colour-cell excites it to respond. Such a view is in strong contrast with the older conception of nervous stimulation by a nerve current. In this more recent interpretation of nerve stimulation, the distinction between it and humoral stimulation tends to disappear, for both types of activation rely for their effectiveness on certain liberated substances which from a near or a far source reach and excite a given colour-cell (Parker, 1932 a).

Substances that are produced in the animal nervous system and its appended glands and that serve as activating agents for other parts of the nervous system or its effectors have been variously called neurohumours (Fredericq, 1927; Parker, 1932 b), transmitters (Dale, 1935), neurohormones (Huxley, 1935), and chemical mediators (Cannon and Rosenblueth, 1937). Such substances may fairly be regarded as hormones even though their region of origin may be very near that of their effective application. Of the several terms that are used for these substances, those of chemical activator and transmitter do not distinguish them from other hormones such, for instance, as the original pancreatic hormone secretin. The other two terms relate them distinctly to nervous activities, and of these the former, neuro-humour, is perhaps to be preferred because of its greater flexibility. It was first proposed for this general purpose by Fredericq (1927) and will be the term commonly used in this survey.

A neurohumour may be defined provisionally as a hormone produced by any type of nerve-cell (receptor cell or neurone) or by a gland controlled by neurones, and effective as an activator or inhibitor for other nerve-cells or for effectors. From this standpoint, neuro-humours are substances that mediate all intercellular relations in the nervous system and its appended receptors and effectors. Whether or not neurohumours will be found to have such a wide application as is implied in this definition is still to be ascertained, but at the moment there appears to be no evidence for the denial of such a possible application. As the present survey will show, neurohumours include a considerable number of substances such as adrenaline, intermedine, acetylcholine, sympathine, other products from chromatophoral nerve-terminals, and whatever of an activating nature occurs in the extract from the eye-stalks or other secretory centres in crustaceans.

Neurohumours are for the most part soluble in water and are consequently open to transportation by blood and lymph. In this way, adrenaline, intermedine, and other like neurohumours are conveyed from place to place. Such water-soluble neurohumours have been called hydrohumours (Parker, 1934h, 1935b). Other neurohumours, such as those that emanate from the nerve terminals in the skins of a number of fishes, appear not to be carried by the blood or lymph, but to diffuse slowly in the skin itself and by means that are not aqueous. They are believed to be soluble in the fat-like materials of the cells, their lipoid constituents, and to diffuse slowly through these from one part of the skin to another. Such neurohumours, which in consequence of their peculiar solubilities are much more limited in range of action than hydrohumours, have been termed lipohumours (Parker, 1934h, 1935b) and constitute a rather unusual and remarkable class of activators. Acetylcholine appears to be a neurohumour of this kind, though it is also soluble in water and may therefore act either as a hydrohumour or as a lipohumour. Probably further study will bring to light still other kinds of neurohumours. From this standpoint, the nervous activation of chromatophores as a special type of stimulation disappears in that it proves to be a form of humoral activation in which the exciting substance comes from the nerve terminal in close proximity to the colour-cell itself.

C. THE OCCURRENCE OF CHROMATOPHORES AND THEIR ASSO-
CIATED COLOUR CHANGES AMONG DIFFERENT ANIMALS

The colour changes in the octopus and in the chameleon were known to Aristotle, who mentions them in his treatise on the *History of Animals*. References to the chromatic activities of both these creatures were made by Pliny in his *Natural History*, where an account of the colour changes of one of the teleost fishes, the red mullet, is also given. The chromatic changes of amphibians and of crustaceans do not seem to have attracted the attention of naturalists till comparatively recent times. The first amphibian whose colour changes were recorded was the common European frog whose chromatic activities were described briefly by Vallisnieri in 1715. The colour changes in crustaceans received no real attention till 1842 when Kröyer gave an account of these responses in the shrimp *Hippolyte*. Thus the five chief animal groups, representatives of which commonly possess chromatophores and show colour changes, were recognized: cephalopods, crustaceans, fishes, amphibians, and reptiles especially lacertilians.

These five groups, however, do not include all the animals that change colour by means of chromatophores. In a number of scattered instances through the animal series, individual species or small groups of species are to be found in which both chromatophores and colour changes are known to occur. Some of these instances are well authenticated, but others remain more or less in doubt. It is questionable whether the colour changes ascribed by von Lendenfeld (1883) to certain sponges such as *Aplysilla* and *Dendrilla* actually occur under natural conditions. Von Uexküll's observations (1896) on the change of tints in sea urchins are quite otherwise. The Mediterranean echinoid *Centrostephanus* was found by this investigator to blanch in darkness and to darken in the light (Fig. 1) through appropriate

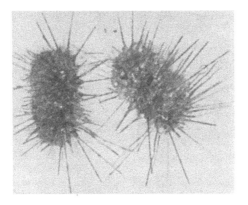

Fig. 1. Photograph of two specimens of *Centrostephanus*, the one to the left after exposure to light, the one to the right after a sojourn in darkness. Von Uexküll, 1896, 327.

changes in its chromatophores. The same was said by von Uexküll to be true of *Arbacia pustulosa*. This observation has been confirmed by Kleinholz (1938 d), though Parker (1931 a) was unable to substantiate it on the allied species *Arbacia punctulata*.

Among annelids, the leeches, most of which have brightly coloured skin-patterns, are known to exhibit colour changes. These have been recorded and in some instances studied in detail by Hachlov (1910), Borrel (1914 a), Stschegolew (1927), Iuga (1931), Wells (1932), Janzen (1932 a, 1932 b), and R. I. Smith (1942). In bright light, these animals become dark, and in darkness pale. The change of tint in the body is well illustrated by Jansen's figures from *Glossosiphonia* in which the general state of the chromatophores in darkness and in light can be seen (Fig. 2). Since leeches are negatively phototactic and live generally in shaded situations, it has been suspected that their dark coloration in bright light is a protective measure (Janzen, 1932 b). Iuga (1931), however, has looked upon the pigmentary cells in *Glossosiphonia* in a very different way and has ascribed to them an excretory function.

Except in cephalopods very little has been done in recent years on the colour changes in molluscs. Weber's study (1923) of the chromatophores in *Limax* show that this slug possesses two types of colour-cells—melanophores and cells containing a reddish brown pigment insoluble in alcohol and hence presumably allophores. *Limax* exhibits a well-marked colour change which persists, though in less degree, after the animal has been blinded. Normal and blinded individuals are said to respond not only to light and to darkness, but

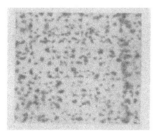 

Fig. 2. Photographs of the same spot on the surface of *Glossosiphonia*; left, in darkness; right, in bright illumination. Janzen, 1932 a, 38.

also to light of different colours, white, red, and blue. The chromatophores in *Limax* are unicellular and relatively simple. In the pteropods such as *Cymbulia* and *Tiedemannia* the colour organs are sacs with coloured contents and surrounded by a circlet of smooth muscle-fibres much like those in the cephalopods. These remarkable chromatophores justify the separation of the pteropods from the gastropods, as originally maintained by the older naturalists, and the establishment of an independent group for these molluscs between gastropods and cephalopods. No work on the pteropod chromatophores appears to have been done for many years.

Very few insects change colour in relation to their environment, and where these changes do take place they are usually slow and relatively permanent. One of the most noted instances of colour change among these animals is that of the phasmid *Dixippus (Carausius) morosus* which has been studied in recent times by a host of investigators (Schleip, 1910, 1915, 1921; Steche, 1911; von Dobkiewicz, 1912; Zacharias, 1913; Schmitt-Auracher, 1921, 1925; Przibram and Brecher, 1922; Hlobil, 1924; Toumanoff, 1926, 1928; Giersberg, 1928; Atzler, 1930; Przibram and Suster, 1931; Priebatsch, 1933; Janda, 1935, 1936; Kalmus, 1938a). *Dixippus* is believed to be without true chromatophores. Within its hypodermal cells are at least four kinds of pigment in the form of granules, green, grey, yellow-red, and sepia-brown. By an appropriate shifting of these coloured particles horizontally, vertically, or both, relatively rapid alterations in the colour of this animal can be brought about. The mechanism for these changes has never been satisfactorily worked out. Enough evidence has been gathered, however, to show that the eyes of *Dixippus* are essential for its chromatic changes, that its central nervous organs are involved as well as some gland of internal secretion, and its blood. How the pigments are shifted without involving chromatophores or elements equivalent to them has not been elucidated. The fact that the colour change is believed to occur in *Dixippus* without the intervention of colour-cells places this instance like the few others among insects beyond the limits of this survey. It is important to note, however, that very recently Hanström (1936, 1937a, 1937b, 1938b, 1940a) has found that extracts of the heads of a number of insects including *Dixippus* are chromatophoral activators for certain crustaceans.

The adaptive coloration of the pupae of certain butterflies, the pierids and vanessids, to their backgrounds has been extensively studied by Brecher (1916–36), who has shown that these adaptations are dependent upon the larval eyes and are prearranged for at this earlier stage to appear only after the larva has changed to a pupa. Apparently, these colour changes are dependent in part on neuro-humours in the insect's blood. Giersberg (1931) asserts that they are due to the deposition of appropriate pigment in or immediately under the insect's skin. When once assumed, the colours of these insect pupae are not subject to alteration. In this respect, they are like the permanent adaptive colours common to so many other animals. Thus, they form what may be regarded as a class in chromatic adaptation midway between that of permanent adaptive chromatic patterns and that in which the colour is open to ready fluctuations from one extreme to the other.

That there are some insects that possess true chromatophores may be inferred from the work of Schmidt (1919d), who found what appear to be structures of this kind in the larvae of certain primitive hexapods. But whether these colour-cells are really motile or not was left undetermined. Schmidt surmised that such colour-cells may have been inherited from ancestral crustaceans. Bodies much more certainly chromatophoral in nature have been described by Martini and Achundow (1929) in the larvae and imagines of mosquitoes and other related dipterans. An intensive survey of the colour conditions in insects generally will undoubtedly bring to light other instances of chromatic activity in a group of animals so rich in functional adaptations.

These sporadic cases of colour changes are significant in indicating that such types of functional responses have probably originated independently many times in the course of animal evolution. The chromatic activities of the echinoids, the leeches, the gastropods, the cephalopods, the crustaceans, the insects, and the vertebrates are types of response which are in no sense genetically connected. They must be regarded as having had independent phyletic origins and to represent, so to speak, separate and wholly unconnected efforts on the part of animals to accomplish a given end. The instances here enumerated are probably only a few of those that have occurred in the past history of the animal kingdom and illustrate the enormous fecundity of the evolutionary process in this respect.

D. STAGES IN THE CHANGES OF CHROMATOPHORES

In describing the conditions of active chromatophores, it is desirable to have terms that designate degrees of concentration or of dispersion of pigment. To this end, the fully concentrated state in melanophores has often been called punctate, the fully dispersed one reticulate, and the intermediate state stellate. Between the three points thus established, two additional ones have been introduced by some workers and designated puncto-stellate and reticulo-stellate. Such a method though clear in itself is obviously cumbersome, as has been pointed out by Waring (1942). It is not surprising that it has failed to gain great favour.

A closely related treatment of the total melanophore range was put forward in 1928 by Slome and Hogben. It has since been several times redescribed and somewhat elaborated (Slome and Hogben, 1929; Hogben and Gordon, 1930; Hogben and Slome, 1931; Waring, 1942), and is now much in use (Abramowitz, 1939). It consists in an arbitrary division of the whole melanophore range into four stretches by five division points which correspond very closely to the five points designated in the older nomenclature as punctate, stellate, reticulate, etc., and in giving to each of these five points a numerical designation from 1 for punctate to 5 for reticulate. The states of concentration or of dispersion of the melanophore pigment for the five points have been illustrated by sketches (Hogben and Gordon, 1930; Hogben and Slome, 1931) and these sketches have served as definitions for the points (Fig. 3). This method at once did away with the cumbersomeness of the older terminology and gave to the work in this field not only greater convenience but a certain quantitative aspect. By means of this system, melanophore indices could be established for the several states of the colour-cells which could then be plotted against time so as to allow a graphic representation of the changing melanophores. Such plottings have been very freely employed by recent students of colour changes (Hogben and Landgrebe, 1940; Waring, 1940; Neill, 1940) and have yielded interesting and important results (see Fig. 49).

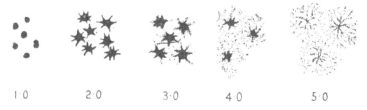

1·0   2·0   3·0   4·0   5·0

Fig. 3. Stages in the changes of chromatophores as defined by Hogben and Slome (1931) and as related to the more usual practice: 1, punctate; 2, punctate-stellate; 3, stellate; 4, stellate-reticulate; 5, reticulate. H. G. Smith, 1938, 251.

Several modifications of this system have been offered. Following the procedure introduced by Hewer (1926), Matsushita (1938) distinguished in the melanophore pigment changes from full concentration to full dispersion of the Japanese catfish *Parasilurus* six instead of five steps. These steps were defined by means of accurately drawn illustrations (Fig. 4). By the use of a simple formula, Matsushita obtained indices of the average conditions of the melanophores in a given fish at different colour phases and plotted these on a scale of one hundred against time. Thus this worker arrived at an exposition of his results much like that employed by Hogben and others, but on the basis of finer gradations. A move in the opposite direction was taken by Sawaya (1939), followed by Mendes (1942), both of whom, like Hogben and his co-workers, distinguished five dividing points in the melanophore scale, but numbered them in reverse order, I for maximum dispersion and V for maximum concentration. In the plottings made by Sawaya, no averages were employed nor were curves drawn as was done by Hogben and his associates. In consequence Sawaya's tables show the coarseness of his original observations and lack much of the detail shown in the plottings by Hogben and his school. If in refinement Matsushita has somewhat overdone Hogben's method, Sawaya has on the whole underdone it.

The replacement of descriptive terms for the states of melanophores from punctate to reticulate by numbers has not only added great flexibility to the treatment of colour changes, but, as already stated, has given the subject a quantitative aspect. This, however, may be its gravest defect, for it has tempted some of the less critical workers in this field into too great a reliance on what may be done with the quantitative statements that it has been brought to yield. The

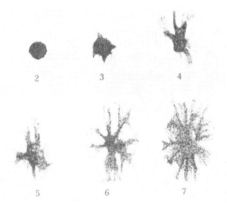

Fig. 4. Stages in the changes of macromelanophores as defined by Matsushita in the catfish *Parasilurus*: 2, contracted; 3, slightly stellate; 4, stellate; 5, slightly expanded; 6, expanded; 7, fully expanded. Matsushita, 1938, 174.

originators of this method repeatedly called the attention of those who might use it to the fact that the subdivisions whereby the steps in the melanophore changes are indicated are made on an arbitrary basis which means that 4 in the scale series is not necessarily twice 2, nor 5 five times 1. Under these circumstances it is very questionable how legitimate are the averages and other mathematical results that have been indulged in and the reliability of the curves based upon these results. It seems possible that to a certain extent the method has run away with its proponents. Undoubtedly it can be made to lead to conclusions of much value, but it must be used with restraint, probably with much more restraint than has been exercised by some of its very recent advocates. No better caution as to its use can be given than that contained in the following passage from the paper in which the method was described by Slome and Hogben (1929). The authors of this paper remark concerning plottings, etc., based upon the use of this method, that "in interpreting these results, which are presented in graphic form, it must be borne in mind that the numerical symbols applied to different configurations of the dermal melanophores are quite arbitrary, and therefore, though some insight may be obtained from a consideration of the intervals which elapse between equilibrium conditions and the intercalation of subnormal or supranormal phases, no significance can legitimately be attached to the gradients of the curves". So clear and understanding a caution as this calls for more conservative estimates of results than those that have been proposed by some of the more recent workers. The temptation seems to have been to use such quantitative results as though they were founded on solid measurements instead of on arbitrary assignments. Because of the tempting ease with which reasonable boundaries in this kind of work can be overstepped, one is led to see greater real security in Sawaya's coarser system or èven in the earlier one of cumbersome adjectives for melanophore gradations which are only in a remote way suggestively quantitative. Possibly such a descriptive system may be as a matter of fact more truthful in portraying what is really observed about colour-cells than one based on arbitrary numerical units not soundly quantitative.

From time to time systems for the recording of chromatophores much more firmly grounded than that introduced by Slome and Hogben (1928) have been suggested. One of these, advanced by Spaeth (1913b, 1916c), much antedates that by Slome and Hogben. Spaeth discovered, by following a line of work initiated by Ballowitz (1913), that the living melanophores in the freshly removed scale of *Fundulus* could be made by an appropriate treatment with barium chloride and sodium chloride alternately to disperse and to concentrate their pigment. This type of response which was rhythmic in character was at a rate essentially the same as that of the normal colour change. Such rhythmic pulsations of the colour-cells reach from complete concentration to complete dispersion and thus reproduce the whole normal melanophore range. In Spaeth's technique, a melanophore was centred in the field of a microscope whose eye-piece contained an automatically adjustable scale. The adjusting mechanism was attached by pulleys to a writing style that was applied to the smoked surface of a kymograph (Fig. 5). As the observer watched the changing chromatophore through the microscope the line at the end of the scale was made to follow the front of the pigment in

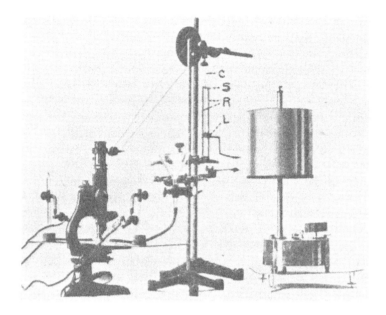

Fig. 5. Recording apparatus for following the movements of a single living melano-phore in a fish scale: *C*, cord which was wound up or unwound following the movements of pigment in the melanophore; *L*, lead block carrying a writing style; *R*, pair of vertical brass rods as guides for the lead block; *S*, perforated brass stop at the upper end of the guides. Spaeth, 1916*c*, 601.

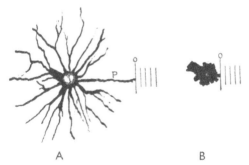

A   B

Fig. 6. Diagrammatic figure of a melanophore before stimulation (*A*), and after contraction is completed (*B*). The line on the eyepiece micrometer marked o is kept tangent to the most distal pigment granules of the process (*P*) during the contraction by turning the adjusting screw which also moves the recording apparatus. Spaeth, 1916*c*, 598.

a given cell process (Fig. 6), and the movement of the pigment was thus transferred to the moving drum of the kymograph. In this way there were recorded, for instance, the effects of increase of strength of tetanic stimulation (Fig. 7) on a colour-cell. With increase of stimulus, there appeared a progressive shortening of the latent period of response and an increase up to a maximum in the amount of contraction. Thus some thirty pulses of a single colour-cell were plotted by Spaeth over a period of about an hour. This gave almost perfect time records of the activity of the melanophore on the basis of absolute measurements. The method seems to have attracted no attention for it appears to have been neither used nor criticized.

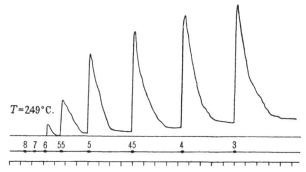

$T=249°C.$

8 7 6 55   5   45   4   3

Fig. 7. Curve of increasing responses with increase of strength of tetanic stimuli. The numbers represent the distances in centimetres of the secondary coil from the primary. Time in minutes. Spaeth, 1916*b*, 585.

A second largely objective technique for measuring melanophore activity was devised by Hill, Parkinson and Solandt (1935). These workers threw a constant beam of light on the back of a restrained *Fundulus*, the surroundings of which could be altered from black to white or the reverse, in order to induce the fish to change colour. The light reflected from the illuminated spot on the back of the fish was focused on a photoelectric cell and the progress of the change read off on a galvanometer. Thus measurements were obtained that could be plotted against time, and in this way curves for the dispersion and the concentration of melanophore pigment could be obtained (Fig. 8). This method agrees with Spaeth's in that it is based upon absolute units. It has been criticized by Wykes (1937) and by Neill (1940), who object to it on the ground that it gives the "sum effect of colour response only and...no information as to the activity of different pigmentary effectors". From the standpoint of its general applicability, this is a serious defect. It must be borne in mind, however, that the fish used by Hill and his associates, *Fundulus heteroclitus*, has on its back, whence the reflected light was taken, very few chromatophores except melanophores. The scattered xanthophores in this part of its body are insignificant in comparison with the dark colour-cells. Consequently the measurements recorded by Hill and his co-workers from this part of the body of *Fundulus* are almost entirely dependent upon melanophores, and the criticisms of Wykes and of Neill fail to apply in this instance.

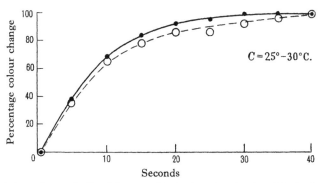

Fig. 8. Colour changes in *Fundulus* on suddenly changing the tint of the background; full circles, black background, fish turning dark; hollow circles, white background, fish turning yellow (pale). Moll micro-galvanometer, period 0·2 sec. Time in seconds. Colour changes in percentage of whole. Hill, Parkinson and Solandt, 1935, 398.

A third distinctly objective method for the study of melanophore changes is that devised by D. C. Smith (1936). This method, like that of Spaeth, depends upon the use of pulsating dark colour-cells in freshly removed scales, in this instance from the fish *Tautoga*. A beam of light is thrown through such a scale under the microscope and the change of intensity in this light as determined by concentration or dispersion of the melanophore pigment is read off by a combination of a photoelectric cell in the microscope and an outside galvanometer. By this means, readings can be taken at 10-second intervals or from ten to fifteen readings for a single chromatic pulse. These readings can be plotted against time and thus made the basis of a curve for chromatophore activity in the same way as in Hill's method. The chief difference between Smith's method and that of Hill is that, whereas in Smith's technique transmitted light is measured, in Hill's it is reflected light. Smith's method like Hill's is based on absolute measurements. It is also open to the same criticism as that urged by Wykes against Hill's procedure. But this has as little force in the case of Smith's records as it has in those of Hill, for in *Tautoga*, the fish used by Smith, the coloration of the scales is due predominantly to melanophores. At the outset of any tests, the melanophores in *Tautoga* commonly beat in phase, which, as Smith pointed out, is essential to good readings. In course of time, however, many of them drop out of step with the result that the records, for instance, of the second quarter of an hour are less regular than those of the first quarter. Notwithstanding this defect, Smith's method has yielded the clearest and most convincing plottings of melanophore responses thus far published (Fig. 9).

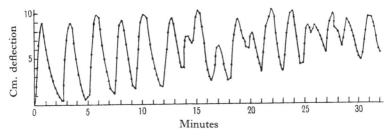

Fig. 9. Pulsations of melanophores as recorded by light transmitted through an isolated scale of *Tautoga* and thrown on a photoelectric cell. Ordinates indicate the amount of galvanometric deflection and upward movement represents pigment concentration. D. C. Smith, 1936 a, 86.

The last three methods here discussed, those of Spaeth, of Hill, and of Smith, are all based on sound physical measurements either of length or of light intensity. In this respect they are much superior to those of Slome and Hogben, whose proposed units are arbitrary and lack real substantiality. As Slome and Hogben themselves remark, "no significance can legitimately be attached to the gradients of the curves" obtained by their technique. Such is not true of the records of the last three methods here described. These, but particularly the methods of Hill and his associates and of Smith, show curves that are consistently uniform and characteristic. These curves are S-shaped, sigmoid in form. This form of curve was noted by Parker (1935 f) in a discussion of the colour changes in *Fundulus*. The colour changes in this fish are almost entirely under nerve control. Blanching begins slowly, owing to the gradual accumulation of a concentrating neuro-humour—probably adrenaline from adrenergic fibres—in the fluids around the melanophores. The later rapid increase of blanching indicates a high concentration of this paling neurohumour, and the following decline in the rate of colour change till it reaches full cessation marks without doubt the limit of responsiveness of the melanophores to the activating agent. Darkening in this fish follows a similar course but in reverse direction, and is probably due to the nervous neurohumour acetylcholine, for intermedine appears to play little or no part in this phase of *Fundulus*. The sigmoid form of the curves for melanophore activity is especially well shown in Smith's plottings, but it is to be inferred clearly and easily from those by Hill and his associates. It can be discerned even in the graphs made by Slome and Hogben's method, though the fact that the plottings based upon this technique usually begin at what is the middle of such a curve disguises the whole reaction measurably. Nevertheless, the elements of such a curve are there discernible. Thus the normal change in the dispersion and the concentration of melanophore pigment in a number of fishes appears to conform, when plotted, to a type of curve, the sigmoid, which is characteristic of the course of many living processes.

In such fishes as *Fundulus*, where the predominant chromatophores are melanophores, or *Ameiurus*, where the colour-cells are exclusively of the dark type, the colour changes conform very exactly to the states of the colour-cells. In dark fishes, the melanophore pigment is greatly dispersed, in pale ones greatly concentrated. This position has been opposed by Neill (1940), who has contended that the colour of a given fish is not closely related to its dark cells and he has tabulated conditions in the eel to substantiate his contention. As the foregoing discussion shows, a determination of this kind depends upon the chromatophoric constitution of the given fish. In the catfish, with only melanophores, the agreement is as near exact as can be measured, but in the eel, with a sexually variable skin background and several classes of diverse chromatophores, it is not to be expected that there would be full agreement between the general tint of the fish and one set of colour-cells, the melanophores. It is surprising indeed that, as the table published by Neill shows, the agreement in the eel is so close. That general colour and states of melanophores are as intimately related as they are in many fishes indicates that of the various types of colour-cells the dark ones commonly predominate, and consequently the colour changes follow in the main this type of chromatophore. In work of this kind, anyone who wished to investigate the activity of xanthophores would not choose a fish whose colour-cells were predominantly melanophores.

Another question in dealing with functioning melanophores has to do with the means by which the momentary state of the changing dark colour-cell is to be recorded. For this purpose photography has been of service. By means of successive photographs of the same living melanophores at different stages, the changes in these colour-cells have been followed in small groups (Spaeth, 1913 a), in a single cell (Perkins, 1928), or in a larger group (Parker, 1935 c; Tomita, 1938 a). This procedure calls for the repeated identification in a living animal after considerable intervals of time of a particular colour-cell or group of such cells and their rephotographing, an exacting exercise at the least. Moreover, the handling of some live fishes induces under certain circumstances changes in the states of their colour-cells that are disturbing in such an operation. Thus *Fundulus* darkens noticeably when taken from the water and handled. It is therefore not surprising that this method is not in common practice, yet it has yielded significant results in the study of the diffusion of neurohumours (Parker, 1935 c).

The great difficulty in determining the exact condition of melanophores in living fishes, as might be inferred from what has been stated, is the ease with which many creatures respond by melanophore changes to handling and the like. This capacity is very different in different species. Thus in the catfish scarcely any change in colour at all is to be seen on reasonably mild manipulation. Flatfishes, on the other hand, are very responsive to the slightest environmental disturbance such as a tap on their container or even the passage of the hand over the aquarium in which they are kept. Sticklebacks, according to Hogben and Landgrebe (1940), are moderately susceptible to such shocks and may thus be brought to shift their tints toward an intermediate phase if in the beginning they are at either extreme of colour. To avoid these disturbing drifts, Hogben and Landgrebe put single sticklebacks each in a small glass vessel supplied with a suitable current of water and with apertures by which the fish could be introduced and through which its tail could project. In taking readings, such a glass with its contained fish was removed from the general aquarium, and with the fish's tail projecting quickly put under the microscope with the tail in the field. Records were then made of the states of the melanophores; whereupon the fish was discarded, for experience showed that it was not favourable material for further work. Much the same technique was followed by Neill (1940) in his study of the colour changes of the eel and other fishes. It has long been the practice in the Harvard Laboratories to treat *Fundulus* in this way, but the colour responses of this fish on handling take place so quickly that only approximate records can be obtained and these can be used only as rough indications of what is transpiring.

To permit of deliberate inspection and measurement of melanophores under the microscope permanent preparations of the tails and fins of fishes have been made. Such preparations were prepared and photographed as early as 1934 by Parker. The method has also been employed by others, especially by Wykes (1937). Much of its success depends upon the way in which the fins have been prepared. The details of this technique have been fully given by Parker (1943 a).

Cinematographic methods for studying living chromatophores have been proposed by Veil, Comandon and Fonbrune (1933), and by Perkins and Cole (1938), but none of these has been brought to great perfection. When all the methods for the study of chromatophore changes are closely compared, it is obvious that each carries with it certain advantages. Nevertheless, a thoroughly reliable quantitative technique is yet to be devised for this field of investigation. Meanwhile, none of the several methods adopted by different workers can well be ignored, for, notwithstanding the broad condemnation issued by such workers as Neill (1940) and Waring (1942) for all methods except their own, no single method has such superiority over others that it can be said to enjoy exclusive possession of the field (Parker, 1943 a).

Chapter II

COLOUR CHANGES IN CEPHALOPODS

THE common Mediterranean devil-fish, a very typical cephalopod, has the historical distinction of having been mentioned by Aristotle as an animal that could change its colour. In his book on the *History of Animals*, the Father of Natural History records the chromatic changes of the devil-fish or octopus and of the chameleon. He notes that the octopus when in pursuit of fishes may change its colour to conform to that of its surroundings and that it does the same when it is alarmed. He thus attributes to this animal colour changes that in common parlance may be said to be voluntary and of a kind to serve in either aggression or protection. The cephalopods are further interesting historically in that their integumentary colour-organs were the first to be designated chromatophores or *cromofori*, to use the exact term first employed by Sangiovanni in 1819 for these parts. Because of the relatively large size of cephalopod chromatophores their structure and activities have been very fully worked out, and in many respects students of this subject are in more general agreement than they are about similar matters in other groups of chromatic animals.

A. CHROMATOPHORES

Chromatophores in cephalopods are to be found as a rule much more abundantly on the dorsal than on the ventral aspects of these creatures. Such colour organs occur in the derma of the skin (Fig. 10), never in its epidermis. In the main, they are limited to this situation, but they are also associated in small numbers with some of the deeper organs, as, for instance, the ink sac. In *Eledone* and *Octopus* they are small, but in *Sepiola*, *Argonauta*, and *Ommastrephes* they are so large that they may be seen even by the unaided eye. In their expanded state, their diameters may vary from twice that of their contracted condition to from fifty to sixty times that dimension.

Cephalopod chromatophores show considerable diversity in colour. In *Loligo*, according to Bozler (1928), they may be brown, red, or yellow. In *Sepia*, Kühn and Heberdey (1929) have noted black, orange, and yellow colour-cells (Fig. 10). In both these instances, the dark chromatophores, either brown or black, are largest and the yellow ones smallest. Some of the older investigators have reported violet or even blue colour organs, but such have not been mentioned by recent workers. From the prevailing tints one would expect carotenoid substances to be present in these colour-cells, but Lönnberg (1936) was unable to identify such materials in the chromatophores of *Eledone*.

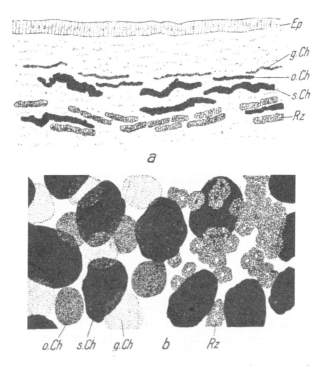

Fig 10 Skin of *Sepia officinalis*. *a*, seen in vertical section, the outer surface of the skin uppermost; *b*, surface view showing outlines of chromatophores and iridophores *Ep* epidermis, *g Ch* yellow chromatophores, *o.Ch* orange chromatophores, *Rz*. iridophores, *s Ch*. black chromatophores. Kuhn and Heberdey, 1929, 236

The structure of cephalopod chromatophores is relatively uniform. Each colour organ consists of a central spherical or spheroidal body (Fig. 11) bounded externally by a firm, highly elastic membrane within which is a fluid carrying a considerable amount of finely divided pigment and a nucleus. This central body constitutes the colour-cell proper. Attached to its periphery and spreading out into what may be called the plane of its equator is a radiating system of filaments, the radial muscle-fibres. These fibres are very long. Their distal ends, which extend far out into the adjacent region, are branched and are firmly attached to the neighbouring tissue. Their proximal ends are blunt and adhere intimately to the elastic membrane of the colour-cell proper. A nucleus for each fibre is imbedded in its proximal end. Each fibre is in reality a single smooth-muscle element. When the system is at rest the colour-cell proper has a spherical or approximately spherical form in consequence of the pressure of its elastic outer membrane on the fluid contents of the cell. Under such conditions, the chromatophore is invisible to the unaided eye and has no noticeable influence on the coloration of the animal. When the radial muscle-fibres of a given chromatophore contract, the coloured, spherical cell in the centre is drawn out into a very thin, flat disc whose diameter as already stated is a number of times that of the

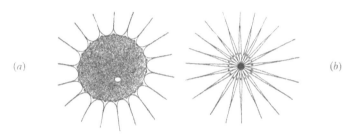

(a)   (b)

Fig. 11. Diagrams of chromatophore from *Loligo* with radial muscle-fibres fully relaxed (a) and maximally contracted (b). Bozler, 1928, 381.

original sphere (Fig. 11b). By this means, the contained pigment is spread over a larger area and thus impresses its tint upon the animal as a whole. When the muscle-fibres relax, the elasticity of the colour-cell membrane asserts itself and the cell returns to its original spherical form and small size and is lost to view. This is the usually accepted opinion of the general structure and action of a cephalopod chromatophore and has been accepted by almost all modern investigators (Bozler, 1928; Sereni, 1930c; Bacq, 1934b).

B. IRIDOPHORES

In addition to chromatophores, many cephalopods possess iridocytes, or better, iridophores. These are cells which contain glistening, often iridescent, material and which form sheets not only in the skin but on many internal organs. In the skin of *Sepia*, for instance, the iridophores occupy a position under the three types of chromatophores (Fig. 12) by which they may be entirely hidden from view. Each iridophore consists of a flat, oval, nucleated cell containing

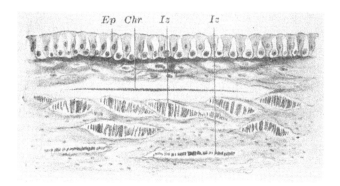

Fig. 12. Transverse section of the skin of *Sepia officinalis*: *Chr*. chromatophore; *Ep*. epidermis; *Iz*. iridophores. Schafer, 1937, 227.

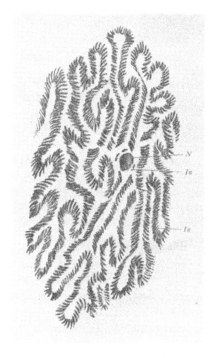

Fig. 13. Surface view of an iridophore from the skin of *Sepia officinalis*: *Is*. reflecting platelets; *N*. nucleus. Schäfer, 1937, 224.

Fig. 14. Transverse section through an iridophore from the skin of *Sepia officinalis*: *Is*. 1, reflecting platelet seen from the edge; *Is*. 2, reflecting platelet seen from the side; *N*. nucleus. Schäfer, 1937, 224.

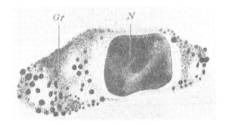

Fig. 15. Developing iridophore from the skin of a *Sepia officinalis* whose body length was 5 mm.: *Gr*. granules from which reflecting platelets will develop; *N*. nucleus. Schäfer, 1937, 233.

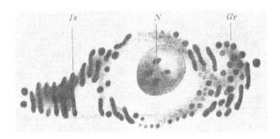

Fig. 16. Developing iridophore from the skin of a *Sepia officinalis* whose body length was 10 mm.: *Gr*. granules from which reflecting platelets will develop; *Is*. reflecting platelets; *N*. nucleus. Schäfer, 1937, 233.

twisted, interlocking chains of platelets (Fig. 13) which, as shown in a transverse section of the cell, can be seen to extend through it from one flat face to the other (Fig. 14). These platelets are formed of reflecting material, guanine-like in character, by which an iridescence is conferred on its possessor. They have been studied recently with great care and fullness by Schafer (1937), who finds that the platelets are lamellated at right angles to their optical axes and are positively doubly refractive. They swell in sodium or potassium hydroxide and thereupon lose their refractive peculiarities. When under normal conditions light falls upon their surfaces a play of colours is seen, due to interference, such as may be noticed in reflections from gratings. These iridophores with their contained platelets are quite motionless during the normal colour changes of cephalopods. They are either exposed to view or cut off from it by the action of the superjacent chromatophores. Their development has been followed by Schafer, who has shown that when an integumentary cell, in *Sepia* for instance, is about to differentiate into an iridophore, it develops first a considerable number of granules (Fig. 15) which sooner or later unite to form the reflecting platelets (Fig. 16).

C. Colour Changes

The colour changes in cephalopods are commonly rapid and often sweep wave-like over the body of the animal so that at one moment a given part may be fully pale and another fully dark. These changes therefore are not easily followed in detail. When cephalopods are

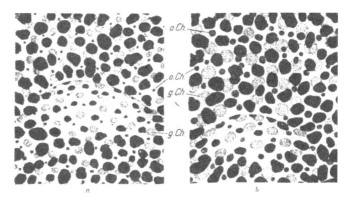

Fig. 17   Surface views of the skin of *Sepia officinalis* when the animal is adapted to a green background (*a*) and when it is adapted to a yellow background (*b*) *g.Ch.* yellow chromatophores, *o Ch* orange chromatophores   Kuhn, 1930, 18.

resting on a coloured background, they may be more constant in tint. Kuhn and Heberdey (1929) have recorded the states of the different chromatophores in *Sepia* on backgrounds of various colours. In green surroundings, the orange chromatophores are fully contracted. The yellow and black elements work together with the deep reflecting iridophores rather greenish in tone and produce a green coloration (Fig. 17 a). In a yellow or red environment, the yellow and orange chromatophores are so completely expanded as to fill the spaces between the black ones. In consequence, the greenish tone of the reflecting iridophores is quite excluded (Fig. 17 b). On white, grey, or black backgrounds, the red and yellow chromatophores are little in evidence and the darker or lighter grey of the animal is determined by the degree of expansion of the black chromatophores. In *Sepia*, tints of pure red or blue are not possible. Nevertheless, this cephalopod can adapt its tint well to the colour of its surroundings. Its chromatic responses according to Holmes (1940) may occur within a second and are believed by this investigator to be chiefly cryptic. In many cephalopods locomotion is quick and abrupt, and under such circumstances the sudden flashing and play of colours is of first significance, not only in pouncing upon prey but also in eluding an enemy. As already stated, these two functions were noted by the earlier observers including Aristotle.

D. Structure of Radial Muscle-fibres

The system of radiating muscle-fibres which spreads out in the plane of the skin from the central pigment cell is the means of stretching this cell from a spherical form to that of a broad, very thin disc. When these muscle-fibres relax, the disc, as already explained, reassumes the form of a sphere in consequence of the elasticity of its membranous cell-wall. The radial muscle-fibres then are the only really active elements in this type of chromatophore.

The radial muscle-fibres attached to cephalopod colour-cells are recorded by Fuchs (1914) as varying from four to twenty-four with a preponderance among the larger numbers. Each fibre is said by Bozler (1928) to be very long, six times the proportional length given in his diagram (Fig. 11). Its distal end is branched and firmly attached to the surrounding tissue. This portion is probably not contractile.

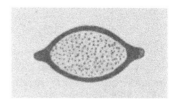

Fig. 18. Transverse section through a radial muscle-fibre: the outer contractile layer is black, the inner sarcoplasm is punctate. Bozler, 1928, 382.

The stouter, proximal portion of the fibre is band-like as can be seen in a transverse section (Fig. 18) where it will be noticed to be composed of a relatively thick outer covering within which is a granular sarco-plasm. Near the proximal end of the fibre, the outer covering may rupture and this covering can then be seen to be composed of coarse longitudinal fibrils (Fig. 19 a), while the plasma within is filled with many minute ones (Fig. 19 b). The proximal end of each radial muscle-fibre is firmly attached to the elastic wall of the colour-cell and carries in its substance a nucleus proper to the smooth muscle-cell. These

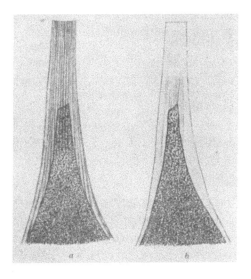

Fig. 19. Radial muscle-fibres from *Loligo*: a, showing the peripheral layer of coarse myofibrils, b, showing the deep layer of fine myofibrils (in the sarcoplasm) Bozler, 1929, 374.

nuclei were mistaken by the earlier workers for the nuclei of a layer of epithelial cells that was supposed to surround the colour-cell proper immediately external to its elastic membrane. It is now generally admitted that no such epithelium exists. When the radial muscle-fibres are relaxed, their nuclei, which are elongated, lie with their long axes in agreement with that of the fibre (Fig. 20 a). When the fibres are contracted, the nuclei come to lie with their long axes at right angles to that of the muscle-fibre (Fig. 20 b). According to Bozler (1928) each radial muscle-fibre is physiologically independent of its neighbour. The few tissue bridges between fibres are not believed to be of functional significance; i.e. the muscle-fibre system is not a syncytium. This opinion is supported by the fact that in fragments of cephalopod skin, single radial muscle-fibres can often be seen undergoing rhythmic contractions without inducing similar activities in neighbouring fibres.

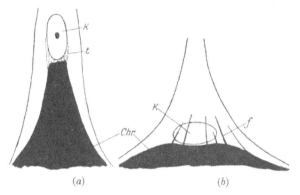

(a)   (b)

Fig. 20. Basal parts of the radial muscle-fibres from *Loligo*: *a*, muscle-fibre relaxed with the long axis of its nucleus corresponding to that of the muscle-fibre; *b*, muscle-fibre contracted with the long axis of its nucleus at right angles to that of the fibre. *Chr.* colour-cell; *f*, grooves in the outer wall of the fibre; *K*, nucleus; *t*, fine fibrils. Bozler, 1929, 375.

E. Development of Chromatophores

The development of cephalopod chromatophores is a subject that excited the interest of the older workers. Their conclusions are well summarized by Fuchs (1914), who points out that the most consistent account of this matter is given by Rabl (1900). Rabl worked on the young of *Loligo* and of *Sepia*. According to him, the colour-cells proper in *Sepia* differentiate from the other mesodermal cells of the skin by a considerable increase in size. This begins when the embryos are some 4 mm. long. When they have attained a length of from 5 to 6 mm. the firm cell-membrane of the colour-cell appears and, still later, the pigment. At 8 mm. in embryonic length, mesodermic cells surround the colour-cell and begin to differentiate into radial muscle-fibres. Thus at this early stage, all the components of the adult chromatophore have made their appearance. As the young cephalopod increases in size, more and more chromatophores are added to those already formed. This account very briefly summarized from Rabl is fully in accord with the general conception of the organization of cephalopod chromatophores as accepted by most modern workers. According to it, such a chromatophore includes at least two classes of cells, a colour-cell and a circlet of radiating muscle-cells. Strictly speaking, such a body is an organ and in this respect the cephalopod chromatophores differ from the chromatophores of most other animals whose bodies are single cells. The chief opponent to this interpretation of the cephalopod chromatophore was Chun (1902), who worked upon the structure of the deep-sea cephalopods collected on the *Valdivia* expedition. Chun (1914) followed the differentiation of the chromatophores in *Bolitaena* and maintained that each colour-cell was formed from an enlarged dermal cell whose cytoplasm differentiated into a peripheral ectoplasm and a central endoplasm. The endoplasm formed the body proper of the colour-cell and the extoplasm differentiated into a system of a dozen or more radiating processes which eventually became the radial fibres. Thus, according to Chun, the whole cephalopod chromatophore, central body and radial processes, was to be regarded as a complex single cell. It is possible, as Fuchs (1914) intimates, that in cephalopods there may be as great differences in the organization of their chromatophores as is implied in the interpretation advanced by Chun and that accepted by most other workers, but it is more probable in consideration of the great uniformity shown in the structure of most cephalopod chromatophores that Chun's view is to be attributed to faulty observation. The work of the last two decades has led modern investigators to accept the view that the chromatophores in cephalopods are simple organs rather than complicated cells.

F. ACTIVITY OF RADIAL MUSCLE-FIBRES

From what has been stated in this survey of cephalopod chromatophores, it is evident that the responses of these colour-organs are due to the activity of their radial muscle-fibres which work against the elastic membrane of the colour-cell proper. An investigation of the responses of cephalopod chromatophores is then an investigation of the physiology of their radial muscle-fibres. This subject has been studied with much detail in recent years chiefly by Fröhlich (1910*b*), Fuchs (1910*b*), Hofmann (1910*a*, 1910*b*), ten Cate (1927, 1928), Sereni (1927–30), and Bozler (1928–31).

It has long been known that, when a mantle nerve in a cephalopod is severed, the area of skin thus denervated becomes pale and ceases to follow the normal colour changes seen on the rest of the animal. In the course of some few days, such areas gradually become dark and retain this deep tint till the death of the animal when they, together with the rest of the creature, blanch completely (Fröhlich, 1910*b*;

Fuchs, 1910*b*; Hofmann, 1910*a*, 1910*b*). Such altered areas indicate of course the general distribution of the controlling nerves. They have been studied in *Octopus* by ten Cate (1927, 1928), who has shown that neighbouring nerves overlap one another to an extent as great as one-fourth or even one-third of the total area of their distribution. This indicates, of course, that there must be a very considerable intermingling of the final branches of large nerve stems. Not only do nerves overlap, but where chromatophores are numerous as on the

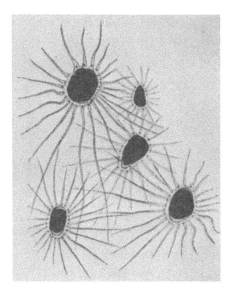

Fig. 21. Five chromatophores from the skin of *Loligo vulgaris*, showing the mutual crossing of their radial muscle-fibres. Hofmann, 1907, Pl. 21.

dorsal aspect of cephalopods the radial muscle-fibres of a given colour-cell overlap with great freedom those of neighbouring cells (Fig. 21). On the ventral aspect, however, where the chromatophores are few and scattered, such overlapping does not occur.

Notwithstanding these complicated relations of nerve and muscle, Bozler (1931) has been able, by means of an ingenious microphotographic apparatus, to record in *Loligo* the movements of the edge of a colour-cell in action and thus to gain an insight into the responses of the attached and activated muscle-fibre. By applying electric stimuli directly to such muscle-fibres, single contractions can be elicited (Fig. 22). These are accomplished in from one-seventh to one-half of a second (21° C.). With a faradic current the muscle-fibre assumes a condition of tetanus of longer or shorter duration. Both these conditions may be superimposed on the muscle-fibre while it is in a state of low tonic contraction. Hence Bozler concludes that the single smooth muscle-cell of the cephalopod chromatophore can exhibit at the same time both tetanic and tonic types of contractions, conditions which are found in separate parts of such a muscle as the adductor of the *Pecten*. In the radial muscle-fibres of the cephalopods,

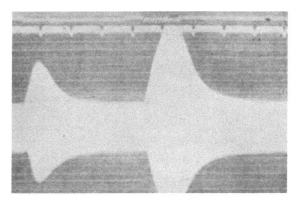

Fig. 22 Single contraction (left) and short tetanus (right) as recorded photographically from the edge of a brown chromatophore of *Loligo* The signal for stimulation is shown at the uppermost edge of the figure, next below this level, time is marked in fifths of a second Temp. 21° C. Bozler, 1931, 764.

the tetanic contractions are believed by Bozler to be carried out by the peripheral coarse myofibrils and the tonic contraction by the finer fibrils in the sarcoplasm. In contraction, the peripheral fibres were observed by Bozler to thicken.

The normal tetanic contractions of the radial muscle-fibres are attributed by Bozler to impulses from the central nervous organs. The occasion of the tonic activities of these muscle-fibres has been much discussed without resulting in great uniformity of opinion. Fuchs (1910b) pointed out that the well-known phenomenon of persistent darkening which follows the cutting of peripheral nerves in cephalopods and which he redemonstrated in a variety of these molluscs, must result from a tonic contraction of the radial chromatophoral muscle-fibres. This type of tonus has since been designated as peripheral autotonus. In Fuchs' opinion this tonus is the main factor in the production of the dark phase in cephalopods. Its inhibition would allow the relaxation of the radial muscle-fibres whereby the colour-cells would contract and the animal blanch. Such an inhibition, in Fuchs' opinion, is mediated by the nerves from the stellar ganglion and these nerves were therefore designated by him as inhibitory nerves, and the function of supplying such inhibitory impulses was attributed to this ganglion. By such means Fuchs believed the blanching in cephalopods to be accomplished. This general view of the action of the chromatophoral system in these molluscs has been accepted, though with some slight additions, by such investigators as ten Cate (1928), Bozler (1928), and Sereni (1930c). Hofmann (1910a), who worked chiefly on *Sepia*, went so far as to attribute the formation of dark spots through denervation to changes strictly peripheral. The normal chromatic banding seen in *Sepia* was, however, in his opinion central in origin. Fröhlich (1910b), whose investigations had to do with the mantle of *Eledone*, *Sepia*, and *Octopus*, and especially with the arms of *Octopus*, expressed the view that tonus played no part in the colour changes of normal cephalopods and that the darkening of denervated areas in these animals was to be attributed, not to the elimination by nerve cutting of tonic impulses, but to the excitation of the peripheral parts of the nerves as a result of the damage and destruction inflicted on them by the cut with which their central connections were severed. This view is very like that advanced by Parker some years later (1936c) to account for the caudal and pectoral bands in fishes and which will be considered subsequently. These several opinions of the way in which the radial muscle-fibres of the cephalopod are activated have never been closely scrutinized and carefully tested by experiment. Until they have undergone some such treatment by a competent investigator an evaluation of them is extremely difficult if not impossible.

G. Nervous Control of Chromatophores

The eyes in cephalopods as in the great majority of chromatic animals are of prime importance as receptors in the excitation of chromatic reflexes. As is well known, when one eye is removed from one of these molluscs, the whole animal still continues to respond with colour changes, but the blinded half less vigorously than the normal half (Sereni, 1930c). The loss of both eyes still further reduces the chromatic possibilities of the animal, but does not eliminate these changes completely. In such colour reflexes, other receptors than the eyes must be involved and these are the sucking cups of the arms.

But even when both sucking cups and eyes are removed, the cephalopod still responds with colour changes to a certain degree. If such a reduced animal be pricked, pinched, or put on its back, it will exhibit reduced chromatic changes, showing that even postural differences may be significant as means of activating colour responses.

The receptor impulses concerned with colour from whatever source they may come impinge upon the brain. Here, according to Sereni (1930c), three sets of chromatic centres may be distinguished. The first are purely chromatic motor centres situated in the suboesophageal ganglia and distinct for the two sides of the body and for different areas of skin. They cannot be substituted one for the other. These centres are overruled by a general colour centre probably located in the so-called central ganglia. Finally there is an inhibitory centre in the cerebral ganglia. Both these centres, the general colour centre and the inhibitory centre, though symmetrically placed, may act for the other side of the body as well as for their own, and the centre on one side may permanently serve in place of the one on the other.

That the chromatophores in cephalopods are provided with activating nerves has been uncritically assumed by almost all the

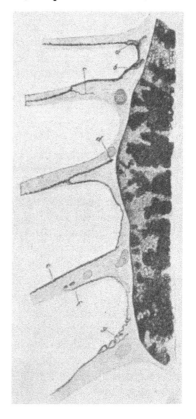

Fig. 23. Nerve-fibres and their branches on the radial muscle-fibres of a chromatophore from *Loligo vulgaris*. Stained with methylene blue: *a, b*, nerve-fibre loops, *c, d, e, f, g*, apparent free-endings of nerve-fibres. Hofmann, 1907, Pl. 21.

older workers and abundantly demonstrated by the more recent ones. The results of nerve cutting, ganglion extirpation, and the like have placed this subject beyond dispute. The histological demonstration of the connection of nerve-fibres with radial muscles has been accomplished by Hofmann (1907) in *Loligo*. In methylene blue preparations, Hofmann has shown that a nerve-fibre extends along the whole length of each radial muscle-fibre from its distal end to its proximal attachment with the colour-cell. Here the nerve-fibre may branch and come into close relation with the fibre from the next radial muscle (Fig. 23). The details of termination have not been fully worked out by Hofmann, but that the nerve-fibre of one radial muscle-fibre is in very intimate relation, if not in actual connection, with that of the next muscle-fibre is suggested. That neighbouring muscle-fibres are physiologically separate has been recently claimed by Bozler (1928), a claim which would imply that the nerve-fibre relations between adjacent muscle-fibres described by Hofmann are not as intimate as he supposed. Furthermore, according to Bozler (1928) each radial muscle-fibre must possess a double innervation, one for tetanic response and the other for tonic inhibition. It may be classed in other words as dineuronic. Evidence of double innervation has been seen by Bozler in the fact that on partial degeneration of the chromatophoral nerve apparatus a stage can be found where the tetanic response is lost, but the tonic still continues. Double innervation of such muscles is likewise conceded by Sereni (1928 b, 1930 c), who describes these innervations as excitatory and inhibitory. Sereni, in commenting on the degeneration of chromatophoral nerves, remarks that though these nerves may require a week or ten days in which to undergo dissolution, in *Loligo* during warm weather they may disintegrate in twenty-four hours.

H. Reactions to Chemicals, Drugs, and Humours

The reactions of cephalopod chromatophores to solutions of various substances when injected into the blood stream of these animals have been studied recently by Sereni (1927–30) and by Bacq (1932–35). Sereni worked upon *Octopus*, *Eledone*, *Sepia*, and *Loligo* and studied the effects on chromatophores of solutions injected into normal animals, animals in which one part of the creature was connected with the rest only by nerves, and by the immersion of pieces of living skin in the solutions themselves. Two sets of substances were distinguished (Sereni, 1927 a, 1927 c, 1928 b, 1930 b, 1930 c), those that expanded the chromatophores and thereby increased the colour of the animal (contracted the radial muscle-fibres) and those that contracted the colour organs and hence blanched the animals (relaxed the radial muscle-fibres). Injections of the following drugs induced an expansion of chromatophores: aconitine, adrenaline, atropine, cocaine, digitalis (strong solution), guanidine, histamine, lobeline, morphine, papaverine, phenol, strychnine, strophanthine, tetrahydronaphthalamine, tyramine, and veratrine. The following induced a contraction of melanophores: acetylcholine, choline, coniine, digitalis (weak solution), ergotamine, caffeine, nicotine, physostigmine, and yohimbine. The action of these drugs was found to be chiefly on the nerve centres, which, as already stated, fall into three categories: motor centres in the suboesophageal ganglion, a general colour centre apparently in the central ganglion, and a colour inhibitory centre in the cerebral ganglion. The general colour centre is excited by strychnine, morphine, and papaverine; the motor centres or endings of the fibres which ran from them to the general colour centre are excited by adrenaline, phenol, tyramine, and veratrine. Atropine paralyses the inhibitory centre or the endings of the fibres that go from this centre to the motor centres. The inhibitory centre is stimulated by acetylcholine, choline, physostigmine, and pilocarpine. Ergotamine, caffeine, and yohimbine injure the general colour centre or the endings of its fibres in the motor centres. The action of veratrine on the chromatophore system has also been studied by Sereni (1927 b, 1929 d).

Salts were found by this worker (1927 a, 1927 c, 1930 c) to act on the muscle-fibres rather than on the nerves. Muscle contraction was induced by the following cations: monovalent, $K > Rb > NH_4 > Cs > Li > Na$ and, bivalent, $Ba > Sr > Ca > Mg$. Mg was found to be almost without contracting power. Mg is the constituent in seawater that inhibits the vigorous contracting action of K; Mg thereby renders seawater neutral. The anions that induce contraction are $SO_4 > HCO_3 > I > SCN > NO_3 > Br > Cl$.

Sereni (1928 b, 1930 c) found that curare was without effect on the radial muscles of cephalopods. There also appeared to be no adrenaline or acetylcholine identifiable in these animals. Tyramine and histamine, on the other hand, were shown to be present in cephalopod blood. Tyramine was identified as coming from the posterior salivary glands of cephalopods (Sereni, 1929 b, 1929 c, 1930 b) and was believed to act as a hormone for the central nervous organs. This humoral aspect of the subject, particularly in relation to the colour changes of cephalopods, was emphasized by Sereni (1928 c, 1929 e) in an ingenious experiment in which the blood circulations in two octopods were united (Figs. 24, 25). Such united pairs of individuals lived some ten to twelve hours after the operation and always became uniform in colour. This was believed to be due to the final balance

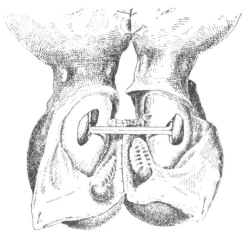

Fig. 24. Sketch of an operation on two octopods whereby their blood vessels are so united as to allow of a full mingling of their blood. Sereni, 1929e, 308.

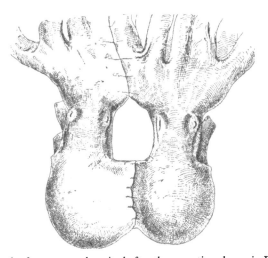

Fig. 25. Sketch of two octopods united after the operation shown in Fig. 24. Such double animals lived ten to twelve hours after having been stitched together. Sereni, 1929e, 309.

of tyramine and histamine in the mixed bloods from the two sources. Another substance that may be of humoral significance in cephalopod blood is betaine, identified in these animals by Henze (1910, 1913) and worked on by Sereni (1930a). These various substances are believed by ten Cate (1928, 1933) and Sereni (1930c) to be answerable for a certain control of central nervous functions such as central tonus. Nadler (1927) has stated that in *Loligo* ephedrine has the same relaxing effect upon the muscles that epinephrine (adrenaline) has.

Bacq's work (1932a, 1932b, 1934b, 1934c) on drugs and salts was done on *Octopus*, *Eledone*, and *Loligo*. In *Loligo*, one part of a solution of KCl 0·59 mol. in five parts of seawater induced a vigorous expansion of the chromatophores. When to five parts of this mixture two parts of a $MgCl_2$ 0·29 mol. were added and the mixture applied, the chromatophore expansion was mild. When finally a mixture of equal parts of the two solutions was used, no chromatophoral change was noticed. The striking activity of K ions as observed by Bacq is thus in agreement with the results of Sereni as were also Bacq's observations on the contraction of chromatophores when they were treated with acetylcholine and ergotamine. Bacq obtained chromatophore contraction in *Loligo* from adrenaline, the reverse of what was reported by Sereni. Otherwise, the two sets of results have much in common.

I. CHROMATOPHORE REACTIONS IN ISOLATED PIECES OF SKIN

When a considerable piece of skin rich in chromatophores is cut from a cephalopod and kept in a favourable environment, a series of striking changes follows. At first, as a rule, the skin is quite pale, but after a few hours it darkens considerably when waves of paler and darker tints may pass over it, till finally as the death of the tissue intervenes it blanches completely. It has already been pointed out that the darkening of the skin has been interpreted by most workers as a resumption of peripheral autotonus. This type of tonus is believed by many workers to be characteristic of the normal skin in place. The initial blanching of freshly excised fragments of skin has received very scant attention. Ten Cate (1928) has interpreted it as due to a reduction in autotonus to be followed eventually by reinforced automatism. The final blanching in such preparations has been generally regarded as indicative of the ultimate collapse of the radial muscle systems. The phenomenon of colour waves on such fragments of skin has aroused considerable interest.

Colour waves may present themselves in several ways. To ten Cate these colour pulsations seemed like the beating of the heart. Bacq (1932 a) has described them in some detail. If a piece of skin from *Eledone* is carefully removed and spread out in seawater, it will remain blanched for some hours. If now a weak solution of KCl is injected under it, in a few minutes it will begin to show signs of colour, after which all the chromatophores over the injected region will contract. From this region now as a centre, waves of paler and darker tints will begin to pass outward over the skin at some such rate as three to eight per minute. This activity may be kept up for from a quarter to half an hour. It is not unusual that several such centres may exist simultaneously on the same piece of skin. This wave response, according to Bacq, is excited with difficulty in skin from *Octopus*, and cannot be aroused at all with solutions of NaCl. It is not open to excitation by the stimulation of nerves. The elements of which it is composed can often be seen in the simultaneous pulsations of small groups of what seem to be semi-isolated chromatophores or in the continued pulsations of single chromatophores or even of a single muscle-fibre on one side of a chromatophore. In all activities in which whole chromatophores respond, the system of radial muscles contracts as a fully coordinated unit, and in the wave these units contract one after another in orderly sequence. According to Sereni (1930c), the forward progress of such a wave is determined by the refractory period which the contraction of the muscles induces and which, so to speak, is left behind by the wave.

The passage of such waves suggests at once a nerve-net as a means of transmission. But the most careful histological (Hofman, 1910a) and physiological investigations have failed to reveal any trace of such a structure. Sereni (1930c) has expressed the view that these waves are due to the stimulus of the pull of the radial muscles of one chromatophore on those of the next and so on across the field. It has already been stated that the neighbouring systems of radial chromatophoral muscles overlap each other to a remarkable degree and such overlapping may well serve as a means of wave transmission. This explanation, which is based on purely muscular relations, seems much more probable than any assumption that nerve responses are concerned with the presence, for instance, of axon reflexes. It must not be forgotten, however, that Bozler (1928) has advanced evidence in favour of the functional independence of radial muscle-fibres, but this independence may be the characteristic of the muscles in the normal intact skin, an independence which may be lost in excised skin. It is by no means certain that the waves so commonly seen in isolated pieces of skin represent processes which have any place in normal wave responses. Waves on skin fragments may be associated only with experimental conditions and may be in no way factors in normal activity. The rhythmicity that underlies these colour waves and which is also seen in the spontaneous contraction and relaxation of a single system of radial muscle-fibres, or even of a single fibre itself, belongs to that category of organic rhythmic movements for which there is at present no very adequate explanation. This type of response will be met again in the description of the reactions of vertebrate chromatophores.

J. Action System of Cephalopod Chromatophores

Although it must be admitted, contrary to the view advanced by Chun, that the cephalopod chromatophore is a simple organ rather than a complicated cell, its organic nature does not necessarily make its physiology as a colour effector more difficult to understand than would be the case if it were unicellular. In fact, as has already been pointed out, the active responses of a cephalopod chromatophore are those of its circlet of radial, smooth muscle-fibres. The blanching of a cephalopod is due to the contraction of the colour-cells in its chromatophores. This contraction depends upon the unopposed elastic response of the firm membranes bounding such a cell, a response which is possible when the radial muscle-fibres relax. This response is purely mechanical, so that blanching in these animals may be said to have nothing especially vital about it. It reflects in a purely physical way the state of the muscle-fibres. The assumption of colour in a cephalopod as contrasted with blanching is the result of the expansion of the colour-cells in its chromatophores. The prevailing view that a peripheral autotonus is exerted by each muscle-fibre whereby it would remain indefinitely contracted, and that this autotonus is reduced or extinguished by inhibitory impulses from the central nervous organs so as to allow muscle relaxation and colour-cell contraction, is not beyond question. It is based upon an interpretation of the expansion shown by chromatophores after their nerves have been cut, an interpretation that implies that cutting frees the muscle-fibres from central inhibition. But it must be remembered that when in cephalopods a chromatophoral nerve is cut, the area thus denervated first blanches and only after a few hours darkens. This darkening, as Fröhlich (1910) has pointed out, may be due to nerve stimulation as a result of the injury from the cut quite as much as from the exclusion of central impulses. Until the exact nature of this response has been ascertained, it is impossible to assert that the usually accepted view is really established. This whole subject has become more complicated in consequence of Bozler's declaration that the radial muscle-fibres of chromatophores not only exhibit a centrally controlled autotonic activity, but that they also show tetanic contractions. The impulses for these contractions are naturally assumed to emanate from central nervous sources and are superimposed on the momentary tonic condition of the muscle; hence the belief entertained by Bozler that chromatophoral muscle-fibres must have a double innervation.

It must be remembered, however, that many of the complications that have been considered in this discussion may be in reality unnecessary assumptions. Vertebrate skeletal muscle exhibits no small degree of tonic activity and at the same time is capable of tetanic response. The fibres of such muscle are innervated singly. If from the usual conception of the neuro-muscular mechanism of cephalopod chromatophores double innervation, for which there are no good histological grounds (Hofmann, 1910a), and central inhibition, which is not a necessary assumption (Fröhlich, 1910b), are omitted, a condition very like that in vertebrate skeletal muscle is arrived at, except that the cephalopod radial muscles are predominantly tonic and the vertebrate muscles predominantly tetanic in action. Assuming this simplification of the cephalopod system to be possible, the responses of its radial muscle-fibres might then be considered as the result of a variable nerve stimulation, chiefly tonic but in part tetanic. The objection that such a supposition violates the all-or-none law can be met by the statement that it is not known that this law applies to invertebrate muscles such as the chromatophoral radial fibres. In fact, from what is known of this law at present, it is very improbable that it should so apply. Chromatophoral radial fibres may then act under direct stimulation (without inhibition) and under single innervation.

Even if the chromatophore system in cephalopods is open to the kind of simplification suggested in the preceding paragraph, it is not to be imagined that this system, when fully disclosed, will prove to be free of complexities. It must be remembered that the chromatophoral complement of any given cephalopod is not a single set of colour-cells and their motor mechanisms as might have been inferred from the foregoing discussion. Most cephalopods carry in their skins three or even more sets of chromatophores differing in colour—red, orange, yellow and the like—and independently responsive. Under such circumstances, it is evident that, even if the action of each set is found to be in reality less complicated than had been previously supposed, the resulting simplification is offset by the fact that the skins of these molluscs contain not one but a number of such sets of organs which, though relatively simple singly, form collectively a complex aggregate. Thus the problems in the field of the cephalopod chromatophoral systems as viewed from this standpoint are far from exhausted, nor is it to be forgotten that there is just beginning to appear, thanks to the work of Sereni and of Bacq, evidence for a humoral participation in these operations that heretofore had scarcely been suspected. Much evidence is accumulating that muscle action is brought about through nerves, not by simple direct stimulation as formerly supposed, but by liberated substances, the neurohumours, and that these agents play a fundamental part in all nervous activity. No one apparently has as yet approached the excitation of the chromatophoral muscles from this standpoint, nor sought in this particular situation for such substances as acetylcholine which may be concerned with this excitation. This field is at present invitingly open.

Chapter III

COLOUR CHANGES IN CRUSTACEANS

THE colour changes in crustaceans did not attract the attention of the early naturalists. They were first recorded by Kröyer (1842) in the prawn *Hippolyte*, and the earliest account of the chromatophores in these animals was by Sars (1867) in his study of the schizopod *Mysis*. The systematic investigation of crustacean colour changes may be said to have been begun by Pouchet (1872–76), and the subject was put on a firm experimental basis by Keeble and Gamble (1900–05) in a succession of monographs issued during the first few years of the present century. The colour changes in crustaceans have recently been very fully reviewed by P. C. Koller (1929), de Lerma (1936), Hanström (1937c, 1939), von der Wense (1938), Abramowitz (1939) and Kleinholz (1942).

Such changes do not appear to occur among the simpler crustaceans, the entomostracans. The scintillating, jewel-like iridescence of male copepods, such as is shown by this sex in *Sapphirina*, has been studied by Schmidt (1926), who has shown that the truly marvellous display of colours in these microscopic animals is produced by guanine-like crystals contained in the integumentary cells. Such cells are immobile so far as their colour-invoking contents are concerned and hence are not to be classed as true chromatophores. The extraordinary and sudden changes of colour which these organisms show are due to their unusual locomotor activity whereby they are continually shifting the relations of their bodies to any source of light; hence arises their varied coloration. Heads of *Artemia salina* were extracted by Ståhl (1938a, b) with seawater and the concentrated extract injected into blind shrimps, *Leander adspersus*. After about a quarter of an hour, the chromatophores in these shrimps showed a slight contraction which led Ståhl to conclude that *Artemia* probably possesses a pigment-controlling gland, the sinus gland of Hanström. None of these observations gives positive evidence of chromatophores in entomostracans. The small size of these crustaceans and their unspecialized receptors appear to present an unfavourable soil for the growth of chromatic effectors.

Crustaceans that exhibit true colour changes are found almost exclusively among the larger forms, the malacostracans, and range from such primitive types as the mysids through the amphipods, isopods, and stomatopods, to the decapods where chromatic responses reach a very high degree of development. In the eye-stalks of the malacostracan *Nebalia*, Ståhl (1938b) has identified an X-organ, a gland that may be associated with colour changes. Chromatophores have been described in the schizopods *Siriella* and *Praunus* by Degner (1912a). From the eye-stalks of *Praunus*, Koller and Meyer (1930) have obtained a contracting agent for colour-cells. This observation has been substantiated by Perkins and Kropp (1933; Kropp and Perkins, 1933a) who extracted from the eye-stalks of *Mysis* a substance that induced contraction in the chromatophores of *Crangon*. An X-organ and a sinus gland have both been identified in mysids by Hanström (1937a). The sinus gland in these animals is the probable source of a concentrating neurohumour. *Diastylus* has been shown by Ståhl (1938b) to contain an activating substance which will bring about a contraction of the chromatophores in *Leander*. Hanström (1937a) has found both X-organs and sinus glands in a number of stomatopods. The eye-stalks of one of these, *Chloridella empusa*, yielded an extract that on injection brought about an expansion of both the black and the red pigment of *Uca*.

Among the amphipods, the common shore forms, such as *Gammarus*, *Orchestia*, and *Talitrus*, are without chromatophores (Tait, 1917), yet they may show any one of a variety of colours, dark green, slate grey, brown, and even dark red, pale yellow or white. These colour differences appear to depend upon the colour of the blood in the different individuals. Notwithstanding the absence of chromatophores from *Gammarus*, extracts from fifty fresh heads of this crustacean yielded a mixture that when injected into dark specimens of *Leander* brought about a significant contraction of their chromatophores (Ståhl, 1938b). Ståhl has also identified a sinus gland in *Gammarus*.

The pelagic amphipod *Hyperia*, associated probably as a symbiont with medusae, was shown by Lehmann (1923) to be provided with chromatophores and to be capable of colour change. When this crustacean is attached to a jellyfish, it is whitish in tint; when it is free, it changes to a saturated, dark reddish brown. In darkness, deeply tinted specimens assume the pale coloration. These changes are the results of chromatophoral activity. Lehmann's observations have been confirmed by Schlieper (1926), who has shown that background plays no real part in the colour changes of *Hyperia*. Brown individuals placed in darkness become colourless in half an hour, and free transparent individuals become brown when exposed to daylight for five minutes. Brown animals become colourless when they are allowed to attach themselves in daylight to a medusa, to a piece of jelly, or to white or black paper. Schlieper is of opinion that the condition of the chromatophores of *Hyperia* is determined under certain circumstances by the eyes and under others by the tactile receptors on the legs by which this crustacean attaches itself to foreign objects.

Isopods have a much richer array of chromatophores than amphipods and in consequence they have attracted the attention of students of colour changes to a proportionately greater degree. Two isopods

Fig. 26. *Ligia oceanica*. Two individuals photographed after illumination on a white and on a black background. H. G. Smith, 1938, Pl. 12.

in particular, *Ligia* (Tait, 1910; Kleinholz, 1937a; H. G. Smith, 1938; Sawaya, 1939; Kleitman, 1940) and *Idotea* (Piéron, 1913, 1914; Tait, 1917; Remane, 1931; Ståhl, 1938a, 1938b; Peabody, 1939) have been subjects of extensive investigation. *Ligia oceanica* is described by Tait (1910) as possessing melanophores. Smith (1938) credits this species in addition with xanthophores, a condition described in *L. exotica* by Sawaya (1939). *Ligia baudiniana*, according to Kleinholz (1937a), is provided not only with melanophores but also with a white non-cellular pigment and with guanophores. These three species show well-marked colour responses (Fig. 26). In *L. baudiniana*, Kleinholz has described a daily colour rhythm, the animals being dark by day and pale by night. This observation has been confirmed on the same species by Kleitman (1940) and on *L. exotica* by Sawaya (1939). *Idotea tricuspidata* was found by Piéron (1913, 1914) to be of different tints in agreement with its particular environment, but when it was tested experimentally it showed no appropriate true colour responses. The animal did, however, become dark in a vessel with black walls and pale in one with white walls. These changes were accomplished through a dispersion and concentration of its chromatophoral pigment. *Idotea*, in Piéron's opinion, responds to differences in light intensity rather than to differences of colour. Different individuals possess different general body colours which appear to result from the colour of the blood or of the tissues and which are more or less permanent. *Idotea* was found by Piéron to show a daily colour rhythm, being pale by night and dark by day, and this rhythm was said to persist for as long as twelve days in darkness. Peabody (1939) found no diurnal rhythm in either *Idotea baltica* or *I. metallica*, though the chromatophores of these species responded very clearly to background conditions and to darkness. Remane (1931) attributed the colour of *Idotea tricuspidata* to three sources: chromatophores, dark, white, and yellow-green; to an internal colour in the tissues and blood; and to the tint of the chitinous shell. Colour changes in this crustacean were accomplished, according to Remane, by the chromatophores or by a change in the tint of the blood perhaps through a change of food. All these changes were of the nature of slow morphological alterations rather than quick adaptive physiological changes. The production of the differently coloured stocks of *Idotea* found in nature are believed by Remane to result from selection. Ståhl (1938a, 1938b), who worked on *Oniscus*, *Porcellio*, and *Mesidotea* as well as on *Idotea*, found that extracts from the heads of the first three of these isopods when injected into *Leander* induced an expansion of its chromatophores, a reaction which failed with similar material from *Idotea*. Ståhl identified sinus glands in the head of *Oniscus*.

All these observations point to the isopods as a group whose organization for colour changes is considerable, and those who have worked on these crustaceans have raised and discussed questions in the course of their investigations which have been more fully elucidated in much that has been written on the decapods. These questions will not be discussed here, but will be taken up in the remainder of this section in which the decapods will be especially considered.

Among the crustaceans, the decapods show the most striking and complicated colour changes. These changes are seen in most members of this group including macrurans, anomurans, and brachyurans, whose colour responses will be described under the following headings.

A. COLOUR CHANGES ADAPTIVE AND RHYTHMIC

The inability of Doflein (1910), who worked on *Leander*, and of Megušar (1912), who tested *Potamobius, Gelasimus, Palaemonetes*, and *Palaemon*, to induce colour adaptation in these crustaceans when they were kept in the laboratory on appropriate backgrounds, led to the general conclusion that, though such forms could change from pale to dark and the reverse, they were incapable of adapting their bodily tints to the colour of their surroundings. This conclusion has not been supported by the observations and experiments of later workers. It has been clearly shown for instance that the colours of crayfishes vary with the regions from which the animals come (André and Lamy, 1935). Moreover, Koller (1927), who tested *Crangon*, found that this shrimp could adapt itself to backgrounds of white, black, grey, yellow, red, or orange. This shrimp, however, was unable to conform to green or to blue. Colour adaptation was said by Hosoi (1934) to occur. in the Japanese pond-shrimp *Paratya*. Brown (1935 *a, b*) showed that *Palaemonetes* could adapt its colour to backgrounds of white, yellow, red, green, blue, dark grey, and black. The states of the red, yellow, and blue pigments of this form on these several backgrounds are given in Table 1.

TABLE 1. Conditions of the pigments, red, yellow, and blue, in the chromatophores of the shrimp, *Palaemonetes vulgaris*, on the backgrounds of seven different colours (Brown, 1935 *b*, p. 319)

Backgrounds	Pigments		
	Red	Yellow	Blue
White	Concentrated	Concentrated	Absent
Yellow	Concentrated	Slightly dispersed	Absent
Red	Partly dispersed	Broadly dispersed	Absent
Green	Partly dispersed	Broadly dispersed	Present
Blue	Concentrated	Concentrated	Present
Dark grey	Partly dispersed	Concentrated	Present
Black	Broadly dispersed	Broadly dispersed	Present

This problem has also been investigated by Abramowitz (1935 *c*) on two Bermudan crabs, the white crab, *Portunus anceps*, and the red crab, *P. ordwayi*. The colour-cells of the white crab contain white, black, or yellow pigments and the general conditions of these pigments after the crab had been for some time on a background of a given colour are recorded in Table 2.

4

The red crab, *P. ordwayi*, has chromatophores with white, blackish brown, yellow, and red pigments. The conditions of these pigments as seen in crabs that have been kept on the same backgrounds as those used for the white crab *P. anceps* are given in Table 3. Although the steps by which the production of the various colour adaptations indicated in the tables 1 to 3 have not been fully worked out, the record in these tables, as well as other observations already

TABLE 2. Conditions of the pigments, white, black, and yellow, in the chromatophores of the white crab, *Portunus anceps*, on backgrounds of seven different colours. C, concentrated; D, dispersed; I, intermediate (Abramowitz, 1935 c, p. 678)

Backgrounds	Pigments		
	White	Black	Yellow
White	D	I	C
Black	C	D	D
Blue	C	D	C
Red	C	D	D
Yellow	C	D	C
Green	D	D	C
Darkness	C	D	D

TABLE 3. Conditions of the pigments, white, black, yellow, and red, in the chromatophores of the red crab, *Portunus ordwayi*, on backgrounds of seven different colours. C, concentrated; D, dispersed; I, intermediate (Abramowitz, 1935 c, p. 679)

Backgrounds	Pigments			
	White	Black	Yellow	Red
White	D	C	C	C
Black	C	D	D	D
Blue	C	D	C	D
Red	C	D	D	D
Yellow	I	C	D	C
Green	D	D	C	I
Darkness	C	D	C	D

referred to, show beyond doubt that the crustaceans concerned can accomplish a reasonable degree of colour adaptation and that the opinion to the contrary held by some of the older workers must be abandoned.

Such a conclusion, however, does not preclude the possibility of a lack of such adaptive colour changes in other crustaceans. Thus Abramowitz (1937 b) has shown that the fiddler crab *Uca*, though able to change its colour, exhibits no chromatic adaptations to its environment. *Uca* is dark by day and pale by night and this daily rhythm repeats itself regardless of background or of light intensity. When the eye-stalks of this crab with their sinus glands are removed, its daily rhythm is permanently abolished and the animal remains pale. Periodicity is therefore believed to be controlled by rhythmic releases from the eye-stalk. Several individuals were kept on an illuminated white background some 25 days, during which time they were black in the daytime and pale at night. The same was true of specimens kept for the same length of time in total darkness. Abramowitz therefore concluded that in *Uca* adaptation to background appears to be lacking. A rhythmic periodicity is the chief factor in the colour changes of this crustacean.

From what has been stated in the preceding paragraphs, it appears that from the standpoint of colour changes at least two sets of decapods may be distinguished. One of these is represented by *Palaemonetes*, which, though exhaustively studied, has yielded no evidence of a daily rhythm in tint, though it exhibits conspicuous chromatic responses to colour differences in its environment, and the other is exemplified by *Uca* in which a daily rhythm is paramount, but no immediate environmental responses are to be observed. These two types, however, do not exhaust the possibilities of crustacean colour changes, for there are other decapods, such for instance as *Hippolyte*, in which both a daily rhythm and an environmental response have been observed (Keeble and Gamble, 1900–05). The same appears to be true of *Potamobius* (Kalmus, 1938 b). Probably, after the colour changes in the higher crustaceans have been fully investigated, a whole range of examples will be found in which these two types of colour response, daily rhythms and direct environmental adaptations, will occur with mutually varying degrees of efficiency, for, though closely related, they appear to be essentially independent activities.

Intimately connected with the rhythmic chromatophoral responses of decapods are their colour conditions in darkness. It has been recorded with great uniformity that when a chromatic decapod is placed for some time in darkness, it regularly blanches. This condition has been noted by many of the older workers and has been confirmed recently on *Palaemonetes* by Perkins (1928), and by Hanström (1937 a), on *Eupagurus* by Stephenson (1932), on *Hippolyte* by Kleinholz and Welsh (1937), and on *Leander* by Kleinholz (1938 c). As already noted, it is likewise true of the amphipod *Hyperia* (Lehmann, 1923 ; Schlieper,

1926) and of the isopod *Ligia* (H. G. Smith, 1938). This blanched condition in darkness reproduces the darkness phase of the decapod chromatic rhythm and might be considered its exact equivalent were it not for the fact that it occurs in such forms as *Palaemonetes* where no rhythm is known to exist. But notwithstanding this difference, the two classes of response, daily rhythm and darkness blanching, may be in some way physiologically connected.

Another important feature with which these environmental responses are associated is the presence in darkness blanching of a hazy blue coloration which appears to float about the chromatophores rather than to be a part of them. This blue pigment was early seen by Pouchet (1872–76) and studied in *Hippolyte* by Keeble and Gamble (1900–05), who recorded its appearance when the pale night phase of the daily rhythm in this prawn came on. The beginning of this phase was marked by the concentration of chromatophore pigment whereby the animal became pale and by the diffusion throughout its transparent body of the blue colour. Prawns in this condition were designated by Keeble and Gamble as nocturnes. A similar blue pigment was identified by Perkins (1928) in *Palaemonetes*, by Beauvallet and Veil (1934*b*) in *Palaemon*, and by Abramowitz (1935*c*) in the crab *Portunus*. Brown (1934, 1935*b*) recorded this hazy blue colour in *Palaemonetes* when this shrimp was on a green, blue, dark grey, or black background and noted its absence over white, yellow, and red. According to Brown, the blue coloration required some 12 to 24 hours for its full appearance. It disappeared in from three to four hours. It was associated with the red chromatophores, out of which it appeared to diffuse into the adjacent body fluids. Abramowitz (1935*c*) observed that in *Portunus* the blue colour was not confined to a special chromatophore but occurred as a haze in the vicinity of the black pigment-cells. Thus, in addition to the ordinary chromatophores, decapods include in their colour systems a blue component which, though delicate and temporary in character as compared with the other pigments, is nevertheless of no small importance in the colour changes of these crustaceans and is particularly characteristic of their nocturnal phase.

Leucophores in several species of *Leander* have been especially studied by Knowles (1939), who has shown that these chromatophores have what has been called a primary type of response in that they are expanded in the light and contracted in darkness. They also show a secondary response, through the eyes, by expanding on a white illuminated background and contracting on a black one. As will be shown later, these types of response are also met with in a number of vertebrates, especially at their very early stages of life.

Some investigators have intimated that rhythmic colour changes in decapods may possibly be adaptive changes in which the dark tint of daytime is as appropriate for that period as the pale one is for night. The fact that in Leander the colour rhythm depends directly upon the environment in a way favours this view. But it is well known that *Uca* will continue to exhibit its colour rhythm for many days even in complete darkness, a condition which points rather to the independence of rhythmic and adaptive coloration changes. True adaptive colour responses in the higher crustaceans commonly result from the rapid adjustment of chromatophores to particular environments. These changes, however, appear at times to become more or less stereotyped and thus lead to a limitation in the chromatic activity of the individual concerned. This loss of colour plasticity, which may be the basis of the formation of coloured races, is not well understood. It may be associated with ageing. Stephenson (1932) has pointed out that in *Carcinus* old individuals often lose much of their initial capacity for colour change and thus come to be permanently stranded in one phase or another. In this connection, Crozier (1918) had discussed an interesting assembly of crabs belonging to the genus *Planes* and found by him on floating dark timber. *Planes* is usually pale-coloured and found on light-tinted gulf-weed. In the assembly studied by Crozier all the individuals were dark and remained so even after they had been for some six days on pale gulf-weed. *Planes* is well known to have several kinds of chromatophores reasonably active in their responses (Hitchcock, 1941), but the crab itself has been described as very sluggish in colour adaptation, as in fact Crozier's tests show. Apparently, this anomalous condition is due to the accumulation of scattered pigment in the maturing shells of this crab so that before a given shell is cast it becomes so impregnated with colouring matter of one kind or another as to affect the whole animal and obscure any adaptive change that may be taking place in the subjacent chromatophores. Thus *Planes* at certain phases in its life assumes and maintains a single phase of its colour system. Such an example indicates the kind of complexity that may arise in any crustacean with either adaptive or rhythmic colour capacities and shows how difficult it would be in our present state of knowledge to reach any sound conclusions concerning the relations of these two modes of response.

B. CHROMATOPHORES

The chromatophores in decapods are found immediately under the skin proper or hypodermis, or among the deeper organs where, in consequence of the transparency of these parts, they can be readily seen from the exterior. Chromatophores are either single cells, chromatophores proper, or groups of colour-cells so intimately united as to deserve the term chromatosomes as defined in the introduction to this survey. Chromatophores, using the term in the broad sense, may contain a single kind of pigment, in which case they are designated as monochromatic, or two or more kinds, polychromatic. It is current opinion that polychromatic colour-organs are always chromatosomes and that any given cell in a chromatosome will contain only one kind of pigment. At least the intermingling of pigments, though looked for, has not been observed (Abramowitz, 1937 d). In the structure and pigmentation of decapod colour-organs, great diversity may be found even among such parts in a single individual.

In *Crangon*, as Koller (1927) has pointed out, one chromatosome may contain as many as four kinds of pigment: sepia-brown, white, yellow, and red. Combinations of smaller numbers of pigments may also occur and, in addition to these polychromatic colour-organs, there are also monochromatic chromatophores in this shrimp. These, however, always contain sepia-brown and this type of pigment likewise occurs in all the polychromatic chromatosomes. According to Perkins (1928), *Palaemonetes* contains, in addition to the filmy blue pigment, chromatophoral red, yellow, and white. As Brown has pointed out (1933, 1934, 1935 b), white appears in the chromatophores proper of *Palaemonetes* and red and yellow in its chromatosomes. Presumably much the same conditions obtain in *Palaemon* (Beauvallet and Veil, 1934 b), in *Paratya* (Hosoi, 1934), in *Leander* (Hanström, 1938 a), and in *Latreutes* (Brown, 1939 c).

In *Portunus ordwayi*, according to Abramowitz (1935 c), there are true monochromatic chromatophores containing either white, blackish brown, yellow, or red pigment. In *P. anceps* the same condition occurs except that this species possesses no red pigment. Apparently the colour systems of these two crabs include no chromatosomes, all the colour elements being true chromatophores. Abramowitz makes the interesting observations that the fine chromatophoral branches of those cells that have a common colour often unite to form extensive networks. *Uca*, according to Carlson (1936), contains four kinds of pigment: a dark melanin-like colouring matter, and red, yellow, and white. Abramowitz (1937b) states that these colours are carried in part in monochromatic and in part in polychromatic colour-organs. Red occurs in monochromatic chromatophores in *Uca* as does white.

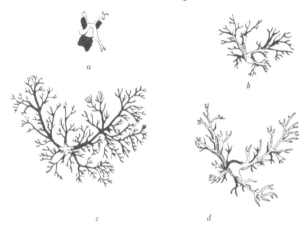

Fig. 27. A chromatosome in *Crangon* after the animal had remained on a white background (*a*), on a medium grey one (*b*), on a black one (*c*) and on a yellow one (*d*). The dark pigment is indicated in black, the red is dotted, and the yellow is in white. Koller, 1938, p. 85.

It is commonly believed that the separate pigments in all crustacean chromatosomes respond independently of each other. This is certainly true in many cases, as for instance in *Crangon* where, as figured by Koller (1928), the white pigment and the dark pigment in the same chromatosome react to different backgrounds with no evidence of uniformity (Fig. 27).

C. PIGMENT MOVEMENT WITHIN CHROMATOPHORES

The method of movement of pigments within chromatophores has been a matter of much dispute. In some chromatophores, the pigments appear to be dissolved in the chromatophoral cytoplasm. This is especially true of the red and yellow colouring matters. In other colour-cells, the pigment has the form of minute granules. This condition is met with in all the colours, but is especially often seen in the dark brown pigment and is the exclusive form in which white occurs (Bauer, 1912; Bauer and Degner, 1913). In living chromatophores, the pigment granules often exhibit Brownian movement (Perkins and Snook, 1932).

Among the older workers, two extreme views were expressed concerning the ways in which chromatophore pigments moved. According to one of these, each chromatophore was like a stationary amoeba which could throw out or retract its pseudopodia. Chromatophoral expansion was accomplished by the formation of many extensive, branching, pseudopodia-like processes which carried with them the pigment characteristic of the given colour-cell. Chromatophoral contraction was the withdrawal of these processes with their contained pigment into the body of the colour-cell. The opposing view regarded the chromatophore as a body with fixed, relatively permanent outlines including its branches, into which the pigment passed in expansion and from which it withdrew in contraction. On comparing, the exact outlines of two successive expansion phases in a given chromatophore as a simple stationary amoeba was quite untenable, for the repeated outlines were in perfect agreement. Such successive outlines have been often sketched (Bauer and Degner, 1913), or still better photographed (Perkins, 1928). A photographic illustration of these changes as shown by an active chromatophore from *Palaemonetes* is given in Fig. 28. It is perfectly evident from these results that the outline of the chromatophore even to its smaller processes is fixed, but it does not follow from this that the chromatophore may not be amoeboid, for it is conceivable that what has been regarded as the outline of the chromatophore in its expanded state may be a system of circumscribed and limited tissue spaces into which the plastic chromatophore flows and from which it retreats. Such an amoeboid chromatophore would then on every expansion repeat its outline (A. Fröhlich, 1910).

As Perkins and Snook (1932) remark, "whether crustacean chromatophore processes with contained pigment move in an amoeboid fashion into pre-formed tissue spaces or whether pigment granules suspended in cytoplasm flow into and out of fixed cell processes is the debatable question". It is this question that these two investigators have worked upon. According to them, the distal movement of the pigment appears to be accomplished by a flowing of cytoplasm and pigment into collapsed tubes which are the walls of the chromatophore branches fixed in position and lying in tissue spaces. When the proximal movement of the pigment takes place, the walls of the processes constrict within the tissue spaces and leave only faint lines behind them. These lines may eventually disappear. In the proximal movement, small amounts of pigment may be left behind on the line of the process. These may become incorporated in the same branch during a subsequent distal streaming. Thus the chromatophore

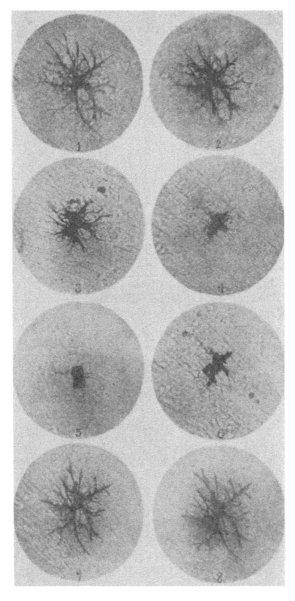

Fig 28 A single chromatophore from the shrimp *Palaemonetes*, showing the changes of form from full dispersion of pigment (1) to full concentration (5) and back to full dispersion (8) Perkins, 1928, 101.

appears to maintain a thin-walled process which expands and contracts within a tissue space and under the inflow or outflow of cytoplasm with its contained pigment. This question will be met again in the account of the colour changes in vertebrates. In these animals much more evidence has accumulated than in the crustaceans and fuller and more convincing conclusions can be reached, though whether these conclusions apply to crustaceans or not remains to be ascertained.

D. Activation of Chromatophores, Nervous and Humoral

Pouchet (1872–76), in his studies on the colour changes in animals, was led to experiment upon crustaceans as he did upon fishes and made the discovery that when the eye-stalks were removed from shrimps, these animals, like blinded fishes, failed to show adaptive chromatic responses. He therefore assumed that the chromatophores of crustaceans were under the direct control of nerves as were the colour-cells of many fishes. He maintained this belief notwithstanding the fact that all his experiments on cutting nerves in crustaceans failed to show any modifications in their colour responses as contrasted with the striking chromatic alterations to be observed in fishes after similar operations. All the early workers on this subject who followed Pouchet confirmed his observations and accepted his conclusions. Thus Keeble and Gamble (1900–05), who worked so extensively on the colour changes in *Hippolyte*, accepted the general view that crustacean chromatophores were under direct nerve control as were those of fishes.

This problem was very exhaustively investigated by Perkins (1928) who showed that, when the ventral ganglionic chain in *Palaemonetes* was severed in the region between thorax and abdomen, no observable effect could be noticed in the subsequent colour changes in the experimental animal. The anterior and posterior parts of the creature continued to change colour in complete unison. Moreover, if in different shrimps various cuts were made across the abdomen near its connection with the thorax and in such ways that the cuts collectively would have severed the abdomen from the thorax, none of these disturbed the colour changes in the shrimp except those that passed through the dorsal blood-vessel. Whenever this vessel was severed, the chromatophores, if they were not already expanded, would pass quickly into that state and remain so indefinitely. If a small side branch from the dorsal vessel were cut, the chromatophores of the region supplied by that branch would expand and remain permanently so, while those on the rest of the abdomen would continue to show the characteristic colour changes. From experiments of this kind it was suspected that chromatophoral nerves might accompany the blood-vessels and be distributed to the colour effectors along the lines of the dorsal vessel and its branches. When, however, these vessels were studied histologically, not the least sign of nerve fibres could be discovered on them, and Perkins finally concluded that nerves were not the controlling agents for the chromatophores of *Palaemonetes*—that these chromatophores were in other words aneuronic.

Meanwhile, Koller (1925, 1927) had made the interesting observation that when blood was drawn from a dark *Crangon* and injected into a pale one, the recipient became distinctly dark even when it was kept on a white background. This observation at once suggested a humoral as contrasted with a nervous control of chromatophoral activities, an idea the significance of which was fully appreciated by von Buddenbrock and was reported by him in a preliminary way in the second part of his *Grundriss der vergleichenden Physiologie* (1928). Perkins (1928) was unable to carry out on *Palaemonetes* successful experiments of this kind, but he attempted to discover in the body of this shrimp the source of a substance that might induce colour changes. It was evident from the experimental results already brought forward that the light that entered the eye of *Palaemonetes* is the external stimulus to colour change, and that what immediately induces this change must be carried in the blood. Perkins endeavoured to discover where such a substance as this was elaborated. He tested watery extracts from most of the organs in the body of *Palaemonetes* but without success. Finally he removed the eye-stalks from several pale *Palaemonetes*, crushed them in seawater, and injected the extract thus obtained into a blinded *Palaemonetes* whose chromatophores, in consequence of the blinding, were expanded. Within an hour, the dark pigment in these chromatophores had contracted and it remained so for about a day. No change was induced by the injection of pure seawater into a dark shrimp. It was therefore concluded by Perkins that when the retina of *Palaemonetes* is stimulated by light from a white background, the eye-stalks of this shrimp produce a substance that passes into the blood and excites a concentration of the dark pigment in those chromatophores to which the blood is distributed. In confirmation of this view, Perkins succeeded by a very simple operation in closing temporarily the dorsal abdominal blood-vessel in

Palaemonetes. When this vessel was thus closed the portion of the shrimp posterior to the region of closure became dark, and after the vessel had been released and the current of blood was re-established, this portion of the animal again became pale. Although it was carefully sought for, no substance that would darken a pale *Palaemonetes* was found by Perkins.

These results were abundantly confirmed on *Crangon* and on *Leander* by Koller (1928), who also showed that if an extract is made from the rostral region of *Crangon* and this extract is injected into an individual whose chromatophores are contracted, these colour-cells will expand. Thus, a so-called black organ was located whose function was to produce a hormone the opposite in effect from that in the eye-stalk. Koller designated the substance produced by the eye-stalk as "contractin" in consequence of its concentrating action on the dark chromatophoral pigment, and he called that from the rostral organ "expantin" because of its dispersing power. Thus, according to Koller, *Crangon* is to be credited with two neurohumours, one for concentrating and the other for dispersing chromatophoral pigments.

If the colour changes in crustaceans are controlled by hormones and not by nerves, as is now universally admitted, it is not surprising that all the experiments on the cutting of nerves with the view of discovering the tracts over which chromatophoral impulses might pass should have resulted negatively. Nor is it surprising that no one has found indubitable nerve terminals for crustacean chromatophores. There is apparently no good reason to suppose that chromatophoral nerve-fibres or terminals exist in crustaceans. In this respect crustacean chromatophores are in the strongest contrast with those in cephalopods. The cephalopod chromatic organ is completely under nerve control, and so far as is known is uninfluenced by hormones. Crustacean chromatophores appear to be excited by purely humoral means and without immediate nervous intervention. It is evidence of this kind, in addition to anatomical differences and the like, that points to the complete independence in origin of the two systems of chromatophores, that in the cephalopods and that in the crustaceans.

E. SOURCES OF EYE-STALK AND OTHER CRUSTACEAN
NEUROHUMOURS

That the eye-stalk extract discovered by Perkins (1928) as the activator of the colour-cells in *Palaemonetes* contains a chromatophoral neurohumour that is common to a considerable number of crustaceans was soon established by numerous workers. Koller (1928) found that on injecting eye-stalk extracts from one species of *Leander* into another, the recipient blanched. The same was true when *Crangon* was used as the donor and *Leander* was the recipient and vice versa. Thus the interspecificity of this neurohumour was established (Koller, 1930). This eye-stalk hormone was reported in *Macrobrachium* by D. C. Smith (1930a), in *Praunus* by Koller and Meyer (1930) and in *Callinectes* by Perkins and Kropp (1932b). Additional evidence of the effectiveness of the crustacean eye-stalk extract was brought forward by Koller and Meyer (1930) and by Meyer (1931) who injected this extract into the fishes *Gobius* and *Pleuronectes*, and produced pallor thereby. Eye-stalk extracts from *Callinectes* and from, *Palaemonetes* were injected by Perkins and Kropp (1932b) into the tadpoles of frogs. In this instance a well-marked darkening took place instead of the usual blanching. Perkins and Kropp (1933) made eye-stalk extracts from six different decapods, *Crangon*, *Pandalus*, *Homarus*, *Pagurus*, *Libinia*, and *Cancer*, and from one schizopod, *Mysis*, and tested these extracts on crustaceans, fishes, and amphibians. All such extracts blanched dark specimens of *Crangon*. This was true of the extracts from *Homarus*, *Libinia*, and *Pagurus*, notwithstanding the facts that these decapods show no obvious colour changes themselves. When these eye-stalk extracts were injected into fishes, some blanched (*Anguilla*, *Fundulus*, *Myoxocephalus*, and *Limanda*) and others changed in tint only slightly or not at all (*Clupea*, *Pollachius*, and *Scomber*). Extracts from the eye-stalks of *Pagurus* and of *Homarus* were injected into the tadpoles of *Rana clamitans* with the result that the melano-phores of these tadpoles underwent a maximum expansion. This result confirmed the earlier one of Perkins and Kropp (1932b) with eye-stalk extracts from *Palaemonetes* and *Callinectes* on tadpoles. Notwithstanding these relatively wide applications of the crustacean eye-stalk extracts, they failed to excite the gonads of growing rats (Kropp, 1932).

Kropp and Crozier (1934) made extracts from the eye-stalks of pale and of dark *Palaemonetes vulgaris* and injected these extracts separately into dark individuals of the same species. The extracts in both instances caused blanching, but those from the eye-stalk of pale shrimps were much more efficient than those from dark individuals. They concluded that the condition of a pale shrimp favoured the production and liberation of the eye-stalk neurohumour. Beauvallet and Veil (1934b) recorded the blanching of *Palaemon* from its own eye-stalk extract. When four eye-stalks were extracted in 2 c.c. of seawater and 0 1 c.c. of this extract was introduced into a shrimp, blanching was immediate. Extracts of the eye-stalks of *Penaeus* were stated by Hosoi (1934) to blanch *Paratya*. Kleinholz (1934) showed that the eye-stalk extract from *Palaemonetes* would induce an inward migration of the distal retinal pigment cells of this shrimp. Thus the question, often discussed, of the interrelation of crustacean retinas was shown to have a possible neurohumoral basis.

In the year following these publications, Carlson (1935, 1936) made the interesting discovery that the eye-stalk responses of the fiddler crab *Uca* were very unlike what had been described for other crustaceans. The loss of the eye-stalks from this crab was followed

Fig 29. Colour changes due to loss of eye-stalks in the crab *Uca pugilator*. The specimen to the left is normal and its legs are dark; that to the right has been devoid of eye-stalks for two hours and its legs are pale. Carlson, 1936, 6.

not by darkening but by blanching (Fig. 29), as had been previously noted by Megušar (1912). When an extract from the eye-stalks of *Uca* was injected into a pale individual, Carlson found that the recipient darkened, and yet when this extract was injected into *Palaemonetes* this shrimp blanched as it did to its own eye-stalk extract. Thus the same extract from the eye-stalks of *Uca* blanched *Palaemonetes* and darkened *Uca*. In this respect, this extract acted upon *Uca* as the *Palaemonetes* and *Callinectes* extracts had acted upon the tadpole (Perkins and Kropp, 1932 b). Subsequently, Abramowitz (1937 d) showed that the eye-stalk extract from *Palaemonetes* when injected into *Uca* caused this crab to darken, although the same extract was the means of blanching *Palaemonetes*. This led Abramowitz to declare that, irrespective of the sources of these eye-stalk extracts, they invariably darken *Uca* and blanch *Palaemonetes*. These observations imply that the chromatophores of *Uca* and of *Palaemonetes* are radically different. This difference, judging from the recent work of Carstam (1942) on a wide range of Swedish crustaceans, may as a matter of fact be a general difference between brachyurans and macrurans.

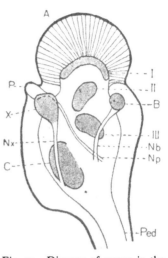

At the same time that Carlson's work was appearing, Hanström published an important series of papers (1935, 1937 *a*, 1937 *b*, 1937 *c*) directed toward the question of the localization of the secretory organ that produces the eye-stalk neurohumour. Hanström had already made important contributions to our understanding of the crustacean nervous system (1931, 1933, 1934 *a*, 1934 *b*), in the course of which he had pointed out two organs, the blood gland, or, as it is now called, the sinus gland, and the *X*-organ, either of which might well be concerned with the elaboration of the eye-stalk neurohumour (Fig. 30). These organs, of which the sinus gland has been especially studied by Sjögren (1934), usually lie in the crustacean eye-stalks and are innervated from the medulla

Fig. 30. Diagram of organs in the eye-stalk of *Parapandalus*: *A*, eye; *B*, blood gland or sinus gland; *C*, corpus hemiellipsoidale; *Nb*, nerve for the sinus gland; *Np*, nerve for the eye papilla; *Nx*, nerve for the *X*-organ; *P*, eye papilla; *Ped*, pedunculus lobi optici; *X*, *X*-organ; I, lamina ganglionaris; II, medulla externa; III, medulla interna. Hanström, 1934, 12.

terminalis of the optic apparatus. To test the possible connection of these organs with the production of the chromatic humour, Hanström tried out the eye-stalk extracts of a large number of higher crustaceans including *Palaemonetes*, *Homarus*, *Cancer*, and *Callinectes*, previously tested by Kropp and Perkins (1933 *a*) and, in addition, *Crangon*, *Leander*, *Cambarus*, *Pagurus*, *Carcinus*, *Libinia*, *Ovalipes*, and *Panopaeus*. The extracts from these various eye-stalks were injected into *Palaemonetes* and *Crangon* and in both instances called forth a concentration of the red and yellow pigments. In two cases, *Gebia* and *Hippa*, the eye-stalk extracts were inactive, but the heads of these two decapods yielded a material which was capable of inducing blanching. It was found that in all the crustaceans enumerated, excepting *Gebia* and *Hippa*, the sinus gland and the *X*-organ were contained in the eye-stalks, but in *Gebia* and *Hippa* these organs were located on the surface of the brain. This evidence points to these two glands as possible sources of the neurohumour.

Further search was made for facts which would allow a discrimination between the two glands. In *Pagurus*, the sinus gland extends through the proximal and middle thirds of the eye-stalks. The *X*-organ in *Pagurus* on the other hand is limited to the proximal third of this organ. In testing this eye-stalk, extracts from the middle third were found to concentrate the pigment as strongly as those from the proximal third, thus showing that the sinus gland was certainly effectively concerned in this process. The evidence, however, does not preclude the possible participation of the *X*-organ in this operation.

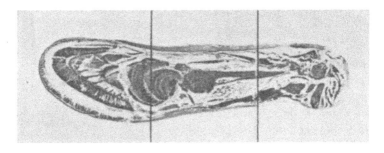

Fig. 31. Dorso-ventral section of the eye-stalk of *Uca* showing the contents of the stalk and by vertical lines its division into thirds. Carlson, 1936, 7.

Carlson (1936) attempted similar tests on *Uca*. The distal third of the stalk of this crab (Fig. 31) yielded an ineffective extract, an observation that substantiated Koller's previous statement (1930) that extracts of the retina of *Crangon* had no effect on the chromatophores. In a similar way, the proximal third of the eye-stalk of *Uca* was shown by Carlson to be inactive. The middle third, however, yielded an extract that was fully efficient in darkening *Uca* and in blanching *Palaemonetes*. Moreover, specimens of *Uca* from which the distal two-thirds of the eye-stalks had been removed blanched as did those from which all of each of these parts had been excised. Evidently in *Uca* the effective organ is located in the middle third of its eye-stalk. This part of the stalk contains, besides ganglionic masses, a well developed sinus gland, a condition which favours the view that the sinus gland is the source of the chromatic neurohumour. Carlson, however, calls attention to a small group of cells in the middle third of the *Uca* eye-stalk which he suspects to be a rudimentary *X*-organ and thus again the evidence is not wholly conclusive. Such as there is, however, favours the view that the chromatic neurohumour is a product of the sinus gland.

The development of the sinus gland and the *X*-organ has been investigated in *Pinnotheres* and in *Homarus* by Pyle (1943). In both these crustaceans, the sinus gland appears earlier than the *X*-organ and shows evidence of cyclic secretion, conditions which, however, leave untouched the question of the relations of these two organs to the activation of colour-cells.

In 1936, Kleinholz, continuing his work on the crustacean retinal pigments, showed that eye-stalk extracts from *Cancer, Libinia, Uca* and *Carcinides*, when injected into *Palaemonetes*, induced migrations of the distal and the reflecting eye pigments, and Kropp (1936) pointed out that eye-stalk hormones had a profound influence on the vertebrate heart. Hanström (1936, 1937*b*, 1938*b*) extended his search for chromatic activators to the insects and found that extracts from the heads of a number of these animals induced responses in the chromatophores of shrimps. In 1937, he published an exhaustive review of the problem of invertebrate hormones including in particular those concerned with colour changes. Kleinholz (1937*a*), in work carried out on the isopod *Ligia*, the eyes of which are of course sessile, found that an extract of this crustacean's head blanched a dark individual and had no effect upon a pale one, acting in this respect as did the eye-stalk extracts from most decapods. Presumably, the sinus glands are imbedded in the head of *Ligia* as they are in *Gebia* and in *Hippa*. Kleinholz and Welsh (1937) noted that eye-stalk extracts from *Leander* and from *Hippolyte*, when injected into experimental *Hippolyte*, caused a concentration of chromatophoral pigment accompanied by a diffusion of blue pigment. No evidence of such a chromatic activator was found by Cooper (1938) in the horseshoe crab *Limulus*, though Brown and Cunningham (1941) have more recently identified a principle from the central nervous organs of this animal, particularly from its circumoesophageal ganglion, that will concentrate the chromatophoral pigment of *Uca*.

Hanström (1937*a*), in his general survey of the sinus gland, pointed out its almost universal occurrence among higher crustaceans as contrasted with the *X*-organ which is absent, for instance, from *Cambarus, Astacus, Sesarma, Aratus*, and probably *Uca*. The fact that the sinus gland is present in crustaceans without colour change (*Homarus, Hippa, Libinia, Cancer*), in colourless forms (*Anchistioides*) and in blind forms (*Lepidopta*), shows that this organ very probably has important functions other than that of chromatic control. According to Hanström, these may be retinal adaptation (Kleinholz, 1934, 1935), calcium regulation (Koller, 1930), and growth control (Navez and Kropp, 1934). Although the sinus gland shows relatively great histological uniformity, its physiology, as Hanström's survey indicates, must be far from simple.

The standardization of the eye-stalk hormone was undertaken by Abramowitz (1937 b). When a dose containing an extract equivalent to that from one-twentieth of an eye-stalk is injected into a blinded *Uca*, colour states appear in the following sequence. Melanophore expansion begins in from 15 to 20 minutes after the injection has been made. Maximum expansion is reached within an hour and the melanophores remain fully expanded for about three and a quarter hours. About four and a half hours after injection, the melanophores begin to contract and in half an hour the crab is again pale. All these periods are independent of the concentration of the extract except that of maximum expansion which appears to be roughly proportional to it. On the basis of such tests the following *Uca* unit was established by Abramowitz. An *Uca* unit is the amount of neurohumour contained in 1 c.c. of solution of which 0·05 c.c. when injected into *Uca* pugilator blinded two days previously produces a chromatophore response over a period of five hours. The response period is the time measured from the moment of injection to that of final blanching. With this as a basis, numerous important observations on the states of the eye-stalk neurohumour under different circumstances were made by Abramowitz.

It was found, for instance, that the amount of the neurohumour in the eye-stalks of different crustaceans varied from about a quarter of an *Uca* unit per stalk to as much as five such units (Table 4).

The amount of neurohumour in the eye-stalks of *Palaemonetes* was the same regardless of whether the neurohumour was being secreted into the circulation (white background) or whether it was absent (black background). In darkness each eye-stalk contains approximately half the amount of neurohumour that it does in the illuminated animal. Light, irrespective of background, seems then to increase the production of the humour and a white background brings about a release of this substance into the circulation. The daily colour-rhythm of *Uca* is believed by Abramowitz to reflect the activity of the eye-stalk: during daylight an increased rate of production of neurohumour and its discharge; during night a slowing of the rate of production and a cessation of discharge.

Observations similar to those briefly stated in the last paragraph were continued by Abramowitz and Abramowitz (1938). The effect of varying the concentration of the eye-stalk neurohumour in *Uca* was tested, and it was found that to a dose of o·oo2 of an eye-stalk about 15 per cent of the crabs responded, but to a dose of o·o4 of an eye-stalk all responded. Thus, the melanophore response in *Uca* appeared to

TABLE 4. Neurohumour contents in *Uca* units of a single eye-stalk from eleven species of decapods (Abramowitz, 1937 *b*, p. 352)

Average weight of animals in grams	Species	*Uca* units
2·46	*Uca pugilator*	1·0
3·3	*U. pugnax*	1·0
11·0	*Pagurus pollicaris*	1·25
309·0	*Homarus americanus*	1·20
2·0	*Hippa talpoidea* (head)	0·25
0·38	*Crangon borealis*	0·25
219·0	*Carcinus maenas*	1·25
80·0	*Cancer irroratus*	5·0
0·4	*Palaemonetes vulgaris*	0·36
15·0	*Uca minax*	1·5
127·0	*Libinia dubia*	4·0

be graded as, for instance, is that of smooth muscle. With a method of testing sensitive to o·ooo,o16 of a gamma (= o·oo1 of a milligram) of chromatophoral neurohumour, it was found that a single eye-stalk contained o·2 of a gamma of this substance. The injection of distilled water was found to induce in blinded *Uca*, in from 15 to 20 minutes, complete melanophore expansion, a reaction which was suspected on further investigation to depend upon possible osmotic relations connected with the resulting hypotonicity of the blood.

Enough has been stated in this survey of the activation of crustacean chromatophores to show that in this operation the eye-stalk neuro-humour, probably a product of the sinus gland, is of first importance. This agent commonly induces a concentration of the dark pigments in crustacean colour-cells and a consequent blanching of these animals, but in *Uca* its effect is the reverse and under its influence these crabs darken. Other sources for activators of crustacean chromatophores than the eye-stalks have been sought, but with very little success.

In 1928 Koller, after having substantiated on *Crangon* Perkins' discovery that the eye-stalk extract from *Palaemonetes* caused blanching, declared that an extract from the rostral region of *Crangon* would induce a darkening of that animal. This investigator (1928, 1929, 1930) thus favoured two chromatic neurohumours from two separate sources as a means of carrying out the chromatic changes in *Crangon*. One of these, as already mentioned, he called "contractin", a product of the "white organ" (sinus gland) in the eye-stalk; and the other, "expantin", from the "black organ" of the rostral region. His belief in the presence of a rostral or "black organ" in *Crangon* rests on four sets of observations (Koller, 1930): melanophore expansion follows the feeding of parts of dark shrimps to a pale one; melanophore expansion results from the injection of rostral extract of a dark shrimp into a pale one; the power to darken is lost after the rostral region of a shrimp has been cautiously cauterized; and melanophore expansion takes place in a given shrimp after a slight pressure has been applied to its rostral region.

Perkins and Snook (1931) sought for evidence of a rostral organ in *Palaemonetes*, but without success. Neither were Kropp and Perkins (1933 a) nor Brown (1935 b) able to discover traces of such an organ. Beauvallet and Veil (1934 b) noted a slight darkening in some instances when *Palaemon* was injected with large doses of rostral extract. Kleinholz (1938 c), who worked upon the same species of *Crangon* that Koller had used, thoroughly retested the question of the presence of a rostral organ. When pale specimens of *Crangon* were injected with blood from a dark individual, only 27 per cent of the recipients darkened. When blood from pale individuals was injected into other pale ones, 22 per cent darkened. Injections of extracts from the rostral region darkened slightly over 50 per cent of the recipients. Superficial cauterization of the rostral region had no effect on the colour changes of shrimps so treated. Deep cauterization of the rostral area caused nine individuals in sixty-nine to become pale. In all these nine, swimming and equilibrium were abnormal, and it was believed by Kleinholz that this treatment injured the brain with consequent disturbances in the sinus gland. The conclusion to be drawn from this work may be stated in Kleinholtz's words: "more critical evidence will be needed before the existence of a melanophore-dispersing hormone, originating in the rostral region, can be unqualifiedly accepted."

Still other parts of decapods than the eye-stalks and the rostral region have been suspected of producing chromatic neurohumours. Brown (1933, 1935 a) noted that extracts of the ventral nerve-cord of *Palaemonetes* would blanch blinded dark individuals, and Brown and

Wulff (1941 b, 1941 c) have stated that extracts of the commissural ganglion in *Crangon* yield a chromatic activator. Hosoi (1934) also declared that extracts of the ventral nerve-cord and male genitalia of *Penaeus* would cause blinded specimens of *Paratya* to blanch. Extracts of the stomach and muscles of *Penaeus* were also slightly active, but extracts of the heart were inactive. In none of these positive instances was the response strong. Hosoi estimated that an eye-stalk contained about one hundred times as much neurohumour as the ventral nerve-cord did. Kleinholz (1936 a) suggested that these traces of chromatic neurohumour found in organs other than the eye-stalk were held there in blood by which the humour was being transported when this fluid coagulated. Of the six suggested sources for chromatic neurohumours in the crustaceans, the sinus gland, the X-organ, the rostral organ, the ventral nerve chain, the commissural ganglia, and the gonads (Brown, 1941), only the first can be said to be an unquestioned organ of origin.

This section cannot be closed better than by a quotation from Abramowitz (1937 b), who has said that "it can be concluded from this discussion that the eye-stalks are the chief source of the production of the chromatophore hormone. It can also be stated that none of the experiments already mentioned shows conclusively that tissues other than the eye-stalks may produce the hormone. The facts that under illumination blinded *Palaemonetes* remains steadily dark, and that blinded *Uca* remains continuously pale, indicate that if other tissues capable of forming the chromatophore hormone are present, they play an insignificant part in the ordinary chromatic physiology of these animals."

F. Properties of Eye-stalk Neurohumour

The eye-stalk neurohumour is not destroyed by boiling in water (Koller, 1930; Perkins and Snook, 1931; Hanström, 1935; Kleinholz, 1937 a) nor by being boiled in weak hydrochloric acid or weak sodium hydrate (Carlson, 1936; Abramowitz, 1937 b). It is destroyed when it is boiled two hours in sodium hydrate (Abramowitz, 1937 b). It is active in extreme dilution, one part in a hundred thousand (Koller, 1930). It is readily soluble in water and fairly soluble in ethyl alcohol and in methyl alcohol, very slightly soluble in petroleum ether, benzene, and pyridine, and insoluble in acetone, ethyl ether, chloroform and ethyl acetate (Abramowitz and Abramowitz, 1938). It is not destroyed by drying (Perkins and Snook, 1931; Hosoi, 1934; Carlson, 1936; Abramowitz, 1937 b). It adsorbs easily to various substances present in an aqueous solution in a refrigerator for some time without appreciable loss of activity, but it is destroyed slowly when kept dissolved in water at room temperature (Abramowitz, 1937 b). The substance or substances are obviously fairly stable.

G. NUMBER OF CRUSTACEAN NEUROHUMOURS

Koller (1928, 1929, 1930, 1938), in his study of the colour changes in *Crangon*, was led to conclude that there must be at least three neuro-humours involved in the chromatophoral responses of this shrimp: melano-contractin from the eye-stalks, melano-expantin from the rostral organ, and a yellow activator from an unknown source. Brown (1933, 1934, 1935 a, 1935 b), from his work on *Palaemonetes*, took a very different view of the required number and kinds of neurohumours in this shrimp. He found no evidence in *Palaemonetes* of humours concerned with the dispersion of the three pigments, red, yellow, and white. This process he regarded as the result of an innate activity on the part of the chromatophore itself. The diffusion of the blue colour out of the red he also regarded as a spontaneous operation. Concentration of the red, the yellow, and the white pigments and suppression of the blue he believed to be independent operations and to call in each instance for a separate neurohumour. Such a general view then necessitated at least four such agents, all concerned with concentration or, in the case of blue, with suppression. Thus, Koller and Brown took radically different views as to the nature of the humoral control of the chromatophores in the two shrimps that they studied though they agreed in assuming that the number of neuro-humours was above one. This general opinion was also entertained by Parker (1932 a) and by Hanström (1935, 1937 a). H. G. Smith (1938), on the basis of the differences in the curves illustrating the darkening and the blanching of the isopod *Ligia*, was led to conclude that the melanophores of this crustacean were acted on by two neurohumours, one for pigment dispersion and the other for con-centration. In this respect, Smith revived in a way Koller's original view of an expanding and a contracting hormone.

Influenced by the fact that the crustacean eye-stalk extract when injected into *Palaemonetes* caused a concentration of pigment and when injected into *Uca* a dispersion of pigment, Abramowitz (1937 d) was led to consider the possibility of a single neurohumour for the control of two or more kinds of colour-cells. Should this prove true, the assumed multiplicity of agents might be avoided. The problem had already been broached by both Carlson (1935) and by Abramowitz

(1936*f*). It had been discussed still earlier for other hormones by Abel (1930), who had designated the opposing conceptions as the unitary and the multiple theories of hormone action. These designations were adopted by Abramowitz, who urged the importance of a determination in favour of one or other view. Of the earlier workers, almost all of whom had favoured in a mild way the multiple theory, Brown (1935 *a*) had advanced what perhaps was the strongest evidence, namely, the mutually independent reactions of the four pigments in *Palaemonetes*. Of the activators concerned, two each are believed to come from the sinus gland and from the commissural ganglion (Brown and Wulff, 1941 *a*, 1941 *b*, 1941 *c*). As Abramowitz remarked, "if this is true the conclusion is almost inescapable that there must be more than one hormone". Brown's evidence was carefully analysed by Abramowitz, who showed that the observations therein recorded were not in his opinion so unequivocally in favour of independence as had been inferred. Nor had any of the earlier workers taken into account the possibility of different concentrations of the neurohumour acting differently on the colour-cells. This matter of threshold, which had been suggested some years before by Przibram (1932 *a*), may have a very important bearing on the extension of neurohumoral activity (Abramowitz, 1937 *b*). Moreover, acidity may play an important part in chromatophore reactions (McLean, 1928). Thus, a simple and direct test of this question is apparently not so obvious. Kleinholz (1938 *c*) approached the problem by determining the threshold values for retinal and integumentary pigments in *Leander* to the eye-stalk hormone. The lower threshold limit for the distal retinal pigment was equivalent to 0·016 mg. (wet weight) of eye-stalk; that for the integumentary pigment was 0·0008 mg. These values do not support the unitary theory which requires that the minimal threshold for the retinal pigment be lower than that for the body chromatophores. Carstam (1942), who has worked recently on a wide range of crustaceans, states that in *Leander* there is good reason for believing in separate hormones for its yellow and brown pigments. The question of the exact number of crustacean chromatophoral neurohumours must, however, be left for the present in much uncertainty, for information on the subject is still too scanty to allow a safe conclusion to be drawn. Attempts at the chemical purification of the eye-stalk hormone (Abramowitz, 1940 *b*) are steps in the right direction. Abramowitz's distinction of unitary and multiple theories is perhaps unfortunate, for the important question is, not whether there are one or many neurohumours, but what the actual number of these agents may be.

H. Drugs

The attempt to excite chromatophoral activity by the injection of drugs into crustaceans has scarcely been made at all. This is rather remarkable considering the large number of trials of this kind that have been carried out on vertebrates. Beauvallet and Veil (1934 *b*) tested adrenaline on shrimps. This agent, which is well known to blanch fishes, was found to darken *Palaemon*. Abramowitz and Abramowitz (1938) could distinguish no chromatic effect from adrenaline when injected into *Uca*. They also tested by injection into the same crab, pale or dark, the following sixteen drugs: atropine, morphine, acetylcholine, histamine, eserine, pilocarpine, cocaine, brucine, veratrine, curare, strychnine, guanidine, chlorbutanol, caffeine, nicotine, and hyoscine. All these drugs, with the exception of hyoscine, failed to produce definite, positive results with blinded specimens (pale) and all were without effect on the melanophores of normal dark individuals. These drugs, with the exception of hyoscine, have therefore no direct or indirect action in causing the melanophores of *Uca* to contract or expand. Hyoscine induced a dispersion of pigment in *Uca*, but whether this drug resembles the crustacean eye-stalk neurohumour in any other respect remains to be seen.

I. Blinded State in Crustaceans and Direct Stimulation of their Chromatophores

The terms direct and indirect as applied to the stimulation of chromatophores have been used in the past with considerable looseness. This question has already been alluded to in the introductory portion of this survey. In the older papers, indirect stimulation of chromatophores meant activation through nerves and direct stimulation by means of agents which reached the colour-cells independently of nerves. This view was held at a time when the nervous activation of colour-cells was believed to be almost, if not quite, universal. When, however, it was discovered that in whole groups of animals, such as the crustaceans for instance, there was no evidence whatever for the indirect activation of chromatophores by nerves and that consequently in these animals the only method of chromatic excitation would be by what had been called direct, this usage of the term was changed in that the word direct was restricted to those exciting agents which, like heat and light, come from the exterior and as such penetrate the substance of the animal to impinge essentially unaltered on the colour-cells. From this standpoint, neurohumours and other internal products, though they come with the blood into immediate contact with the chromatophores, are no longer classed as direct agents. Using direct in this more limited sense, it may be asked, to take a specific case, does light exert a direct influence on crustacean chromatophoral activity?

This problem can best be approached by subjecting the colour-cells in crustaceans to varying degrees of illumination after the eyes of these animals and all other means of exciting their chromatophores have been excluded. From an operative standpoint, it is a fortunate circumstance that the eye-stalks of crustaceans contain not only the retina but also what appears to be the only significant endocrinal chromatic gland in these animals. Hence, a removal of the eye-stalks not only blinds the animal, but eliminates at the same time the chief internal source of chromatophoral hormone. In past experimentation in which blinding has been resorted to, it has not always been explicitly stated whether in carrying out this operation only the retina or the whole stalk was removed. For the present purpose, it is of course desirable to remove the whole eye-stalk.

The results of such an operation as performed on a number of crustaceans by earlier workers are well given in Table 5 from the work of Carlson (1936).

Carlson remarks that with few exceptions all species under the Natantia and the Reptantia expand their dark chromatophores on loss of their eye-stalks. The exceptions are *Hippolyte* and *Crangon*; *Hippolyte* with contracted dark chromatophores and *Crangon* with contracted (Degner) or intermediate (Koller) dark chromatophores. All the Brachyura tested showed a chromatophoral contraction and blanched on losing the eye-stalks. Under the Isopods, *Ligia* was uniformly dark and *Idotea* either dark (Piéron) or intermediate (Bauer) on the elimination of the eyes.

In addition to the species of crustaceans whose colour responses on loss of their eye-stalks are recorded in Carlson's table, the following have been noted: under the Natantia, *Palaemonetes vulgaris* (Perkins and Snook, 1932; Brown, 1935 a) and *Paratya compressa* (Hosoi, 1934), both dark; under the Brachyura, *Portunus anceps* (Abramowitz, 1935 c) and *Uca minax, U. pugilator, U. pugnax* (Abramowitz, 1937 d), all pale; and under the Isopods, *Ligia baudiniana* (Kleinholz, 1937 a) and *Ligia oceanica* (H. G. Smith, 1938), both dark. All these additional records support Carlson's general conclusion that the majority of the Natantia darken on loss of their eye-stalks and that all Brachyura thus far tested blanch under the same conditions. These peculiarities, as pointed out previously, probably depend upon special responses of the dark colour-cells in the several sets of crustaceans.

TABLE 5. States of the dark chromatophores, contracted, expanded, or intermediate, and the colours of blinded crustaceans with a list of the species tested and the authorities (Carlson, 1936, p. 12)

Date	Authority	Crustacean	Colour of blinded animal	State of dark chromatophore
		DECAPODA		
		Natantia		
1900	Keeble and Gamble	*Hippolyte varians*	Pale	Contr.
1908	Minkiewicz	*H. varians*	—	Contr.
1872	Pouchet	*Palaemon serratus*	Dark	Expand.
1912	Megušar	*Palaemonetes varians*	Red brown	Expand.
1912	Degner	*Crangon*	—	Contra.
1928	Koller	*C. vulgaris*	Interm.	Interm.
1878	Jourdain	*Nike edulis*	Dark	Expand.
1930	Smith	*Macrobrachium*	—	Expand.
		Reptantia		
1912	Megušar	*Potamobius astacus*	Dark	Expand.
		Brachyura		
1908	Megušar	*Uca pugnax*	Brown-yellow	Contr.
1936	Carlson	*U. pugilator*	Pale	Contr.
1936	Carlson	*Ovalipes ocellatus*	Orange	Contr.
1936	Carlson	*Ocypoda albicans*	White	Contr.
		ISOPODA		
1905	Bauer	*Idotea*	Interm.	Interm.
1913	Piéron	*I. tricuspidata*	Dark	Expand.
1910	Tait	*Ligia oceanica*	Dark	Expand.

To ascertain whether crustacean chromatophores are open to direct stimulation by light, blinded animals must be tested. Thus it can be discovered if there are any colour differences in these creatures when they are exposed to darkness or bright light. This test has been attempted by Stephenson (1932, 1934) on *Leander*, by Kleinholz and Welsh (1937) on *Hippolyte*, and by H. G. Smith (1938) on *Ligia*. Gamble and Keeble (1900) declared that *Hippolyte* showed a daily colour-rhythm which continued in phase even after these shrimps had been blinded. This observation was not confirmed by Kleinholz and Welsh, who found that after the destruction of the retinas or the removal of the eye-stalks from *Hippolyte* it lost its daily rhythm, though it would still blanch in darkness and darken in daylight. As they believed the eyes to be the only receptors concerned with colour changes in this crustacean, they concluded that, as its chromatophores were contracted in darkness and expanded in light, these colour organs were open to direct stimulation. Stephenson, whose work was done on *Leander*, had come to a similar conclusion. Smith's tests were carried out on *Ligia oceanica*. In this isopod, the eyes are sessile and very conveniently arranged for occlusion with varnish followed by black enamel. The responses of normal and blinded individuals under bright light, dim light, and in darkness are well shown in Table 6, where the chromatophoral records are based on a conventional rating in which 1 stands for full contraction and 5 for full expansion of the colour-cells.

TABLE 6. Chromatic behaviour of *Ligia oceanica* under three conditions of illumination and either normal or blinded. The rate is based on a scheme in which 1 stands for full contraction of the chromatophores and 5 for their full expansion (H. G. Smith, 1938, p. 252)

Animal	Background	Bright light	Dim light	Total darkness
Normal	Black	$5\cdot0\pm0\cdot00$	$4\cdot6\pm0\cdot08$	$2\cdot7\pm0\cdot10$
Blinded	Black	$4\cdot2\pm0\cdot07$	$3\cdot9\pm0\cdot07$	$2\cdot7\pm0\cdot10$
Normal	White	$1\cdot7\pm0\cdot08$	$1\cdot4\pm0\cdot06$	$2\cdot7\pm0\cdot10$
Blinded	White	$4\cdot2\pm0\cdot06$	$3\cdot9\pm0\cdot09$	$2\cdot7\pm0\cdot10$

For the question at hand, the significant part of Table 6 is that which indicates the responses of the blinded isopods. Such individuals naturally show no differences to black or white backgrounds, but exhibit a graded condition in accordance with the degrees of their illumination. In bright light, their chromatophores are well expanded ($4\cdot2$), in dim light partly expanded ($3\cdot9$), and in darkness on the contracted side of the intermediate state between expanded and contracted. These differences fall in line with the degree of illumination and have been assumed by Smith to result from a direct stimulation of the chromatophores. Although it may be maintained that in *Hippolyte* a light receptor other than the eye has been overlooked and that in *Ligia* the elimination of the eye has not been complete, it seems much more probable that the chromatophoral changes herein described are due to the direct action of light on the colour-cells, a view that will be again discussed under the vertebrates.

J. RETINAL FIELDS

The retinal field in crustaceans, composed of a roughly radial system of many ommatidia, is by no means homogeneous as a receptive area for impulses to colour changes. The study of this subject was initiated by Hanström in his work on *Palaemonetes* (1937a) and *Leander* (1938a) and by H. G. Smith who investigated *Ligia* (1938). Hanström occluded parts of the retina or the whole of this organ in *Palaemonetes* by covering its chitinous outer shell with a lacquer impervious to light. As is well known, the red and yellow chromatophores in a normal *Palaemonetes* contract when it is placed on an illuminated white background. This response was found also to occur when the shrimp was in complete darkness, and its retinas were destroyed, but its eye-stalks otherwise left intact, as well as when its eyes were fully covered with lacquer. When only the dorsal halves of its eyes were covered, the ventral halves being left open, and the animal was put on an illuminated white background, its chromatophores contracted. When such a shrimp was placed on an illuminated black background, its chromatophores expanded. Thus the ventral half of the eye is capable of serving as a receptor for the animal's normal colour responses. When, however, only the dorsal halves of the eyes were left open and the shrimp was put on a white or a black illuminated background, the chromatophores under both conditions expanded, showing that the determining influences for full colour change came from the ventral part of the retina. Put in another way, it may be stated that impulses toward the discharge of the eye-stalk neuro-humour (blanching) come only from the ventral half of the retina; impulses toward the inhibition of this operation (darkening) may come from the dorsal as well as from the ventral portion of this organ. Much the same condition was found by Hanström (1938a) in the eyes of *Leander*.

In *Ligia*, the retinal field was investigated by H. G. Smith (1938), who eliminated parts of the retina by paint or by restricting the direction of the light which fell upon this organ. In *Ligia* as in the other crustaceans studied, two retinal regions could be distinguished, one dorsal and the other latero-ventral. Stimulation of the dorsal region was found to result in melanophore expansion, while stimulation of the latero-ventral area induced melanophore contraction. Although Smith assumes an explanation of these changes very unlike that given by Hanström, the facts of chromatic response so far as they are recorded by him, namely that blanching is the property of the ventral part of the eye and darkening is associated with the dorsal part, are in agreement in the two cases. The work of both Hanström and Smith must be regarded as distinctly preliminary. These two investigators have shown that the subject of retinal fields in crustaceans is an important one and deserves active cultivation. It has been scarcely more than touched upon.

K. CRUSTACEAN COLOUR SYSTEMS

The elements that enter into the colour systems of crustaceans are the four pigments yellow, red, dark brown, or black, and white, and in addition to these the hazy, evanescent, blue element which, though it may originate in chromatophores, is not limited to them. Apparently, no one crustacean possesses all five of these colours, but not a few, as for instance *Palaemonetes* and *Crangon*, may have four of them, though not necessarily the same four. In some crustaceans they may be reduced to two or possibly even to one, as in *Ligia*. How are these elements organized into systems for the harmonious interplay of crustacean colour responses?

At present there is no conclusive evidence in favour of crustacean chromatic activators from any source other than the eye-stalks or adjacent parts and in most instances they appear to originate in the sinus gland. The only other sources to be seriously considered in this respect are parts of the central nervous organs and the rostral organ, which would seem to be limited to *Crangon vulgaris*, but even here their presence is very doubtful.

It has already been pointed out that the sinus gland is relatively simple in structure. If there are not other chromatic organs of internal secretion in the crustacean eye-stalk, a multiplicity of function falls upon this gland. That crustacean chromatophores are differentiated in a remarkable way is to be concluded from the fact that the eye-stalk extracts from two such sources as *Palaemonetes* and *Uca*, if injected separately into each of these crustaceans, will invariably call forth concentration of pigment in *Palaemonetes* and dispersion of pigment in *Uca*. This may be a clue to some present difficulties, for, assuming that this extract contains only one neurohumour, the instances just cited point to the probability of profound differences in the chromatophores themselves. In addition to these differences, crustacean colour-cells may also have thresholds at different levels, and, if chromatophoral differentiation and threshold differences are combined, a simpler explanation for the control of chromatophores may be possible than at first seemed likely. This, however, has not yet been worked out, for the unitary theory as advocated by Abramowitz has not been elaborated to such a degree as would lead to the understanding of the colour changes of so complex a form as *Palaemonetes* or *Crangon*. To chromatic activities such as are represented in these shrimps, a larger number of neurohumours than one would seem to be called for, as has been suggested by several workers. Even here, however, a simplifying factor may be involved, for it is possible, as has been suggested for *Palaemonetes*, that in this and closely allied forms chromatic expansion may be an inherent property of the colour-cell itself; chromatic concentration would then be the only phase requiring a neurohumour. This condition appears to be reversed in such brachyurans as have thus far been studied, for in these forms the presence of the neurohumour darkens and its absence blanches the individual concerned. But in one way or another, it may still be assumed that one of these two operations is in the hands of the colour-cell and the other in those of the neurohumour. That two neurohumours are present for these operations in *Ligia* has been recently maintained by H. G. Smith (1938). The evidence presented by this worker for the assumption of a double humoral control depends upon the difference in the opposing curves for chromatophoral activity as the pigments concentrate and expand. These curves, however, may be interpreted as indicative of the rates of appearance and disappearance of one neurohumour as well as of the interaction of two opposing elements. Hence, the assumption of a duplication of agents does not necessarily follow from them.

The conclusions that can be drawn from this discussion as to a system of control for the colour mechanism in crustaceans may be stated thus. This control is not due to direct nerve action, but to one or more neurohumours liberated from organs such as the sinus gland which is connected on one hand by nerves with the optic apparatus and on the other by blood with the body at large. The number of neurohumours involved in these operations is not yet determined, but it is probably several rather than one. These agents excite only one of the two types of chromatophoral responses, concentration or dispersion of pigment, and production or elimination of the diffuse blue coloration. The chromatophores, which include red, yellow, white, and very dark pigments, are highly differentiated and may respond in radically different ways to what may be the same neurohumour. Crustacean chromatophores appear to be open also to direct stimulation by light in that under bright illumination their pigment is dispersed and in darkness it is concentrated. Such a summarized statement of the chromatophoral system in these animals shows that in this field there still remains much that is to be worked out.

Chapter IV

COLOUR CHANGES IN VERTEBRATES

1. *INTRODUCTION*

THE vertebrates are the third large animal group whose representatives exhibit colour changes. Such changes are shown by only certain lower members of this group; namely, the cyclostomes, the elasmobranchs, the teleost fishes, the amphibians, and the reptiles. These forms are all cold-blooded and possess true chromatophores which are absent from the birds and mammals, the warm-blooded members of the vertebrate phylum.

Most chromatic vertebrates exhibit fully formed chromatophores before they hatch and these embryonic colour-cells are often responsive in remarkable and peculiar ways. These early colour activities have been designated primary colour responses and are in strong contrast to those of adult life, which have been called secondary responses (Slome and Hogben, 1928; Parker, 1943 b).

A. PRIMARY COLOUR RESPONSES

These colour responses were first studied by Babák (1910 b) in the Mexican axolotl. They made their appearance, according to this worker, when the young animals were about 1·5 cm. long. In complete darkness such small creatures were pale, but in bright light they became dark. When they had attained a length of 5 cm. they were almost the reverse of this in that in complete darkness they were dark and in bright light they were either pale or dark depending upon their surroundings, being pale on a white background, dark on a black one. Of the two general conditions, the early larval one is the primary colour phase, the succeeding one the secondary phase. When a young larva in the secondary phase was blinded, it was found to revert to the primary phase. When a young larva in the primary phase was deprived of its eyes, it retained its capacity for that colour phase without alteration. It was Babák's belief that the primary phase was due to the direct stimulation of the colour-cells, melanophores, by light and that the secondary one resulted from the indirect stimulation of these cells through the eyes. Although Babák's explanation of the primary phase has been rejected in certain instances (chameleon, Zoond and Eyre, 1934; ammocete, Young, 1935), his idea of the sequence of a primary by a secondary one has been shown to occur in many animals as, for instance, in pleuronectids (Wenckebach, 1886), in axolotls (Pernitzsch, 1913), in *Rana pipiens* (Hooker, 1914 c), in several species of *Amblystoma* (Laurens, 1914 c), and in several other amphibians (Fischel, 1920), in *Perca* (Duspira, 1931 a), in *Macropodus* (Tomita, 1936), in two species of *Salmo*, *S. salvelinus* (Duspira, 1931 a) and *S. trutta* (Neill, 1940), and in *Hoplias* (Mendes, 1942). As Babák himself pointed out, some lower vertebrates appear to have abbreviated the development of their colour sequence by the suppression of the primary phase. Examples of this condition have been noted in *Bombinator* and in *Hyla* (Babák, 1910 b), in *Fundulus* (Bancroft, 1912; Spaeth, 1913 b; Wyman, 1924 b; Gilson, 1926 a), in *Lebistes*, *Gambusia* and *Xiphophorus* (Tomita, 1936), in *Mustelus* (Parker, 1936 a), in *Gasterosteus* (Hogben and Landgrebe, 1940), in *Scyllium* (Waring, 1942), and in *Xenopus* (Landgrebe, quoted by Waring, 1942). *Onchorhynchus* has been relegated to this group by Tuge (1937), but it is more probable that this fish is a member of the first group whose response to darkness develops late in ontogeny as compared with its response to light. Laurens reports *Amblystoma tigrinum* as a species that remains throughout life in the primary phase. It is interesting to note that no instance is known among chromatic vertebrates in which there is a secondary phase followed by a primary one. This is in accordance with what is implied in Babák's general conception of the relation of these two states.

As Babák first pointed out, when a chromatic vertebrate has passed into its secondary phase and is blinded it commonly lapses into the primary one. This condition has been identified in a considerable number of forms: in *Rana esculenta*, *R. fusca* (Babák, 1910 b), in several species of *Amblystoma* (Babák, 1910 b, 1912; Laurens, 1914, 1915), in *Triton* and *Salamandra* (Fischel, 1920), in *Xenopus* (Slome and Hogben, 1928, 1929), in *Macropodus*, *Lebistes*, *Gambusia* and *Xiphophorus* (Tomita, 1936) and in *Phrynosoma* (Parker, 1938 a). Thus a considerable body of evidence, direct and indirect, supports Babák's original claim of primary and secondary phases of colour response in the lower vertebrates. Whether the changes in the primary phase are induced by the direct stimulation of the colour-cells by light, as is believed by a number of workers (see Parker, 1938 a), or are due to other forms of activation (Zoond and Eyre, 1934; Zoond and Bokenham, 1935; Kleinholz, 1938 b), remains uncertain. Nor has any clear view been expressed as to the biological significance of the primary colour phase, which as the examples quoted show, presents itself most strikingly just before or after the young animal has hatched. The possible intervention of neurohumours in these changes has not even been suggested.

B. Secondary Colour Responses

Secondary colour responses characterize larval, especially late larval, life and the whole adult period. They are often lost in creatures in full maturity or in old age. They are dependent upon the functional activity of the eye as can be shown by their disappearance in completely blinded animals. Their immediate excitation may result from nervous or from humoral agents or from a combination of these. Before 1915, all such colour responses were believed to be under exclusively nervous control. This belief was in accordance with the findings of such early investigators as Brücke, Lister and others, nor is it entirely without truth to-day. From the standpoint of innervation taken exclusively, three classes of chromatophores have been distinguished (see chapter I). They are termed dineuronic, mononeuronic, and aneuronic (Parker, 1943 b). Among vertebrates, dineuronic chromatophores, those with two kinds of nerve-fibres, are almost entirely limited to bony fishes, though they may also possibly occur in chameleons (Parker, 1938 a), where in fact they were originally suspected by the French physiologist, Bert (1875). Mononeuronic chromatophores are known from only two vertebrates, the smooth dogfish *Mustelus* (Parker and Porter, 1934) and the horned toad *Phrynosoma* (Parker, 1938 a). In both these forms, the single kind of nerve-fibre is of the concentrating type. No animal has been recorded whose mononeuronic chromatophores are provided with dispersing fibres. Aneuronic chromatophores are found among vertebrates in the cyclostomes (Young, 1935), many elasmobranchs (Young, 1933; Waring, 1936 a; Wykes, 1936; Parker, 1937 b), amphibians (Hogben and Winton, 1922 a) and some lizards such as *Anolis* (Kleinholz, 1938 a). Such chromatophores might also be designated from their means of activation as humoral.

C. Fossil Chromatophores

Colour-cells, because of their small size and delicate nature, would scarcely be expected to occur as fossils. Yet Voigt (1934, 1935) has described and figured what are undoubtedly fossil melanophores in the skins of three fishes and one frog from the brown-coal deposits

in the Geiselthal, Germany. The scales and the bones of the heads of three fishes belonging to the genera *Thaumaturus*, *Palaeoesox*, and *Anthracoperca* were found to contain well preserved remains of melanophores (Fig. 32). A reddish tint observable in the same parts was interpreted as evidence of erythrophores. In other portions of these fossil fishes, good examples of muscle-fibres with cross-striations easily visible were discovered, showing how unusually well preserved the material was. In the skin of a fossil frog from the same deposits, remains of melanophores 0·05 to 0·10 mm. in diameter were easily seen as well as what appeared to be xanthophores surrounded by melanophores. All the melanophores showed pigment in a dispersed state. These preparations came from deposits in the middle Eocene and

Fig. 32. Four fossil melanophores from the preoperculum of *Thaumaturus spannuthi*, a teleost from the middle Eocene deposits of brown coal in the Geiselthal, Germany. Voigt, 1934, Pl. 12.

may therefore be set down as having an approximate age of fifty million years.

A brief general account of vertebrate chromatophores and their activities has been given recently by Walls (1942) in his volume on the vertebrate eye.

2. *COLOUR CHANGES IN CYCLOSTOMES*

Comparatively little work has been done on the colour changes in cyclostomes. In addition to Young's comprehensive account (1935) of these changes in *Lampetra planeri*, there appear to be only three other references in the literature to the chromatophoral activities in these animals. The first of these is the statement by Wild (1903) that *Petromyzon marinus* may change from a bright yellow with dark marbling to a dark blue with white patches, a change which according to him is exceeded in degree only by the colour responses in such animals as the chameleon and the octopus. The second is from Young and Bellerby's account (1935) of *Lampetra planeri*, wherein it is stated that when an extract of the anterior pituitary lobe of the ox is repeatedly injected into this cyclostome in its dark phase, it blanches noticeably. Using the arbitrary scale proposed by Hogben and Slome (1931), in which 5 represents full dispersion of melanophore pigment and 1 full concentration, *Lampetra* was found to change from 4·65 ± 0·08 as shown by the control animals to $3 \cdot 45 \pm 0 \cdot 12$ in the injected individuals. The third of these accounts is that by Coonfield (1940) on *Myxine glutinosa*. This cyclostome, when kept in a black-walled illuminated tank, became uniformly dark in two to three days. When in a white receptacle, it blanched in about the same time. In darkness, the creature assumed a light grey tint. It blanched gradually to injections of adrenaline. Apparently its colour changes are entirely under neurohumoral control as has been shown to be true of *Lampetra* by Young (1935).

According to Young (1935), both the larvae and the adults of *Lampetra* are dark dorsally and yellow or silvery white ventrally. The dark colour is due to melanophores, which are the only kind of chromatophore possessed by this cyclostome. *Lampetra* shows a well-marked daily colour rhythm, becoming pale during the night and dark during the day. The changes here involved often extend from complete pallor to maximal darkening. The pale state assumed at night fluctuates more than the dark one taken on during the day. This led Young to conclude that the pale state was the active one and the dark state the resting one. In the allied species *Lampetra fluviatilis*, these changes, though present, were much less marked. When placed upon an illuminated light or dark background, *L. planeri* remained under both circumstances indefinitely dark, showing that it was not responsive to such differences. When the larvae of this species were kept in continuous light, their melanophores failed to contract as evening came on and remained in almost complete expansion as long as the light continued. Retention of the animals in total darkness produced less constant results. In some instances, the diurnal rhythm was completely arrested, the melanophores remaining in the expanded state. In others, the rhythm continued over many days of total darkness though diminished in extent. Young concluded that the rhythm was controlled primarily by the effect of the light on the animal, but that this effect was dependent upon an internal mechanism which was capable of a certain degree of activity even in total darkness.

Young tested both nerves and endocrines as a means of exciting melanophore responses. Groups of spinal nerves were cut, and spinal nerves and the spinal cord were stimulated electrically without inducing any observable change in the melanophores of the regions concerned. It was concluded, therefore, that nerves were not immediate factors in the colour changes of this animal; in other words, their chromatophores may be classed as aneuronic.

Total removal of the pituitary glands from the ammocoete larvae was followed in an hour or so by a contraction of melanophores and after a few hours by the complete blanching of the animals. This state persisted when the larvae were kept in either bright light or in total darkness and was still to be seen in individuals that had remained alive eleven months after the operation. Further experimentation showed that the parts of the pituitary complex whose loss was necessary to induce permanent blanching were the pars intermedia and the pars nervosa. The injection of mammalian pituitary extract into hypophysectomized lampreys was followed by melanophore expansion, which reached a maximum after two to four hours and disappeared within twenty-four hours. After death lampreys blanch, which was assumed by Young to result from the absence from the dead animal's blood of the pituitary principle. This type of blanching could be retarded in pieces of lamprey skin by the application of pituitrin to the fragment and could be accelerated by a similar use of adrenaline. All the evidence accumulated by Young went to show that in *Lampetra* a secretion from its pituitary gland was the normal agent by which its dark colour was maintained. That adrenaline was concerned in a similar way with its pale phase was doubtful. When the blood supply of a given region was obstructed the region blanched, probably because of the interference in the transfer of the blood-borne pituitary hormone.

Search was made by Young for the afferent pathway by which the colour changes in *Lampetra* were controlled. On the removal of the pineal and parapineal organs from lampreys, they became dark and their daily rhythm of colour change ceased. This cessation was also characteristic of the metamorphosed adults, though in this case it was necessary to remove the lateral eyes as well as the pineal organs. It is thus clear that the exteroceptor by which the colour changes in these ammocoete larvae are regulated is the pineal (and parapineal?) eye and, in the adult, this eye supplemented by the lateral eyes. It is probable that these parts affect the melanophores through nerves which check the secretion of the melanophore-expanding substance of the pituitary gland. The paling of an ammocoete would then be due to an inhibition of pituitary secretion, a secretion which in itself would induce darkening. From this standpoint *Lampetra planeri* appears to be an animal whose chromatophoral activities are controlled through the presence or absence in its blood of a single pituitary neurohumour (unihumoral type).

3. COLOUR CHANGES IN ELASMOBRANCH FISHES

The normal colour changes of elasmobranch fishes are relatively slow and require for their completion hours and even days as compared with the briefer periods of a few minutes as in the teleost *Fundulus*, or even seconds as in the squirrel fish *Holocentrus*. In fact, *Raja clavata* and *R. batis*, two dark-coloured rays, are said by Schaefer (1921) not to change colour at all on being transferred from a black to a white background. Veil and May (1937) in a recent paper have stated that *Torpedo marmorata*, likewise a dark species, shows no response to changes in the background. This statement, however, has been questioned by Vilter (1937 b), who was able to observe such changes in the torpedo when it was kept under inspection over a period of some weeks. Possibly the same might have been true of the species of rays studied by Schaefer had they been under observation over a similarly long period. According to Parker (1936 e, 1937 b) *Squalus acanthias*, a dark dogfish, when introduced into an illuminated aquarium with white walls, showed ordinarily little or no change of tint, though some individuals under such circumstances blanched somewhat in the course of two days or so. Such moderately pale fishes on being transferred to a tank with black walls slowly reassumed their original dark colour. Waring (1938), who also worked on *Squalus*, has confirmed these statements. He found, however, in the fishes at his disposal a larger proportion of responsive individuals than were reported by Parker. *Scyllium catulus* and *S. canicula*, like *Squalus acanthias*, appear to be relatively sluggish in their colour changes (Young, 1933; Hogben, 1936; Wykes, 1936; Waring, 1938; Waring, Landgrebe and Bruce, 1942). According to Waring, *S. canicula* will change from pale to dark or the reverse in from eighty to one hundred hours. Such periods are also required by this dogfish to change from an intermediate tint to either extreme as well as from one extreme to the other. Newly hatched individuals of this species take about the same periods of time for their colour changes as do adults (Waring, 1938 a).

Hogben observed (1936) that after *Raja clavata* had been deprived of its pituitary gland, it would blanch, showing that its chromatic system was not absolutely inactive as might have been inferred from Schaefer's account. *Scyllium catulus* and *S. canicula* also become pale on the loss of their pituitary glands (Hogben, 1936). The same is true of *Squalus acanthias* (Parker, 1936 e). This fish, however, does not lighten till some three or four days after hypophysectomy. Notwithstanding the fact that all these relatively inactive dark forms show a sluggish response to changes in their surroundings, their dark colour cannot be attributed to a complete lack of ability to change their tint, for on the loss of their pituitary glands they all blanch noticeably. Their more or less constant dark coloration is therefore probably due to a steady production of a dispersing pituitary neurohumour, a production which in this instance is apparently not influenced to any considerable extent by environmental changes.

In contrast with these relatively inactive elasmobranchs are others whose colour responses are accomplished in shorter periods. Thus *Mustelus canis*, whose colour changes were first noted by Lundstrom and Bard (1932) in their experimental study of the pituitary gland in

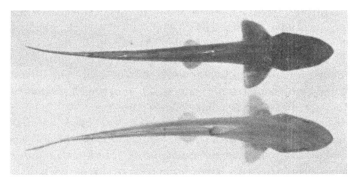

Fig. 33. Dorsal views of two newly born *Mustelus canis*, the upper one in the dark phase, the lower one in the pale phase. Preserved in formaldehyde-alcohol. Parker, 1936b, 5.

this species, assumes a deep slate grey on a black background and becomes ashen pale on a white one (Parker and Porter, 1934). Adult fishes change from pale to dark in from half an hour to two hours and from dark to pale in some two days. The newly born young of this species are fully responsive to the colour changes in their environment immediately after birth (Fig. 33), and react to these changes rather more quickly than do the adults (Parker, 1936b, 1938b). Parker observed (1933c) that *Raja erinacea* on a white background blanched in about twelve hours and on a black one darkened in some nine hours. Griffiths (1936) noted that a moderately dark *Trigonorrhina fasciata* when transferred to an illuminated tank with white walls became noticeably pale in seventeen hours. Other elasmobranchs which are reputed to be relatively active in their colour responses, but whose exact times of change have not been recorded, are *Raja brachyura*, *R. maculata*, and *Rhina squatina* (Hogben, 1936; Wykes, 1936; Budker, 1936). This distinction of two classes of elasmobranchs, one relatively slow and the other more rapid in their colour responses, though not one that can be rigidly carried through, has already been suggested by Vilter (1937 b).

In all the elasmobranchs whose colour changes have been considered thus far, these changes have been described as limited almost exclusively to the dorsal side. Weidenreich (1927) has recorded in *Raja oxyrhyncha* a change of tint that affects the ventral surface of this species, a remarkable and decidedly puzzling phenomenon.

A. CHROMATOPHORES

The chromatophores of *Scyllium* (*Scyliorhinus*) *canicula* and *S. stellaris* were described by Ford (1921) as of two kinds: lemon yellow and brownish black to black, both said to be found all over the body. According to Hogben (1936), elasmobranchs may possess three kinds of chromatophores: epidermal melanophores, larger and more richly branched dermal melanophores, and xanthophores. These occur in *Raja brachiura*, *R. clavata*, *R. maculata*, *R. microcelatus*, *Scyllium canicula*, *S. catulus* and *Rhina squatina* from European waters and in the American *Mustelus canis*. Waring (1936a) has described in much detail these three types of colour cells in *Scyllium canicula* as superficial, probably epidermal melanophores containing very dark pigment, deep-seated, probably dermal melanophores with light brown pigment, and yellow cells, xanthophores, at approximately the same level as the deep-seated melanophores. In the newly hatched young of this species, Waring (1936a) has recorded in addition to the kinds of chromatophores already mentioned a class of greyish-brown cells not hitherto noticed. Their pigment, like that of the xanthophores, is soluble in alcohol. In a later publication concerned largely with *Scyllium canicula*, Waring (1938) has recorded the degree of expansion and contraction of the two classes of melanophores shown during their changes and has demonstrated that in general the epidermal melanophores at almost all stages exhibit a greater amount of pigment concentration than do the dermal ones (Fig. 34). Otherwise the two sets respond in much the same way. Waring (1942), using the system of arbitrary units devised by Slome and Hogben, has measured in *Scyllium canicula* the diameters of its melanophores and found that when the fish is on a black background they average 4·9 units in diameter, on a white one 1·4, and in darkness 2·9. According to Hogben (1936), the two classes of melanophores and the xanthophores in the elasmobranchs studied by him expand and contract in unison. The list of elasmobranch chromatophores was somewhat extended by Vilter (1937d), who worked on *Trigon pastinaca*, *Raja undulata* and *Torpedo marmorata*. In these species, five classes of chromatophores could be distinguished: dermal guanophores, dermal and epidermal

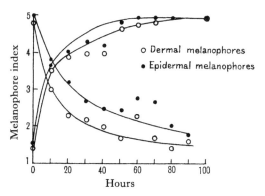

Fig. 34. Plottings to show the rates of change of *Scyllium canicula* from pale to dark and from dark to pale when on black or white backgrounds. Waring, 1938, 269.

xanthophores, and dermal and epidermal melanophores. Vilter pointed out, however, that of these the dermal melanophores were the dominant group in the colour changes. Most elasmobranchs are rather uniform in coloration, but some dogfishes, such as *Squalus acanthias*, and many skates of the genus *Raja*, are mottled and spotted. The relations of these special markings to the general colour changes in these fishes have been commented upon by no student of the subject except Vilter (1937d).

B. The Dark Phase

Lundstrom and Bard (1932), in their study of the pituitary gland in the smooth dogfish, *Mustelus canis*, discovered that the removal of this gland from the fish was followed by a profound, general, and permanent blanching (Fig. 35). Slight paling was first noticed about half an hour after the loss of the gland and maximal pallor was attained in about twelve hours. This pallor could be temporarily overcome by injecting into the fish a small amount of pituitary extract. The injection of all of the extract from a single gland was followed in three minutes by a distinct general darkening of the fish and after an hour by its maximum deep coloration. "Pituitrin" (Parke, Davis and Company) and "infundin" (Burroughs, Wellcome and Company), both preparations of mammalian pituitary glands, invariably produced darkening when injected into pale hypophysectomized dogfishes. A microscopic examination of the skin of these animals showed that in the pale condition the melanophores were contracted and in the dark one they were expanded. It was therefore concluded by Lundstrom and Bard that the dark phase of *Mustelus* was brought about by a hormone produced by the pituitary gland and carried in the blood to the melanophores which were thereby induced to expand.

Fig. 35. Two smooth dog-fishes, *Mustelus canis*, originally of the same colour, 24 hours after the removal of the hypophysis from the fish on the right. They now show extreme differences in tint. Lundstrom and Bard, 1932, 5.

The conclusion thus reached was confirmed by Parker and Porter (1934) who showed in addition that when the defibrinated blood from a dark dogfish was injected into a pale one, a dark area due to the expansion of the melanophores of the given region was induced. No colour change was called forth when such blood was injected into a dark fish. Furthermore, when blood from a pale fish was introduced into either a pale or a dark one, no change of colour occurred in the recipient.

The various tests made upon *Mustelus* were repeated by Parker (1936e) on *Squalus acanthias* and with confirmatory results, except that *Squalus* was found to be a less favourable species in which to excite the pale phase than *Mustelus* had been. A repetition in many respects of the procedure initiated by Lundstrom and Bard for *Mustelus* was carried out by Hogben (1936) on several species of *Raja*, on *Scyllium* and on *Rhina*, by Waring (1936a, 1936b, 1936c) on *Scyllium*, by Wykes (1936) on *Raja*, *Rhina* and *Scyllium*, and by Griffiths (1936) on *Urolophus* and *Trygonorrhina*, and in all instances the conclusion that the dark coloration in elasmobranchs is due to a secretion of the pituitary gland was fully supported. This secretion was designated by Hogben as the B-substance; it has since been commonly called intermedine. Further evidence in favour of it as the exciting agent of the dark phase of elasmobranchs has come from the work of Parker on *Squalus* and *Mustelus* both adult (1937*b*) and young (1933), from that of Veil and May (1937) on *Torpedo*, from Vilter's extended studies (1937*b*; 1937*d*) on *Trigon pastinaca, Raja undulata,* and *Torpedo marmorata,* and from Waring's work (1938) on *Scyllium canicula* and *Squalus acanthias.* Few statements are more generally assented to by investigators in this field than that the dark phase of elasmobranchs is induced by the pituitary secretion intermedine. To this there seems to be no ground for objection.

In this connection considerable attention has been given to the problem of the part of the pituitary complex from which intermedine is secreted. According to Lewis and Butcher (1936*a*), who appear to be the most recent workers on the structure of the pituitary gland in elasmobranchs, this organ in *Squalus acanthias* and in *Raja stabuliforis* is composed of the following six parts: pars distalis, pars medialis, pars intermedia, pars neuralis, saccus vasculosus, and pars ventralis. Of these, the pars distalis and the pars medialis are together equivalent to the anterior lobe of other workers and the pars intermedia and the pars neuralis to the posterior lobe. The histology of these parts has been described in a brief, preliminary way by Butcher (1936).

Lundstrom and Bard (1932) tested certain portions of the pituitary complex for the materials concerned with the dark colour change. They found that when the posterior (neuro-intermediate) lobe was removed, the fish blanched as it did after complete hypophysectomy. Extirpation of the anterior lobe (pars distalis and pars medialis) did not result in paling. Injection into a hypophysectomized dogfish of extracts of the posterior lobe resulted in a darkening of the fish. No effect upon pigmentation was observed when extracts of cerebellum, skeletal muscle, or pancreas were injected. Administration of a suspension of anterior lobe caused a relatively weak darkening of a pale fish. It was concluded by Lundstrom and Bard that the posterior lobe of the hypophysis was responsible for the considerable melanophore expansion seen in the dark skin of normal dogfishes.

This conclusion, which is in agreement with the earlier work of Hogben and his associates on amphibians, has been substantiated by the investigations of most later students. Hogben (1936), who has done extended work not only on amphibians but also on elasmobranchs, observed that the loss of the neuro-intermediate lobe from *Raja* was followed by extreme pallor which did not result from the ablation of the ventral lobe. Removal of the anterior lobe alone appeared merely to abolish the white background response. Waring's studies on this general question (1936a, 1936b, 1936c, 1938), which included detailed experimentation chiefly on *Scyllium* and *Squalus*, led to essentially similar results: the removal of the neuro-intermediate lobe was always associated with maximum pallor. Whether the loss of the ventral lobe interfered with the black background response was not certain. The white background response disappeared with the removal of the anterior lobe. Lewis and Butcher (1936a), whose work on the parts of the pituitary gland in *Squalus acanthias* and *Raja stabuliforis* has already been referred to, tested the expanding capacity of extracts of these several parts on the melanophores of the frog (1936b). They obtained positive results with extracts of the pars intermedia and the pars distalis, but negative ones from those of the pars ventralis, pars medialis and the saccus vasculosus. They did not test the pars neuralis. Geiling and Le Messurier (1936) in similar tests obtained evidence of an expanding principle in the posterior lobes of the pituitary from the dogfish and the skate, but not in the saccus vasculosus of these fishes. It is interesting to note that in tissue cultures of the parts of the pituitary, extracts of the pars intermedia and of the pars distalis darken frogs, showing that even after cultivation *in vitro* the cells of these lobes contain an active melanophore-expanding principle (Lewis, 1936). All these results lead to the general conclusion that the dispersion of the melanophore pigment in elasmobranchs is brought about by a water-soluble neurohumour, intermedine, produced chiefly, if not entirely, in the neuro-intermediate lobe of the pituitary gland and circulated by the blood (Waring, 1936b). This view was assented to by Veil and May (1937) in their study of the *Torpedo* notwithstanding the fact that, as noted by them, blood from a normal *Torpedo* when injected into a hypophysectomized one did not seem to alter the coloration of the recipient.

C. THE PALE PHASE

The way in which the pale phase is excited in elasmobranchs is a problem of considerable uncertainty. Much work has been done on it and numerous views have been expressed about it, but no real uniformity of opinion concerning it has been reached. Lundstrom and Bard (1932), in their tests of *Mustelus canis* with adrenaline, found nothing to favour the idea of an adrenal or sympathetic control. They intimate the possibility that blanching in *Mustelus* may depend upon the simple disappearance of the pituitary secretion from the blood of the fish. This uncertainty led Parker and Porter (1934) to undertake the study of the pale phase in this dogfish. When cutaneous nerves are cut anywhere on the body, but especially on the pectoral fins of *Mustelus*, a pale band appears covering the denervated area. Such a band will extend from the initiating cut to the edge of the fin (Fig. 36).

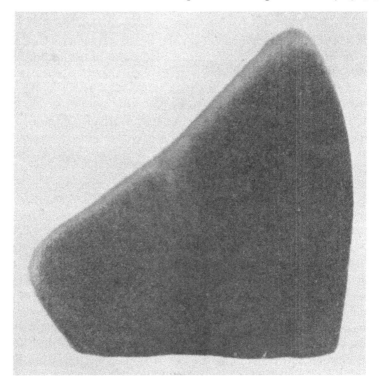

Fig. 36. A pectoral fin from a dark dogfish, *Mustelus canis*, showing a pale band which appeared as the result of an incision transverse to the direction of the fin rays and about 1½ cm. from the edge of the fin. Parker and Porter, 1934, 33.

It will begin to appear in from ten to fifteen minutes after the cut has been made, will reach its maximum of distinctness in about a day and will disappear in two to three or more days. This band was believed by Parker and Porter to be induced by the excitation of concentrating melanophore nerve-fibres due to the cut and the irritation induced by it. When the melanophores in such a band are inspected under the microscope, their pigment is found to be fully concentrated, a condition not always realized by the melanophores in a dogfish normally, and apparently fully blanched by exposure to white surroundings. Thus this method may induce an extreme concentration in melanophore pigment. It was the opinion of Parker and Porter that the normal blanching of *Mustelus* was due to the action of concentrating nerve-fibres as indicated by this experimental outcome and that these fibres acted in opposition to the pituitary secretion in the colour changes of *Mustelus*.

Other tests in this matter showed that the formation of pale bands by cuts was by no means common in elasmobranchs. Parker (1936e) showed that in *Squalus* the bands were at best faint and at times failed entirely to appear. No trace of them could be seen in *Raja erinacea* (Parker, 1936e). Young (1933) had previously reported them as absent from *Scyllium*, an observation subsequently confirmed by Waring (1936a) and by Wykes (1936), who also showed that nerve section in *Raja brachyura*, *R. maculata*, and nerve stimulation in *R. brachyura*, *Rhina squatina* and *Scyllium catulus*, failed to induce colour changes. In *Squalus*, however, Waring (1938) succeeded in producing better pale bands than Parker had been able to elicit, though these bands were apparently not so pronounced as in *Mustelus*. Pale bands, therefore, have been identified in only two species of dogfishes and have not been observed in some half a dozen other elasmobranchs where search has been made for them.

What interpretation is to be placed on such bands where they do occur is by no means certain: Parker and Porter's view that the pale bands described by them are due to the direct action of concentrating nerve-fibres excited by the initiating cut has been criticized by Hogben (1936), who has pointed out that such blanching may result from vascular or vaso-motor disturbances concomitant on the cut. This opinion has also been voiced by Wykes (1936), and by Waring (1938). Parker (1937d, 1938c) has pointed out three conditions opposed to the vascular interpretation of pale bands. The first has to do with the circulation of the blood itself. When a cut is made, the flow of blood immediately around the cut is of course disturbed, but at a short distance distal to the cut in consequence of collateral connections the flow of blood appears in all respects normal. This is particularly true of the capillary circulation next to the melanophores as can be seen under the microscope. Another feature opposed to the vascular interpretation is seen in the occurrence of the bands. They are found in only a few dogfishes; other elasmobranchs when cut do not exhibit them and yet the cuts in these fishes must disturb the circulation as much as they do in those forms that show bands.

Finally, when steps are taken to block completely the circulation of blood in a fin, as for instance by the use of a tourniquet at the base of this structure, a band may still be excited by a cut. Hence, Parker has been led to conclude that the pale bands in *Mustelus* and probably in *Squalus* are not due to vascular disturbances. The fact that in *Mustelus* they may be induced by the faradic stimulation of integumentary nerves (Parker, 1935*a*, 1938*b*), and may be revived after they have faded by a new cut distal to the old one (Parker, 1936*f*), has supported the conclusion that such pale bands result from the direct action of concentrating nerves (Figs. 37, 38).

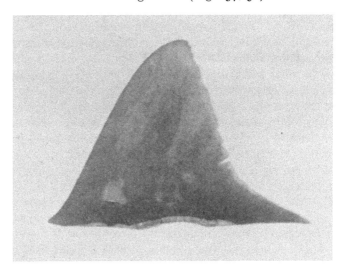

Fig. 37. Side view of an anterior dorsal fin of a dark smooth dogfish, *Mustelus canis*, which had been stimulated to the formation of pale bands by a transverse cut near the anterior edge of the fin and by electric stimulation posterior to this cut. The resultant pale areas can be seen in the fin peripheral to the two regions of stimulation. Parker, 1935*a*, 2.

In a paper published a few years ago, Abramowitz (1939*a*) has offered renewed criticism of Parker and Porter's view that *Mustelus* possesses concentrating chromatic nerves. Without denying the easy production of pale bands produced by cuts in the pectoral fins of this dogfish, Abramowitz states that when the brachial nerve-plexus or one of the ophthalmic nerves is severed the region of skin thus denervated does not blanch as would be expected but remains dark. Abramowitz, in agreement with earlier workers, regards the pale bands resulting from cuts in the pectoral fins as due to disturbances in the blood supply. Abramowitz's results from the cutting of the brachial plexus and the ophthalmic nerves led Parker (1942*a*) to a re-examination of the question of innervation with an outcome quite different from that reached by Abramowitz. When the ophthalmic nerve of one side of the head of a *Mustelus* is cut in the region of the orbit, the skin of that side of the head not only blanches but blanches to an extreme of paleness as does a band produced by a cut in the pectoral fin. This blanching of half the head by the severance of the appropriate ophthalmic nerve can be obtained with great certainty and regularity. It is not commonly produced in fishes that are in a decidedly dark

Fig. 38. Right pectoral fin of *Mustelus canis*. The pale band from the initial cut in the fin had darkened before the second cut distal to the first one was made. This second cut excited the formation of the distinct distal pale patch within the area of the former pale band. Parker, 1936*f*, 257.

phase and whose blood, as indicated by their colour, is rich in the chromatic dispersing agent intermedine. Under such circumstances, it is extremely difficult to produce blanching not only on the head but on the pectoral fin. This antagonism between the blanching agent and intermedine was pointed out some time ago (Parker, 1937*g*), and inattention to it is the probable cause of the inability of Abramowitz to obtain a pale area on cutting the ophthalmic nerve. In testing for blanching, it is natural to test with a dark fish but such a fish is the poorest material possible for such operations. Animals should be selected that are moderately pale and yet dark enough to show a really pale area. Such fishes must have a very meagre supply of the darkening agent intermedine in their blood. Apparently this condition escaped the attention of Abramowitz; hence his misleading results.

In the course of Parker's reinvestigation of this subject, he was led to look into the interrelation of nerve and blood in experimental colour tests. For this purpose, the pups of *Mustelus* are very convenient. In these young fishes, the majority of the nerves in the brachial plexus are easily accessible through about one millimetre of flesh in a region dorsal to the anterior edge of the pectoral fin and immediately posterior to the gill-clefts. By an incision about half a centimetre in length and appropriate in position, all these nerves can be severed and without the loss of blood, for the nerves are at a considerable distance from the brachial artery and its branches. Soon after such a nerve section has been accomplished, the pectoral fin of the appropriate side begins to blanch and finally takes on full paleness. An inspection of the blanched fin under a low power of the microscope shows a full and complete flow of blood in its superficial capillaries which is in no way different from that in the opposite normal fin. Hence, Parker was led to conclude that blanching of the type shown in pectoral fin bands and other like conditions in *Mustelus* was not due to disturbance in its blood supply but was a nervous phenomenon most readily understood on the assumption of the presence of concentrating chromatic nerve fibres in the organization of this dogfish.

Griffiths (1936), without giving any new experimental evidence, appears to favour a nervous interpretation of elasmobranch blanching. Such a view is fully advocated by Vilter (1937*b*, 1937*d*) as a result of his experiments with ergotamine on *Trigon pastinaca, Raja undulata*, and *Torpedo marmorata*. According to Vilter, blanching in elasmobranchs is due to the tonic action of the sympathetic nervous system on melanophores. This tonic action is in opposition to the dispersing activity of the hypophysial agent. It can be excited by the electric stimulation of a section of isolated spinal cord whereby the corresponding section of the skin becomes pale, and it can be eliminated by an injection of ergotamine into a pale fish, whereby the melanophores can be relieved of sympathetic tonus and are thus free to expand under the action of intermedine. That the darkening of a fish under ergotamine is not due to the direct action of this substance on the melanophores is shown, according to Vilter, by the absence of melanophore expansion in fragments of pale skin subjected to ergotamine. The fact that ergotamine can induce darkening in hypophysectomized elasmobranchs shows that this substance does not induce expansion by an indirect action on the pituitary gland. For these reasons Vilter believes that ergotamine acts by the exclusion of concentrating nerve-fibres.

Dittus (1937) has attacked the problem of blanching in elasmo-branchs from an entirely novel standpoint. His work was done chiefly on *Torpedo ocellata* and *T. marmorata* and was concerned with the hormone from the interrenal bodies. After the extirpation of their interrenal organs, torpedoes breathed more slowly and less intensely, and there followed a contraction of their melanophores with the result that the fishes took on a dirty grey tint. This Dittus interpreted as due to a deficiency of oxygen in the blood, an interpretation in accord with the early work of von Frisch (1911c, 1912b), and the more recent studies of Wykes (1936). An injection of an extract of the interrenal cortex intensified breathing with the result that the melanophores expanded. Thus the blanching of these fishes was attributed to a low oxygen supply in the blood.

Hogben (1936), in line with the conclusions arrived at by Hogben and Slome (1931, 1936) on amphibians, has presented evidence in favour of the anterior pituitary lobe in elasmobranchs as a source of a blanching hormone for these fishes. Hogben's work was done on *Scyllium canicula*, *S. catulus*, and *Raja brachiura*, and the particular hormone concerned has been designated by him as W-substance. When in a dark fish the anterior pituitary lobe is removed, no change in colour is noted. If this lobe is removed from a pale fish, the creature becomes dark even though it is kept on a white background, an observation confirmed by Waring (1936b). Thus the loss of the anterior lobe abolishes the pale response. Implanting an anterior lobe into a pale fish produced, however, negative or relatively slight results. Implanting such a lobe into a dark fish produced so slight a change as to justify no definite conclusion. Waring (1938) tested the effect of the anterior lobe on blanching in *Scyllium canicula* by other means. Hypophysectomized fishes were prepared, some with the anterior lobes intact and others with these lobes removed. When both kinds of fishes, all of which were of course pale, were injected with inter-medine, the two kinds darkened and at about the same rate, but as the effect of the intermedine wore off, those with intact anterior lobes blanched more fully and more quickly than those from which these lobes had been removed. Thus the presence of the anterior lobe facilitated recovery from the darkening effect of intermedine (Fig. 39).

Parker (1937b) has made the following comments on the work of Hogben and Waring. If a concentrating pituitary neurohumour, such as the W-substance is believed to be, is present in elasmobranchs, it ought to be identifiable in the blood of these fishes when they are in the pale states. When blood from pale *Mustelus canis* (Parker and

Porter, 1934) or from a similar *Squalus acanthias* (Parker, 1936*e*) is injected into moderately dark individuals of the appropriate species, these individuals neither develop pale spots nor change noticeably in tint. The same results were obtained in blood transfers from pale skates, *Raja erinacea*, to dark ones (Parker, 1937*b*), though when blood from a dark skate was injected into a pale one a darkening of the recipient occurred. These results led Parker to question the general validity of the conclusions reached by Hogben and by Waring.

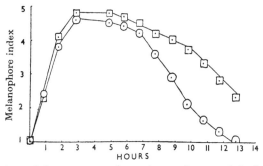

Fig. 39. Plottings of the responses of two classes of operated dogfishes, *Scyllium canicula*, to injections of equal doses of intermedine (B-hormone). Each curve represents the average of several injections with five sets of dogfishes. The records represented by squares are from fishes whose neuro-intermediate and anterior pituitary lobes had been removed; those shown by circles are from fishes whose neuro-intermediate lobes only had been removed. Waring, 1938, 273.

Another way in which elasmobranchs may blanch is by the loss of intermedine from their blood. This is what occurs experimentally in all instances of simple hypophysectomy. A direct experimental test of this question can be carried out conveniently on the young or pups of *Mustelus*. If a strong cord is tied firmly around the head of a dark pup at the level of the posterior margins of its eyes and another at the level of the vent, the pup will be divided into three regions: an anterior one which in consequence of the ligating cord is without circulation, a posterior one also devoid of circulation, and a middle one containing the heart and connecting blood-vessels and in which a circulation of blood still goes on. Such a pup will remain dark for a full hour, though, as death supervenes, it will eventually blanch. If into the ventral aorta of a pup freshly ligated in the manner described an irrigation cannula is tied, the middle region can be washed free of blood in some five minutes by a flow of Ringer's solution. During this washing, the middle region will blanch very completely and will remain so indefinitely (Fig. 40). Under the microscope, the melanophores of the middle region can be seen to be reduced almost to the punctate state, while those of the anterior and posterior regions remain with dispersed pigment (Fig. 41). That this blanching of the middle region is not due to the death of its tissues including the melanophores

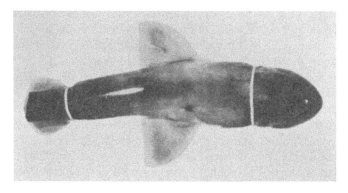

Fig. 40. Dorsal view of a young *Mustelus canis* originally dark and divided by firm ligatures into three regions, anterior, posterior, and central. The central region has been fully irrigated with Ringer's solution from the ventral aorta and has in consequence largely blanched, while the anterior and posterior regions have remained dark. Parker, 1938c, Pl. 2.

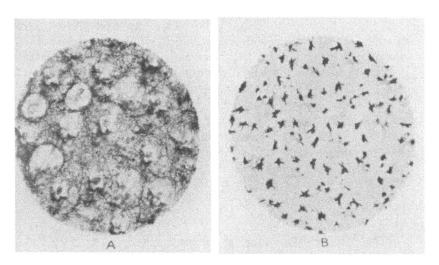

Fig. 41. Melanophores from (A) the anterior region with pigment dispersed and consequently dark, and from (B) the middle region with pigment concentrated and consequently pale, of the dogfish shown in Fig. 40. Parker, 1938c, Pl. 2.

is seen in the fact that when this region is further irrigated with a mixture of one volume of Ringer's solution and two volumes of commercial pituitrin, the middle region will again darken through the expansion of its melanophores. *Mustelus*, then, may be made to blanch by the simple loss of intermedine from its body fluids. This type of blanching appears to occur naturally in *Raja erinacea* (Parker, 1937 b), for this skate gives no evidence of possessing concentrating nerves, nor, when in the pale state, of carrying a concentrating neurohumour in its blood, and yet it blanches under appropriate circumstances. It is difficult to explain blanching in this fish except on the assumption of the loss of intermedine, an assumption which also implies that when such melanophores are unstimulated they lapse into a state of concentrated pigment.

This survey of the means of blanching in elasmobranchs leads to no certain and simple conclusion such as that which obtains for darkening. It is improbable that oxygen deficiency, as advocated by Dittus, plays any part in normal blanching. Apparently it is only concerned with the pallor of death. It is also unlikely that vaso-motor or other vascular disturbances, as were originally intimated by Lundstrom and Bard, have to do with normal blanching. Lack of intermedine in the blood of the fish, as was originally maintained for amphibians by Hogben and Winton, is probably a subordinate factor in elasmo-branch blanching, though possibly in such instances as *Raja erinacea* it may be of prime importance. Concentrating nerves appear to play an unquestionable part in the blanching of *Mustelus* and probably *Squalus*, but that they occur with the universality that is attributed to them by Vilter is extremely doubtful. That many elasmobranchs blanch in consequence of a neurohumour from the anterior pituitary lobe as first advocated by Hogben is probable, though conclusive evidence is still lacking. The process of blanching in elasmobranchs, viewed as a whole, seems to be carried out by different means in different species; hence, the diversity of opinions concerning it. This lack of uniformity already pointed out by Parker (1936e) has been reiterated by Waring (1936b, 1942). It is opposed to the unihormonal type of chromatic organization as advocated by Abramowitz (1939a).

D. The Blinded State

After having been blinded, an adult *Mustelus canis* follows the general rule for the majority of chromatic vertebrates and assumes a dark tint in consequence of the expansion of its melanophores (Parker and

Porter, 1934). In this respect, it differs somewhat from other elasmo-branchs. In eyeless *Raja brachiura*, according to Hogben (1936), the melanophores are neither fully expanded nor fully contracted, and in the related batoid *Urolophus* no colour change could be discerned by Griffiths (1936) during the four hours that followed enucleation. In *Torpedo marmorata*, Veil and May (1937) observed that the usual brownish tint of this species was maintained unchanged as well without the eyes as with them. Possibly some of these discordant results may disappear when better means are provided for working on these relatively unwieldy fishes. In this respect, the newly born young or pups of *Mustelus* are much more satisfactory as experimental material than are the adults (Parker, 1938 b).

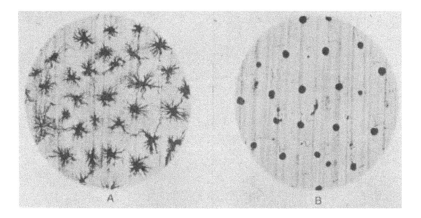

Fig 42. Melanophores from the pectoral fins of *Mustelus canis* A, with fully dispersed pigment (dark phase of the fish), B, with fully concentrated pigment (pale phase of the fish). Parker, 1938 c, Pl. 1.

Normal pups of *Mustelus* were found to change fully pale or fully dark in response to an appropriately illuminated background (Fig. 42). They could be readily blinded by cutting their optic nerves through a buccal opening. The blinded pups soon became relatively dark. When some were placed in an illuminated aquarium with black walls and others in one with white walls, the two sets remained dark and indistinguishable. When some were placed in a white-walled illu-minated tank and others in a tank from which all light had been excluded and were kept there half a day or so, those in darkness were obviously paler than those in the light. This difference was not only visible to the unaided eye, but clearly discernible under the microscope

(Fig. 43). As a test of responsiveness the pectoral fins of both sets of pups were cut. In half an hour a band of maximum paleness

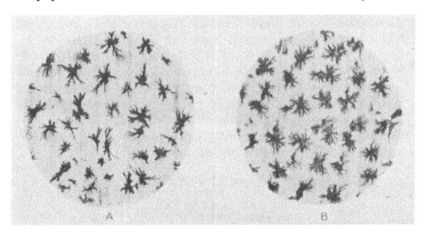

Fig. 43. Melanophores from a blinded young *Mustelus canis*: *A*, in darkness; *B*, in light. The pigment is slightly more concentrated in the dark than in the light. Parker, 1938 *c*, Pl. 1.

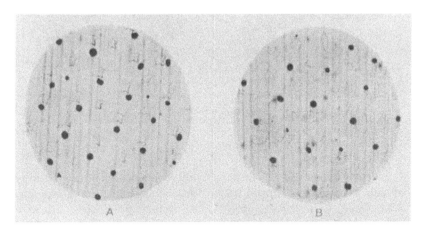

Fig. 44. Melanophores from a blinded young *Mustelus canis*: *A*, on an illuminated black background; *B*, on an illuminated white background. In both, the pigment is fully concentrated and the two conditions are indistinguishable. Parker, 1938 *c*, Pl. 1.

had appeared in each fin. To ascertain the influence of the pituitary secretion, pups were hypophysectomized and some were put in an illuminated tank with white walls and others in one with black walls.

After twenty hours both sets had reached maximum paleness. The melanophores of these two sets of individuals were seen under the microscope to be fully punctate (Fig. 44). Some were now put in complete darkness. After seven hours, they were found to be still fully pale. Cutting the optic nerves of these fishes made no alterations in their condition. From these results, Parker concluded that the darkening of blinded *Mustelus* must be due to intermedine and not to a direct effect of the light, and that the intermedine is present in slightly greater amounts when the fishes are in the light than when they are in darkness. Thus there appears to be no evidence in *Mustelus* of a direct or primary response of its melanophores as is believed to be the case in *Squalus* by Waring (1938).

E. Elasmobranch Neurohumours

Of the elasmobranch neurohumours, intermedine is the most commonly recognized and appears to be the only one concerned with the dark phase of these fishes. Its universal occurrence in this group is attested by all investigators of elasmobranch colour changes. It is the B-substance of Hogben.

Adrenaline, a very active agent in the concentration of dermal melanophore pigment, has long been known to be effective in this respect when injected into elasmobranchs. Lundstrom and Bard (1932) showed that when 1 c.c. of a 1 : 1000 solution of adrenaline chloride was injected into *Mustelus canis*, the fish began to pale in some 10 minutes. This blanching reached a maximum in about an hour and persisted approximately two hours. The dose was admittedly enormous. Parker and Porter (1934) found that 0·5 c.c. of 1 : 10,000 of this agent would induce temporary blanching in an adult smooth dogfish. *Mustelus* pups reacted to his hormone in much the same way as did adults (Parker, 1936 b). Young (1933) was unable to induce pallor in *Scyllium canicula*, *S. catulus* and *Torpedo ocellata* by injecting 1 mg. of adrenaline chloride into these fishes, though this dose was amply sufficient to blanch *Pleuronectes*. Wykes (1936) succeeded in producing only a slight degree of pallor in *Raja brachyura* and *Rhina squatina* on injecting 1 c.c. of a 1 : 10,000 adrenaline hydrochloride. Possibly in Young's tests and certainly in those of Wykes, the concentrations used were very near the effective limits of this reagent. Veil and May (1937) found it necessary to inject 2 mg. of adrenaline into a torpedo of 200 g. weight to induce a noticeable blanching. In all these instances the effective doses were truly prodigious. That the blood of such pale elasmobranchs as *Mustelus*, *Squalus*, and *Raja* has been found to have no effect in blanching, even to a small degree, other dark or partly dark elasmobranchs, has led Parker (1937 b) to agree with Waring (1936 a) that adrenaline very probably plays no part in the normal pale reactions of these fishes.

Because of the nature of the response and particularly of its remoteness, the extract of the interrenal organs advocated as a chromatic agent by Dittus is hardly to be classed as containing a chromatophoral neurohumour. It is quite otherwise with the W-substance of Hogben. This neurohumour, a product of the anterior pituitary lobe, appears to be the effective agent in the blanching of many elasmobranchs. Although its presence in these fishes is not established beyond dispute, it seems probable that this neurohumour may be the means of bringing about a contraction of their melanophores.

All the chromatic hormones thus far mentioned, intermedine, adrenaline, and the W-substance of Hogben, are soluble in water and consequently may be carried in the blood. They have been designated by Parker (1935 b) as hydrohumours and have been set off against another group of neurohumours which appear to be soluble not in the aqueous components of an animal's body but in its lipoids. These are the so-called lipohumours and are represented in such elasmobranchs as *Mustelus* by the material that appears to emanate from those nerve terminals that bring about a concentration of melanophore pigment. Such material can be extracted from the skin, and especially from the fins, of a pale *Mustelus* by the use of oil or of ether (Parker, 1935 b). In carrying out this operation, the fins of a pale fish were cut off and ground to a fine pulp and this pulp was then mixed with about 2 c.c. of Italian olive oil. This mixture was further ground by hand for about half an hour in a rough porcelain mortar till it reached the consistency of a thick paste. It was then set aside to undergo extraction. In most instances, it was sterilized by heat before it was extracted. Its extraction was carried on at the low temperature of an ordinary ice-box. After the paste had stood some fifteen hours, it was mixed with its own volume of seawater and the thick liquid that resulted was set aside to allow the oil to rise to the top. In this way there was collected a crude water-and-oil emulsion which, after having been roughly filtered through sterile cheesecloth, was vigorously agitated and injected subcutaneously in appropriate amount into a dark dogfish. Very soon after the injection had been made, there commonly appeared on the skin of the dogfish and a little in front of the point of insertion a few small pale spots which, however, soon disappeared. As these spots were to be seen when small amounts of indifferent fluids were injected as controls, they were regarded as of purely operative significance. In from one to two days after the injection, relatively large pale areas made their appearance in the skin immediately over the region into which the fin extract had been introduced (Fig. 45). These large areas were very persistent and, as could be seen under a low power of the microscope, were produced

Fig. 45. Left side of the trunk of a *Mustelus canis*, showing a secondary pale spot due to an injection of 0·5 c.c. of an emulsion of olive oil extract of blanched fins and seawater made a little over a day previously. Parker, 1935b, 840.

by the concentration of melanophore pigment. That the pale skin included in these spots was essentially normal was demonstrated by the injection of pituitrin into a fish with such a spot. Shortly after the injection of this reagent had been made, particularly if the region of injection was close to the pale spot, it disappeared by the expansion of its melanophores, only to return after a few hours as the effect of the pituitrin wore off.

These large spots were not produced by injections of seawater, oil, oil extracts of dark fins or of muscle, seawater extracts of pale fins or defibrinated blood from pale or dark fishes (Fig. 46). They were produced by oil extracts, sterilized or not sterilized, and from cold ether extracts and Soxhlet ether extracts of pale fins from the same.

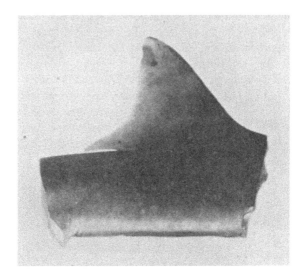

Fig. 46. Right side of the trunk of the same dogfish as that depicted in Fig. 45 showing no change of colour after the injection of 0·5 c.c. of an emulsion of olive oil and seawater. Parker, 1935*b*, 840.

Fig. 47. Dorsal view of a right pectoral fin of *Mustelus canis* in the process of darkening and showing a restricted and irregular pale band formed from a transverse cut. Parker, 1937*g*, 597.

These various tests led to the conclusion that the induced pale areas in *Mustelus* are due to the action of a lipohumour from the pale fins of this fish. The few known properties of this lipohumour, apart from its solubility in olive oil and in ether and its insolubility in water, are its resistance to dry heat up to 110° C., to treatment with 2 per cent sodium hydroxide and with 2 per cent hydrochloric acid (Parker, 1937 *b*). This lipohumour may be extracted from fins that have been

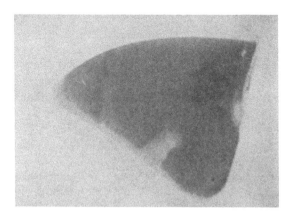

Fig. 48. Dorsal view of a left pectoral fin of *Mustelus canis* in the process of blanching and showing a full and regular pale band formed from a transverse cut. Parker, 1937*g*, 597.

kept in a dry state in the laboratory for a year (Parker, 1938 *b*). It must, therefore, be a relatively stable material. What it is, is unknown. It is certainly not adrenaline. It may be conveniently designated selachine (Parker, 1942*g*). Thus the dark phase of elasmobranchs is excited by the hydrohumour intermedine and the pale one either by another hydrohumour, the W-substance of Hogben, or by selachine. This is probably only the beginning of a list of elasmobranch neuro-humours. Parker (1937*a*) has pointed out two types of pale fin-bands in *Mustelus*, one with blurred edges (Fig. 47) and the other with relatively sharp edges (Fig. 48). These two types apparently depend upon the presence or absence of an antagonism in this fish between its hydrohumour, intermedine, and its lipohumour, selachine.

Chapter V

COLOUR CHANGES IN VERTEBRATES (*continued*)

4. *COLOUR CHANGES IN TELEOST FISHES*

T HE colour changes of the ordinary bony fishes did not escape the attention of the ancients. Thus Pliny in his *Natural History*, book IX, chapter 30, was led to remark that "masters in gastronomy inform us that the mullet while dying assumes a variety of colours and a succession of shades, and that the hue of the red scales growing paler and paler gradually changes, more especially if the fish is looked at enclosed in glass". Seneca, in book III, chapter 100, of his *Physical Investigations*, observed to the same effect that a "mullet even if just caught is thought little of unless it is allowed to die in the hands of your guest. Such fishes are carried about enclosed in globes of glass and their colour is watched as they die which is altered by the struggles of death into various shades and hues". Thus the chromatic changes of the mullet excited at least the social interests of the Romans.

It was not until 1830, however, that Stark noticed that several British fishes, *Leuciscus phoxinus*, *Gasterosteus aculeatus*, *Cobitus barbatula*, and *Perca fluviatilis*, as he named them, became dark on a dark background and pale on a light one, conditions which he described as protective in nature. According to Van Rynberk (1906), the chromatophores of teleosts were first observed independently by von Siebold and by Buchholz in 1863. Fuchs (1914), however, attributes their discovery to Vogt (1842), who in his study of the embryology of the salmon figures the chromatophores of this fish. The control of the colour cells in teleosts by nerves, which was the early opinion of the way in which these cells were made to contract and expand, was first fully established by Pouchet (1871–76) in his experimental work on turbots. This noted French investigator pointed out further that the particular part of the nervous system concerned with chromatophore control was the sympathetic or, as it would be designated to-day, the autonomic system. These conclusions were confirmed in the masterly studies on teleost colour changes by von Frisch (1910–12), who first clearly traced the courses of the chromatophoral nerve-fibres and located the regions in the medulla oblongata from which they emanated.

The idea that neurohumours had a part in the control of teleost colour changes was a gradual growth. Already in the work of von Frisch (1911c) there were observations on the chromatophores of denervated areas that were difficult to understand except on the assumption of humoral agencies. Hogben's suggestion that such changes were to be accounted for by the photo-receptivity of the regions of skin concerned was shown to be inapplicable by Scharrer (1928), who together with Giersberg (1930) inclined to an endocrinal interpretation of the facts. That the melanophores of fishes were really influenced by hormones was shown beyond a doubt by Spaeth (1913 b) in his experiments with adrenaline on the isolated scales of *Fundulus*. Thus was initiated the neurohumoral study of the colour changes in bony fishes, a study much of which has been based upon the technique of caudal bands as originally employed by Wyman (1924 a).

Teleosts as a group show great variations in the possibilities of their colour changes. Some are extremely limited in this respect; others are highly diversified. Thus the catfish *Ameiurus* ranges from a pale olive to an almost full black with no other additional tints (Bray, 1918; Pearson, 1930; Parker, 1934e; Matsushita, 1938). The squirrel fish *Holocentrus* has also a limited colour scheme between the two extremes of clear red and of white (Smith and Smith, 1935; Parker, 1937c). Although *Fundulus* usually exhibits a simple series of tints between steel-grey and pearl-white, it may also assume subordinate tones of yellow, red, green, and blue (Connolly, 1925). *Crenilabrus* has a range of pronounced colours including red, yellow, green, and blue (von Frisch, 1912c), and *Macropodus* shows responses to black, white, red, yellow, and blue (Dalton and Goodrich, 1937). Probably the most remarkable array of colour changes are those seen in the flatfishes where, beside the extremes of ashen grey and almost black, the fishes may assume blue, green, yellow, orange, pink or brown tints (Sumner, 1911; Mast, 1916; Hewer, 1927, 1931). These fishes can change not only their tints, but also their patterns, as was first clearly shown by Sumner and subsequently fully confirmed by Mast. In this respect, flatfishes exceed in all probability even the far-famed chameleon. Teleosts thus range from those with very simple and limited colour schemes to those which possess the most remarkable and diversified chromatic possibilities.

The times required for these changes in different teleosts are as various as are the changes themselves. In a recent publication, Parker (1943 d) has tabulated the majority of time-records thus far published for the colour changes in teleosts. They are reproduced in

Table 7 where the names of the fishes are in the order of time magnitudes beginning with the shortest.

As an inspection of Table 7 will show, some teleosts accomplish their colour changes with extreme rapidity, especially where these changes depend upon erythrophores. Thus, according to von Frisch

TABLE 7. Recorded instances of times for melanophore and erythrophore changes from pale to dark and dark to pale as shown by teleosts (Parker, 1943 d)

MELANOPHORES			
Animal species	Authors	Time	
		Pale to dark	Dark to pale
Mollienisia latipinna	Pierce, 1941 b	10–25 sec.	20–50 sec.
Fundulus heteroclitus	Pierce, 1941 a	45 sec.	1 min.
Fundulus heteroclitus	Parker and Brower, 1937	1 min.	2 min.
Salmo salar	Neill, 1940	30 min.	10 hr.
Lebistes reticulatus	Neill, 1940	35 min.	7 min.
Tautogolabrus adspersus	Hunter and Wasserman, 1941	50 min.	50 min.
Gasterosteus aculeatus	Hogben and Landgrebe, 1940	1 hr.	1 hr.
Ameiurus nebulosus	Parker, 1934 e	1 hr.	3·5 hr.
Parasilurus asotus	Matsushita, 1938	1 hr.	3·5–4 hr.
Lophopsetta maculata	Osborn, 1939 b	1 day	2 hr.–2 days
Phoxinus laevis	Healey, 1940	1–2 days	5–6 days
Paralichthys dentatus	Osborn, 1939 a	1–3 days	2–4 days
Pseudopleuronectes americanus	Osborn, 1939 b	3–5 days	3–7 days
Anguilla vulgaris	Neill, 1940	20 days	20 days
ERYTHROPHORES			
Prionotus strigatus	Lee, 1942 b (exc. pallor)	10–15 sec.	2–4 sec.
Holocentrus ascensionis	Smith and Smith, 1935	10 sec.	5 sec.
Holocentrus ascensionis	Parker, 1937 c	16·5– 22 sec.	4·5– 8·5 sec.

(1912 c), *Crenilabrus*, whose colour changes are chiefly in the reds and yellows, carries out these changes in the course of a few seconds. *Holocentrus*, also a red fish, is said by Smith and Smith (1935) to change from red to white in some five seconds and from white to red in about ten seconds. Parker (1937 c) recorded the red to white change in this fish as ranging from 4·5 to 8·5 seconds with an average from thirty trials on sixteen fishes of 6·38 seconds, and from white to red in from 16·5 to 22·0 seconds with an average of 18·7 seconds (19° C.), thus confirming in the main Smith and Smith's statements. These are astoundingly rapid changes as compared with those of hours or days as noted in the slower species. In *Fundulus*, the true colour changes require several days or even a week or more, but the changes from dark to pale and the reverse are affairs of a few minutes (Connolly, 1925). Parker and Brower (1937) found that *Fundulus heteroclitus* blanched in about two minutes (average 121·8 sec.) and darkened in a minute (average 60·0 sec.). According to the observations of Hill, Parkinson and Solandt (1935), made by means of a photoelectric cell, the period for blanching and that for darkening in this fish were found to be nearly equal. Cole and Schaeffer (1936, 1937), who also worked on *Fundulus heteroclitus*, state that blanching occurs in this species in a shorter time than darkening. Blanching, according to these workers, takes place 2·8 times more rapidly than darkening. The occasion of these discrepancies is not known.

Colour changes in other fishes are relatively slow. Thus *Ameiurus*, according to Parker (1934e), may require as much as 36 hours to blanch and 19 hours to darken, periods that on repetition of the responses are soon reduced to approximately three hours and one hour respectively (Parker, 1934e; Abramowitz, 1936c). These observations have been substantiated by Matsushita (1938) on the oriental catfish *Parasilurus*. *Phoxinus*, according to Healey (1940), changes from dark to pale in some five to six days and from pale to dark in from one to two days. These periods agree fairly well with the times recorded by Osborn (1939b) for the two flatfishes, *Paralichthys* and *Pseudopleuronectes*, studied by him. These intervals of days are in strong contrast with those of seconds shown by such a fish as *Mollienisia* (Pierce, 1941b). Considering this striking range of times, it is not surprising that Hogben (1924) should have felt called upon to direct the attention of investigators to this aspect of the colour problem.

The most lengthy interval ascribed to these changes for any of the fishes included in Table 7 is that of about 20 days recorded by Neill (1940) for both the darkening and the blanching of the European eel. These truly remarkable records stand in strong contrast with all others in Table 7. As the following discussion will show, they may not be in reality fully comparable with the other tabulated times. As reported by Neill, they were taken in water at a temperature of approximately 8° C. It is well known that at this degree of cold, melanophore activity falls off very considerably. Wykes (1938) states that at 6° C. the colour activities of *Ameiurus* are at almost full abeyance. It is, therefore, probable that the long time intervals for the colour changes in the eel as reported by Neill are due in part at least to the temperature at which his work was done. At 8° C. the responses of melanophores must be greatly slowed down as compared with what they are at 18° to 20° C.—the temperatures at which many of the other readings in Table 7 were made. It must also be confessed that when Neill's plottings of these changes in the eel are inspected (Fig. 49), one is tempted to surmise that he has in some way failed to distinguish between what students of this subject have been accustomed

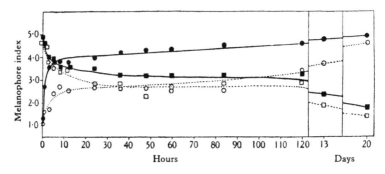

Fig. 49. European eel. Black-background and white-background responses in light. Temperature about 8° C. Melanophore index; 1, concentrated pigment; 5, dispersed pigment. Each curve represents a group of six animals. Dermal melanophores in black circles and squares. Epidermal melanophores in clear circles and squares. Neill, 1940, 79.

to call physiological and morphological colour changes. This surmise is aroused by the fact that each of Neill's curves illustrating this part of his work is made up of a short interval of rapid change (about 10 hours), which may well be the physiological change, followed by a very long one of very slow change (some 20 days), the morphological change. Such a separation in Neill's plots would bring his results more in harmony with what was originally given for the eel by Waring (1940) and by Waring and Landgrebe (1941), whose records cover what would ordinarily be regarded as the period of colour change. However, the difference in this respect between the work of Neill and of the two later investigators leaves the whole problem more or less obscured, notwithstanding the qualifying comments recently made by Waring (1942).

Hogben's insistence (1924) on the importance of the time intervals in colour changes with their possible relation to types of chromatophoral activation led Neill (1940) to the following specific statement: "When the total time taken for colour change exceeded two hours hormonal co-ordination is indicated. On the other hand, a total change time of the order of 10 minutes or less may be safely held consistent with predominantly nervous co-ordination, through the direct innervation of the melanophores themselves." This general statement was based by Neill upon a very meagre array of examples. When a reasonably full list of instances, such as is given in Table 7, is inspected, it is obvious that these do not fall from the standpoint of time intervals into two such classes, one slow and humoral and the other rapid and nervous, but that there are all gradations from the extremely slow to the extremely rapid with great diversity in the methods of activation (Parker, 1943 d). This leads to the conclusion that although humoral activation is usually associated with slow chromatic response and nervous with rapid reaction, there must be other important factors than the types of activation concerned with these temporal aspects of colour responses. One of these may well be the nature of the special chromatophore itself (Parker, 1943 d).

In concluding this section on teleost chromatophores, attention must be called to the fact that the concentration and dispersion of pigment in these colour-cells as worked out by the use of critical methods (Spaeth, 1916 c; Hill, Parkinson and Solandt, 1935; D. C. Smith, 1936 a) follow the course of a sigmoid curve. This has been discussed in Section D on the stages in the changes of chromatophores in the first chapter of this volume.

A. CHROMATOPHORES

The chromatophore system in the teleosts so far as variety and combination of colour-cells is concerned is a fair rival to that in the crustaceans (Becher, 1924; Schnakenbeck, 1925; Ballowitz, 1931). It includes melanophores with dark brown or black granular pigment, melanin, lipophores with yellow (xanthophores) or red (erythrophores) pigment soluble in alcohol, allophores with red or reddish brown pigment insoluble in alcohol, guanophores (leucophores or iridocytes) with granular or crystalline inclusions of guanine, and possibly still other types of chromatophores. Some tropical fishes possess blue skin pigment, though this seems to be a free pigment and not carried in colour-cells (Goodrich and Hedenburg, 1941). These various elements, which may occur singly or in a variety of combinations, add greatly to the complexity of the teleost colour system as a whole.

Melanophores are the most conspicuous and commonly studied type of chromatophore in the teleosts. They contain melanin in the form of either blackish or brownish granules. In the embryos of *Fundulus* the first melanophores, particularly those of the yolk sac, are brown; the later ones more intimately associated with the body of the growing fish are black (Stockard, 1915 b). Melanophores may assume a variety of forms depending upon the shapes of their processes; thus they may be coarsely stellate as in *Gobius*, or finely radiate as in *Caranx*, or many-branched as in *Gadus*. The expanded stellate melanophore of *Gobius* as figured and described by Ballowitz (1914 c, 1931) consists of a cell centre with over a dozen broad spatula-like blunt processes through which the pigment is scattered in roughly radiating lines (Fig. 50 A). There are commonly two eccentric nuclei situated in the roots of the processes. In the centre of the cell is the pigment-free centrosphere discovered by Solger in 1889. From this

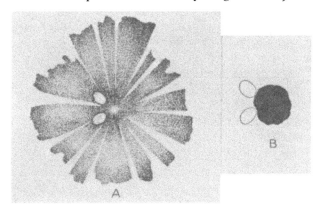

Fig. 50. Melanophores from the membranes in *Gobius*: *A*, pigment fully dispersed; *B*, pigment fully concentrated. Ballowitz, 1931, 507. *a*, *f*.

centrosphere radiate the lines of melanin granules that extend into and fill the cell processes. When the pigment granules begin to concentrate in the contracted state of the melanophore, they migrate centrally over apparently prescribed courses as though they were within radiating tubes. After full concentration is accomplished, the mass of pigment is closely packed around the centrosphere and the two nuclei are left undisturbed in what is now a transparent part of the chromatophore and outside the pigment area (Fig. 50 B). The dispersion of the pigment in the expanded state of the chromatophore is accomplished by an outward radial migration of the melanin granules again on rather prescribed lines. According to Franz (1939) these lines are determined by a gelatinous endoskeleton which forms, dissolves more or less, and reforms with the outward and inward migration of the pigment.

In the branched type of melanophore such as occurs in *Gadus* (Fig. 51), *Fundulus* and *Ameiurus*, the cell-processes stretch out like the roots of a tree, becoming more divided and usually smaller as they near their peripheral ends. These may be so narrow at times as to contain only a single line of melanin granules. In the outward and inward migration of these much irregularity is shown in that one or more granules may move now inward, now outward even though the pigment mass as a whole is moving toward central concentration or peripheral dispersion. All moving melanin particles exhibit Brownian motion. In the type of melanophore possessed by *Gadus*, *Fundulus* and the like there appear to be no prescribed courses for the pigment granules to follow, for they do not move in well-defined lines as they do in the radiating types of melanophores. As a rule, melanophores without a radial arrangement of their melanin particles show no evidence of a centrosphere. Nuclei usually eccentric are present. In the contracted state of the melanophore the pigment is concentrated about the centre of the cell. In the expanded state, it is uniformly dispersed throughout the whole extent of the cell.

Fig. 51. Melanophore from *Gadus*. Schnakenbeck, 1925, 243.

Fig. 52. Transverse section of the integument from the tail of *Ameiurus*, showing small melanophores in the epidermis and large ones below the derma. Parker, 1934*e*, 231.

The integumentary melanophores in teleosts may occur in any of three positions: immediately below the derma, in the derma but next to the epidermis, and in the epidermis. In *Ameiurus* (Parker, 1934*e*) and in *Parasilurus* (Matsushita, 1938), there are both dermal and epidermal melanophores (Fig. 52), but in *Fundulus* there are only dermal ones (Parker, 1935*c*). No teleosts appear to have exclusively epidermal melanophores. As a rule dermal melanophores are large, macro-melanophores, and epidermal ones small, micro-melanophores, to use terms proposed by Gordon (1927). A comparative study of these two types of dark cells has been made in the catfish by Enami (1939). All transitions between macro- and micro-melanophores have been seen in bastards from *Xiphophorus* and *Platypoecilus* by Breider and Seeliger (1938), who doubt the importance of this size distinction. In colour changes, dermal and subdermal, melanophores dominate (Schnakenbeck, 1925). A pathological and excessive growth of melanophores around encysted parasites in the skin of the cod has been described by Hsiao (1941). The occurrence and activities of these dark cells in the sarcomas and other pathological growths in teleosts have been described by Gordon (1937), Gordon and Smith (1938*a*, 1938*b*), Breider (1939*a*, 1939*b*), Grand, Gordon, and Cameron (1941), Gordon and Flathman (1943), Gordon and Lansing (1943) and others. The effect of radiations from a mercury lamp on the common types of goldfishes have been recorded by Goodrich and Trinkaus (1939) and the production of pigmentation in the skin of the goldfish by roentgen rays has been described by Ellinger (1940). The absence of melanophores from fishes, a relatively rare phenomenon, is the occasion of those strains known as albinotic such as occur in *Macropodus*, *Xiphophorus*, *Platypoecilus* and *Carassius* (Kosswig, 1935; Goodrich and Smith, 1937; Dalton and Goodrich, 1937).

Lipophores are chromatophores which contain a yellow or a red colouring matter, a lipochrome, soluble in alcohol, ether, and other like reagents. As already stated, they are known as xanthophores and erythrophores respectively. It is not always possible to distinguish these two kinds of colour cells one from the other. In fact they may be genetically related in that xanthophores may give rise to erythrophores. In many xanthophores and erythrophores there are special concentrations of pigment near the centres of the cells; these are known as xanthoms or erythroms. The significance of these bodies is quite obscure. In *Fundulus*, the xanthophores are more numerous than the melanophores, but because of the inconspicuousness of the yellow cells they play a much less important part in the colour changes of this fish than the dark cells (Fries, 1931). In the expanded state, according to Warren (1932), the yellowish orange pigment is scattered rather uniformly throughout the cell (Fig. 53). In the contracted state it is gathered into a small central sphere (Fig. 54). In some teleosts, such for instance as *Mullus*, the erythrophores have much the form of the stellate melanophores of *Gobius*, and the red particles in these erythrophores move in and out on relatively fixed lines much as melanin particles sometimes do (Ballowitz, 1913 *h*). In *Holocentrus*, according to Parker (1937 *c*), the erythrophores in their expanded state are small disc-shaped bodies with slightly irregular edges and in their contracted condition are smaller spheres about one-fourth the diameter of the discs (Fig. 55). In the sea-robin *Prionotus*, the erythrophores disperse their pigment on a white background and partly concentrate it on black, conditions almost the reverse of those of the melanophores in this fish (Lee, 1942 *b*).

Fig. 53. Xanthophore from *Fundulus* normally and fully expanded. The pigment, represented by coarse and fine stippling, is rather uniformly distributed throughout the cell. Warren, 1932, 634.

Allophores are chromatophores much like lipophores except that their red or reddish brown pigment is not soluble in alcohol. They were first clearly distinguished by Ballowitz (1913 *e*, 1917 *b*) in the Gobiids and Blennids.

The term guanophore has been employed in a variety of ways (D. C. Smith, 1933). It is here used to designate those chromatophores that contain guanine in one form or another. When the guanine is in

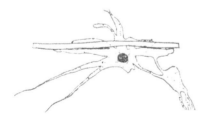

Fig. 54. Xanthophore from *Fundulus* with normally and fully concentrated pigment forming a central sphere. A small blood vessel passes over the xanthophore. Warren, 1932, 636.

Fig. 55. Outline of an expanded *Holocentrus* erythrophore within which is an outline crosshatched of its contracted state. Both outlines are drawn to the same scale. Parker, 1937 *c*, 207.

relatively small granules and is motile (Jost, 1926), the cells may be called leucophores. Such cells occur in *Fundulus heteroclitus* (Odiorne, 1933 *c*; Fries, 1942 *d*). When this material is in plate-like crystals and non-motile, the containing cell may be named an iridophore in conformity with the old term iridocyte employed by Pouchet (1876) for these bodies. The contents of iridophores are as a rule brilliantly scintillating when viewed with reflected light under a low power of the microscope, and have been shown to be birefringent (Shanes and Nigrelli, 1941 a, b). Iridophores give rise to much of the metallic iridescence seen on the flanks and backs of many fishes. Guanophores may combine to form a continuous sheet of silvery tissue, which, as the stratum argenteum, characterizes particular regions in the skins of many teleosts (Millot, 1922 a).

B. CHROMATOSOMES

The chromatic bodies in teleosts are not only unicellular chromatophores such as have just been described, but combinations of these chromatophores, thus giving rise to multicellular elements, veritable diminutive colour combinations, chromatosomes whose range of chromatic change is much beyond that of the unicellular type (Ballowitz, 1914 d, 1930). The most usual type of teleost chromatosome is the melaniridosome. In Trachinus, these bodies may consist of a group of iridophores in the midst of which is a melanophore (Ballowitz, 1914 d). On the expansion of the melanophore, the iridophores become covered and disappear from sight; on the contraction of the black cell, the iridophores shine out with full brilliance. In Fundulus (Foster, 1933, 1937), the iridophore is in the centre of the mass and contains a group of large guanine crystals. This is surrounded by the processes from the melanophores. When the black pigment of these processes retreats, it comes to lie almost completely under the iridophore, whose crystals on illumination may then shine with fascinating and jewel-like brilliancy (Fig. 56). When the pigment moves forward, it covers the iridophore almost completely, so that its scintillating crystals are nearly hidden from view (Fig. 57). These changes, which are wholly dependent upon the melanophore components, are under the control of the nervous system which, however, is not involved in the changes shown by the iridophore. These changes pertain to the iridophore crystals, which upon experimental illumination pass from a bluish green, through yellow and orange, to a wine red. In recovery, the crystals change in colour through the same series, but in the reverse order. In these changes, Foster observed no movement of the guanin crystals. He suggests that the very thin protoplasmic layers between the plate-like crystals may increase or decrease in thickness and thus call forth the colour changes by light interference. The change from blue to red takes place in approximately 5 seconds; that for recovery in from 40 seconds to 2 minutes. There is a refractory period of 15 to 20 minutes. The iridophores do not respond to heat, but when a preparation is warmed the total number of responding iridophores in the field is greater than otherwise. Adrenaline causes a concentration of melanophore pigment and a pinkish colour in the iridophores. Pituitrin is followed by no change in the melanophore component, but an orange or wine red colour in

Fig. 56. Melaniridosomes from *Fundulus* with melanin concentrated for the most part under the iridosome. Fish injected with adrenaline. Foster, 1937, 209.

Fig. 57. Melaniridosomes from *Fundulus* with melanin dispersed. These are the same colour organs as those shown in Fig. 56. Fish anaesthetized with ether. Foster, 1937, 209.

the crystals. These colours are probably influenced by neurohumours. The pale blue coloration seen at times in *Fundulus* as a whole is fully discussed by Foster.

In addition to melaniridosomes, the skins of teleosts may contain other groups of chromatophores so combined as to form colour bodies. Thus in the skin of *Hemichromis*, erythrophores are found combined with iridophores to form what Ballowitz (1917a) has called erythro-iridosomes. In *Gobius*, melanophores and erythrophores are known to combine to form black-red colour groups (Ballowitz, 1913a, 1913g) and in the same fish Ballowitz has described a union of a melanophore with xanthophores, and guanophores. Thus an almost endless variety of combinations may occur in what may be called the multicellular colour elements in the skins of teleosts.

C. Caudal Bands

As early as 1852, Brücke recorded the observation that when a cutaneous nerve in a chameleon is cut the melanophores of the skin thereby denervated disperse their pigment and the tint of the skin darkens. This is likewise true of fishes, as was shown by Pouchet (1876) in the turbots and Fukui (1923) in the file-fish. Von Frisch (1910) employed this technique in tracing the course of the autonomic nerve-fibres concerned with the colour changes in fishes. In from half a minute to a minute after the autonomic tracts have been cut, the area thus denervated begins to darken. In 3 minutes, the dark area is very obvious and in 5 to 10 minutes it has reached a maximum deep tint. After 8 or 9 hours, it may begin to lose its depth of colour. In *Phoxinus*, according to von Frisch (1911 c), denervated areas, at first fully dark, begin to lose in tint after 8 days and finally at 13 days they may be as pale as the rest of the fish. The final blanching of these darkened denervated areas which appears to have been first recognized by von Frisch was a source of concern to him, for he was unable to explain it.

Caudal bands are denervated areas of this kind formed in the tails of teleosts by cutting a bundle of radial nerves in this organ and thereby producing an elongated darkened area that will reach from the cut to the free edge of the fin. Such bands are extremely convenient in the study of the relation of nerves to chromatophores, for they enable the investigator to compare denervated and innervated colour-cells at close range, in fact one next the other. Preparations of this kind were used by Wyman (1924 a) in his study of the relation of blood and nerve as chromatophore excitants. When care is exercised in selecting the position of the initiating cut, a band may be formed without interfering in any serious way with the blood supply to the denervated region. The relation of a successful cut to the blood vessels of the tail has been clearly shown in a diagram (Fig. 58) by Fries (1931), who employed caudal bands in his study of the xanthophores in *Fundulus*.

When an appropriately placed transverse cut about 1 millimetre in length is made near the base of the tail in a pale *Fundulus*, a dark band will begin to appear in about half a minute and will grow in intensity of tint for some 5 minutes, after which it will maintain itself for some time. In 6 hours or so, the band will begin to fade and in about 2 to 3 days it will have become indistinguishable in the general pale field of the rest of the tail (Parker, 1935 c). If at this stage the fish is placed on a black background, it will darken except over the area of the caudal band which will stand out as a pale stripe in an otherwise dark tail (Parker, 1934 a). If such a fish is kept on a dark background for a day or so, the pale band at first easily discernible will darken and be lost to view. If a caudal band is induced in a dark *Fundulus*, it will

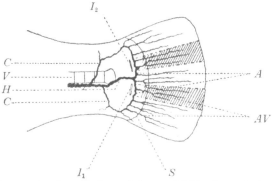

Fig. 58. Diagram showing relations of the caudal blood supply in *Fundulus* to incisions for denervating bands in the caudal fin. The parallel arterial and venous systems are shown as one. *A*, arc of main blood vessels; *AV*, blood vessels of fin; *C*, subsidiary connecting blood vessels; *H*, caudal blood vessel from haemal canal; I_1 and I_2, incisions through tail; *S*, distal limit of scales; *V*, vertebral column; shaded areas represent denervated caudal bands as a result of the two incisions. Fries, 1931, 400.

not be visible as long as the fish remains fully dark. If at any time the fish is made to blanch in consequence of its being put on a white background, the caudal band as a dark area will immediately stand out conspicuously, to disappear in course of time by gradual blanching. Thus, caudal bands after their initial formation may be said to follow the tint of the fish on which they are formed, but with some considerable lag in time.

Bands of this kind have been much used by recent workers in the field of fish chromatophores. They have been employed by Wyman (1924 a), Abolin (1925 b), D. C. Smith (1928), Parker (1931–38), Fries (1931), Mills (1932 d), Matthews (1933), Parker and Porter (1933), Abramowitz (1935–38), Kleinholz (1935), Foster (1937), Dalton and Goodrich (1937), Kamada (1937), Odiorne (1937), Wykes (1938),

Osborn (1938*b*, 1939*b*), Matsushita (1938), Tomita (1938), Vilter (1939*a*) and others. In compliment to the writer of this survey, Kamada (1937) has called the response shown in the formation of a caudal band the Parker effect. As this effect was first demonstrated by Brücke (1852) in his study of the colour changes in the chameleon, it would be more appropriate to designate it the Brücke effect.

D. INNERVATION OF CHROMATOPHORES

In the study of vertebrate chromatophores, it has been assumed from very early times that these colour-cells are under the control of the nervous system. So far as the teleosts are concerned, the work of two distinguished investigators, Pouchet (1872–76) and von Frisch (1910–12), demonstrated by experimental means the correctness of this view. The foundations for this opinion were laid by Brücke (1852), whose studies were made not on fishes but on the chameleon. Brücke aptly compared the melanophoral nerve and its effectors, the colour-cells, to a nerve-muscle preparation. According to Brücke when such a chromatic nerve acts, the melanophores contract as the skeletal muscles do under the influence of their motor nerves. When the chromatic nerve ceases to act, the colour-cells expand which is the equivalent of the relaxation of skeletal muscle. From this standpoint, melanophore contraction would represent the active state of the colour-cell and expansion the resting state. On stimulating melano-phore nerves electrically, as was done especially by von Frisch, the colour-cells were seen to contract precisely as muscle-cells do when their nerves are thus stimulated. On cutting a melanophore nerve, as is done in making a caudal band, the colour-cells expand just as a muscle relaxes when its motor nerve is cut. In the muscle, such a condition is due to paralysis, and this interpretation of dark patches or bands produced by cutting chromatic nerves has been accepted by a host of investigators from the time of Brücke to the present. Almost all the important workers of the past, including Pouchet, von Frisch, and Sand, have accepted this interpretation. This view of the nature of the caudal band and other like conditions has been designated by Parker (1936*f*) as the paralysis hypothesis.

The chief change that Brücke's view on the control of melano-phores has undergone of recent years came from Pouchet's discovery that the nerve-fibres that control chromatophores are autonomic and not cerebro-spinal as originally implied by Brücke. As cerebro-spinal nerves control skeletal or cross-striped muscle and autonomic fibres smooth muscle, it follows that the comparison of chromatophores with muscles is more truthful when the muscle chosen is not a cross-striped one but a smooth one. Smooth muscle, unlike skeletal muscle, is essentially a tonus muscle and chromatophores exhibit in their activities unusual degrees of tonicity; hence the appropriateness of a comparison of a chromatophore with a smooth-muscle cell. The details of this comparison so impressed Spaeth (1916a) that he was led to declare that a melanophore was a disguised type of smooth-muscle cell. Excepting this change from cross-striped muscle to smooth muscle, Brücke's original comparison of the neuro-chromatophore system in vertebrates with their neuro-muscular system remains intact to-day.

The chief departure from Brücke's rather conservative scheme for chromatophore activation was that proposed by Zoond and Eyre (1934) in their study of the South African chameleon, and reaffirmed by Zoond and Bokenham (1935) as well as by Sand in his general account (1935) of the colour responses in reptiles and fishes. The view proposed by these workers was to the effect that when a given chromatic fish or lizard is pale, it is so in consequence of a tonic contraction of its melanophores due to the action of autonomic pigmento-motor fibres. The paralysis that in the opinion of this group of investigators resulted when these fibres were cut was due they believed to the release of the melanophores from this tonus. On such a release these colour-cells would expand and thus the region of skin concerned would become dark. This expansion is believed to occur normally in animals in consequence of the inhibition of the ordinary tonic influences, an inhibition which could be called forth through either the direct illumination of the animal's dermal photo-receptors or the stimulation of its retina by light from a light-absorbing background. Stimulation of the retinal elements by light from a light scattering background was believed to cause an inhibition of the original inhibition, thus allowing the tonic state to reassert itself. As a result the melanophore pigment would become concentrated and the animal would blanch. In commenting on this scheme, Sand (1935) was led to remark that it was by no means simple and that certain parts of it, such as the inhibition of an inhibition, was theoretically cumbersome.

A third view of the way in which nerves may control chromatophores, counting Zoond and Eyre's departure from the older scheme of Brücke as a second one, is to the effect that chromatophores are supplied with two sets of nerves, one concerned with the concentration of pigment and the other with its dispersion. This view was first advanced by the French physiologist, Bert, in 1875 and has recently been revived by Parker (1932 a). According to this view, the concentrating and the dispersing nerve-fibres are antagonists and work one against the other. When for instance the eye of a fish is illuminated by light from a white background, the concentrating fibres are especially activated, the melanophores concentrate their pigment, and the fish blanches. When the eye is illuminated from a black surface, the dispersing fibres become active and the melanophores disperse their pigment, thus darkening the fish. Concentrating fibres are believed to be especially sensitive though not exclusively so to electric stimulation as compared with cutting. Dispersing fibres are more sensitive to cutting than to the electric stimulus. Thus cutting a bundle of melanophore nerve-fibres excites the dispersing elements greatly and the concentrating ones at most very little, and hence a dark band is formed. Ordinary faradic stimulation on the other hand excites especially the concentrating fibres and hence produces a pale band or area. The two sets of chromatophoral nerve-fibres thus set off one against the other are supposed to act in opposition, as the sympathetic and parasympathetic fibres are believed to do, on the heart muscle of the higher vertebrates.

The three views thus described concerning the ways in which chromatophores and especially melanophores in teleosts may be controlled differ as to the kinds of nerve-fibres involved and as to the way in which these fibres act. In attempting to reach a decision as to the validity of these views, it may be well at the outset to discuss the question of the kinds of chromatic nerve-fibres in bony fishes. According to the first and second of these views, only one kind of nerve-fibre is hypothesized. According to the third, two kinds should be present, one concentrating and the other dispersing.

That bony fishes possess concentrating chromatic nerve-fibres is perhaps one of the most generally accepted views in colour physiology. The evidence for this belief rests on the results of electric stimulation of nerves and nerve-tracts as practised by even the early investigators (Brücke, 1852; Pouchet, 1876). When appropriate nerves, nerve-tracts, or nerve-centres are stimulated electrically, the part of the animal tested, or under certain conditions the whole of the creature, blanches with great regularity. If previous to such a test, a part of the skin of the experimental animal is denervated, the blanching will in no way effect the part thus devoid of nerves. This general type of response has been recorded repeatedly in fishes and reptiles by numerous recent workers: in fishes by von Frisch (1911 c), Spaeth

(1913 b), Schaefer (1921), Wyman (1924), Parker (1935 c), Abramowitz (1936 b), Osborn (1938), Wykes (1938), Parker and Rosenblueth (1941), and Hunter and Wasserman (1941), and in reptiles by Redfield (1918), Hogben and Mirvisch (1928 a, 1928 b), Sand (1935), and Parker (1938 a). Since in this type of blanching the colour change never spreads into a denervated area, it cannot be due to a concentrating hormone carried in the blood but must result from nerves acting locally on the colour-cells. The evidence for this view is now so complete that no result in chromatic physiology is more generally accepted than the presence in many lower vertebrates and particularly in the teleosts of the concentrating type of nerve-fibre (Sand, 1935).

Evidence for dispersing chromatic nerve-fibres in bony fishes and other lower vertebrates has not been so easily obtained, nor has what has been discovered proved so generally convincing to workers in this field (Waring, 1942). Much of the evidence for dispersing fibres has come from work on caudal bands, which were discussed in a preceding section of this volume. The nature of caudal bands and other like areas so far as the present account is concerned can well be illustrated by what is to be seen in these bands in the killifish. If a caudal band is induced in the tail of a pale killifish, *Fundulus heteroclitus*, by cutting a bundle of radiating caudal nerves, and the fish is kept in a white, illuminated vessel, the band will fade in a few days (Parker, 1934 a). This kind of blanching in fishes was first pointed out by Pouchet (1876). It was also noticed by von Frisch (1911), but neither Pouchet nor von Frisch gave it any special consideration. If after such a caudal band has blanched a new transverse cut is made within the area of the old band and slightly distal to the original cut, a second

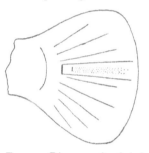

Fig. 59. Diagram of a faded band in the tail of *Fundulus* within which a new short cut has been made with the result that this cut has induced the formation of a new small band within the limits of the larger one. Such a new band shows that the severed dispersing nerve-fibres and their appended melanophores are not paralysed, but are fully active. Parker, 1934 f, 308.

band will appear reaching from the new cut over a part of the original band to the edge of the tail (Fig. 59). Such a revival of the greater part of the old band is a very significant and novel fact and leads to two important conclusions. The first is that that portion of the chromatic apparatus which has been severed from the central nervous system by the first cut is shown by the second cut to be not paralysed, as was claimed by Brücke, but capable of full, even of excessive activity. And the second conclusion is that since the original cut must have obstructed completely all central influences such as tonus and the like the revival of the band by the second cut must be due to a local stimulation of fibres in the severed nerve. Since under this kind of stimulation the melanophores of the band disperse their pigment, such fibres must be dispersing fibres. Hence the general conclusion has been reached from tests of this kind that the chromatic nerves of *Fundulus* contained dispersing as well as concentrating fibres; that is, the melanophores in this fish are dineuronic.

This interpretation of the work on the caudal bands of *Fundulus* has roused considerable opposition. It must be admitted, however, that since the attention of workers was called to the revival of faded caudal bands by recutting, this phenomenon has been recorded in a number of other fishes. It has been noted in the melanophore system of *Parasilurus* by Matsushita (1938), of *Pterophyllum* by Tomita (1938*a*, 1940), and of *Ameiurus* by Parker (1941*d*), and in the erythrophore and xanthophore systems of *Holocentrus* by Parker (1937*c*) and of *Gobius* and of two species of *Fundulus* by Fries (1942*a*, 1942*b*, 1943). The inability of Wykes (1938), of Osborn (1938*b*), and of Vilter (1938*a*, 1939*a*, 1939*b*) to revive by recutting the blanched bands in the teleosts studied by them appears to have been due either to the low temperatures at which they worked or the late application of their tests (Parker, 1941*a*), for the fishes which these investigators studied unsuccessfully have since yielded positive results in the hands of other workers. The revival of caudal bands and other similar dark areas in teleosts by the recutting of their nerves may therefore be said to rest on a sound observational basis.

Although this revival of faded caudal bands in teleosts is now admitted by many workers, the view that such a revival necessarily implies the presence of dispersing nerve-fibres has not proved generally acceptable. Several objections to this view have been raised by recent workers. It has been pointed out that if caudal bands are due to the activity of dispersing fibres, this activity must continue as long as the band lasts, often several days, a period much too long for the continued activity of a nerve excited by a single cut. But it is by no means certain that the nerve is so excited. It is much more probable that its continued activity, like that of nerves of pain, is due to continuous stimulation from the cut itself. If this is true, then a more or less regular flow of chromatic impulses must be passing over such nerves from the region where the impulses are believed to originate, the cut, to the colour-cells themselves. Is there ground for assuming that such a flow is taking place whereby the caudal band as a dark band is maintained?

The presence of such a flow of chromatic impulses can be tested by some form of nerve block. Such technique cannot well be carried out on short nerves, such as caudal nerves, by drugs or other like agents because of the diffusibility of these agents. It may be accomplished, however, by the application of cold. If a well localized cold-block is applied to a caudal band midway between the cut from which it originated and the distal end of the band, it ought to be possible in the course of time to obliterate the band distal to the block, leaving it undisturbed in its proximal extent, provided the band is maintained by chromatic impulses flowing from the cut. A cold block can be made by the application of a cold point at an appropriate position on the band with its contained chromatic nerve. For this purpose, a cold point was made from a sharply bent capillary glass-tube carrying a chilling mixture of alcohol and water from a reservoir cooled by dry ice to some 10° or more below zero centigrade. The cooling liquid, which because of its alcohol

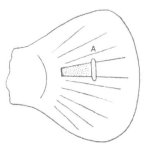

Fig. 60. Diagram of a fully formed band in the tail of *Fundulus* to which a cold block (*A*) has been applied with the result that the distal half of the band has gradually faded. Parker, 1934*f*, 309.

content did not freeze itself, passed through the cold point at a temperature some degrees below zero, as could be seen by the formation of a small amount of ice on the cold point itself, and emerged from the escape tube at a temperature of about 0° C. When this cold point was applied near the middle of a caudal band on a catfish's tail, an interesting phenomenon was to be observed. The band proximal to the region of the application of the cold point remained dark and unchanged; that distal to this region, however, faded out in the course of half an hour (Fig. 60). This is exactly what would be expected if the band were maintained by a continuous flow of impulses from the cut toward the free edge of the tail. Thus this test supports the view that dark caudal bands are preserved as such not by a single initial impulse from the cut but by a succession of such impulses which emerging from the cut pass down the nerve and keep up the dispersed condition of the pigment in the colour-cells. A cold block may be used in still another way to test this question. If the cold point is applied to a spot near the centre of the tail of *Fundulus* and a denervating cut is made some distance proximal to the region of application, a dark band will form from the cut to the region of the cold block, but will not pass beyond this block. If now a cut is made immediately distal to the block and in line with the first one, an additional band will form from the new cut to the edge of the tail (Fig. 61). From this experiment, two important conclusions may be drawn: first, that cold in the neighbourhood of 0° C. serves as an effective block to nerve impulses over chromatophoral fibres; and, second, that what is transmitted from the central organs over these fibres is not an inhibitory influence that is checked when the fibres are cut, but a true activating influence that has been excited locally by severing the nerve.

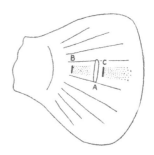

Fig. 61. Diagrams of the caudal fin of *Fundulus* across a part of which a capillary tube (*A*) carrying a chilling mixture has been placed. The cut (*B*) proximal to the tube has been followed by the formation of a dark band which reaches from the cut to the chilled area, but does not enter it. The cut (*C*) distal to the tube has been followed by the formation of a band which reaches from the cut to the edge of the fin. Parker, 1934*f*, 307.

When cuts such as would initiate a caudal band are made, they of course introduce a disturbance in the circulation. Could this be the occasion of such a dark band as has been intimated recently by several investigators (Hogben, 1924; Lundstrom and Bard, 1932; Young, 1933; Wykes, 1936; Vilter, 1939*b*)? Evidence against this view has been recently advanced by Kamada (1937) in his experiments on the melanophore reactions of *Macropodus*. It has also been pointed out by Fries (1931) that the position of the exciting cut may be so chosen that the larger blood vessels are left uninjured and the smaller ones interfered with in no serious manner. Moreover, inspection under a low power of the microscope shows that where these cuts have been carefully made, the capillary circulation in the band, the circulation that is most essential for the chromatophores, is to all appearances quite normal (Parker, 1938*c*). Were the circulation in any way obstructed by the cut and the blood-supply with its oxygen reduced in a given region, that region would be expected to blanch rather than to darken as a caudal band in teleosts always does. Parker has, therefore, concluded that caudal bands are not due to circulatory disturbances, but probably to stimulating materials that result from the injury at the cut and that act continuously upon the severed nerve-fibres. The disturbance appears not to originate from the mechanical abrasion of the surfaces of the wound, for a band will form from a window cut in the tail of a fish as well as from a simple slit itself (Fig. 62).

Although Spaeth and Barbour (1917) published results from a study of the responses of melanophores to ergotoxine and to adrenaline that were indicative of the double inner-vation of these colour-cells, it remained for Giersberg (1930) and for D. C. Smith (1931 b) to attempt a direct experimental attack on this problem. Giersberg found that under the influence of such drugs as ergotamine for sympathetic fibres and pilocarpine, choline, and other such re-agents for the parasympathetic, the re-sponses of the melanophores in *Phoxinus* are such as to give evidence for a double innervation. Similar conclusions were drawn by Smith from his study of the effects of autonomic drugs on the melano-phores of *Fundulus*.

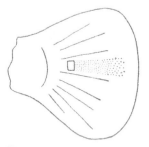

Fig. 62. Diagram of a tail of *Fundulus* in which a window has been cut, from the distal margin of which a dark band has formed. Parker, 1934 *f*, 308.

This problem was then attacked with a somewhat more incisive technique by Mills (1932 a, c), who studied the melanophores on the edges of the caudal bands in *Fundulus*. After these bands are well formed, it is easy to cause the fish to change colour by changing its backgrounds or stimulating the nerve tracts electrically. When the edge of a caudal band is closely observed with the fish first pale and then dark, the limits under the two conditions do not coincide exactly. Moreover, on this edge melanophores will be found that can expand but not fully contract and others that can contract but not fully expand. It is very difficult to explain these conditions on the assumption of a single set of nerve-fibres, but it is very easy to understand them if double innervation is admitted, for the margins of the band would be regions where through the accident of the cut certain melanophores would be deprived of concentrating fibres, but not of dispersing ones and vice versa. It appears to be necessary to assume double innervation to understand these otherwise peculiar states.

Another line of evidence also in favour of the idea of double in-nervation was developed by Abramowitz (1936 a) and is concerned with the regeneration of the melanophore nerves in *Fundulus*. When these nerves are cut so as to produce a caudal band, the portion of the nerve distal to the cut degenerates completely in about two weeks, after which regeneration begins and progresses at a rate of approximately half a millimetre a day (Parker and Porter, 1933; Abramowitz, 1935 b). When regeneration is well under way, the extent to which it has progressed can be seen on darkening a blanched fish by placing it on a black background. The caudal band will then darken as far as its nerve-fibres have regenerated; the rest of the band will remain pale. If now the distal area of the regenerating band is photographed under the dark and the pale conditions of the fish and the individual melanophores in this area are studied critically, four kinds may be identified. Some melanophores will be found that are capable of full contraction, but not of expansion; others with full expansion but incomplete contraction; still others with full expansion and full contraction; and finally some responding neither by expansion nor by contraction. Those responses which go to completion in one direction but not in the other are difficult, if not impossible, to explain except on the assumption that the concentrated phase of the melanophore is dependent upon one set of nerve-fibres and the dispersed phase upon another.

The evidence for the double innervation of the melanophore in *Fundulus* and in *Ameiurus* and for the erythrophores in *Holocentrus*, as presented in the preceding paragraphs, has been recently substantiated on other species of fishes by a number of important investigators in this field (Dalton and Goodrich, 1937; Tuge, 1937; Chang, Hsieh and Lu, 1939; Healey, 1940; Tomita, 1940; Gelei, 1942; Fries, 1942 a, 1942 b; Lee, 1942 b). The insufficiency of this evidence seems to have impressed few but Waring (1942). Since the recent work here referred to covers a considerable number of teleost species, it may well be that double innervation will prove to be the prevailing type of innervation among bony fishes. Whether the chromatophores of the chameleon are supplied with two sets of nerves, as was originally maintained by Bert (1875), is an open question and must await further investigation, but in the instances of the fishes already alluded to there appears to be no doubt that their chromatophores are dineuronic and that their caudal bands and other like dark areas are due to the excitation of dispersing nerve-fibres and not to paralysis, the blocking of central tonic impulses or other like influences.

That the degeneration and regeneration of chromatophoral nerves are processes not involved in the formation of caudal bands so far as the present discussion is concerned, is made clear by the work of Parker and Porter (1933) and of Abramowitz (1935 b) on these activities in *Fundulus*. Parker and Porter studied regeneration in the dispersing fibres of this fish by preparing caudal bands on a number of individuals and then from time to time exciting the newly formed fibres by cutting the nerve proximal to the original cut and determining the extent of regeneration beyond this cut by observing the length of band darkened by the second cut. It was found that at about 5 days after the first cut had been made, degeneration was well advanced. Regeneration began at about 18 days after the initial cut and was completed in about 25 days after that event. Thus about 7 days were consumed by the dispersing nerve-fibres in growing over some 6 mm. of distance. The rate of growth for these fibres was therefore approximately 0·86 mm. per day. As already stated, Abramowitz (1935 b), with a much improved method, carried out observations on both concentrating and dispersing nerve-fibres in *Fundulus* and in *Ameiurus*. Darkened fishes with dark, partly regenerated caudal bands were made to blanch from time to time, and the added length of the blanched band could thus be measured at regular intervals on the same individual fish. In a similar way, pale fishes with pale, partly regenerated bands could be darkened temporarily and the length of the regenerated part of the band then measured. In this way, it was found that both dispersing and concentrating nerve-fibres in *Fundulus* regenerated in the summer months at the average rate of some 0·55 mm. per day. Similar tests on *Ameiurus* showed that both sets of fibres in this fish regenerate at about the same rate as those in *Fundulus* did. In general then, this process appears to go on in these two fishes at about half a millimetre a day. None of these nerves showed evidence of degeneration before some 5 days after they had been cut. Since the formation of a dark caudal band, its blanching and its revival are operations all of which may be carried out fully before the fourth or fifth day after the initial cut, the results of these operations are plainly not dependent upon the destruction or reconstitution of the nerve, and must be regarded as in all respects independent of nerve degeneration and regeneration.

The final conclusion to be drawn from this extended discussion of the innervation of teleost chromatophores may be stated in the following words. Such teleost colour-cells, mainly melanophores and xanthophores, as have been adequately investigated, give indisputable evidence of double innervation. They are in other words dineuronic.

The evidence upon which this conclusion is based is not to be explained away by insufficiency of observations nor by adjacent maladjustments such as vasomotor or other circulation disturbances. Although this opinion has been opposed by some recent workers in animal colour changes as though it were a radical biological novelty, it is by no means such. The vertebrate heart is a good example of a muscle with two opposing sets of nerves. These nerves, moreover, like those to the teleost colour-cells, are both from the autonomic system. Those in the higher vertebrates have been designated sympathetic and parasympathetic, but it is by no means certain that the two chromatic elements in the teleosts can be so classed. At least for the present, it seems best simply to name them in accordance with their functions as concentrating and dispersing fibres and to leave the question of their relation to the two categories of autonomic nerves in the higher forms to future inquiry.

As has been intimated from time to time in the preceding account, the two kinds of chromatic nerve-fibres in teleosts differ in their ease of stimulation; concentrating fibres are easily stimulated electrically but not by cutting, and dispersing fibres are readily stimulated by cutting and wounding but not electrically. The common failure to stimulate dispersing fibres by means of the faradic current has been especially puzzling and led Parker and Rosenblueth (1941) to re-investigate this subject. By using an electric outfit in which the stimulating current could be appropriately varied, these workers found that dispersing fibres as well as concentrating ones were open to electric activation. When the chromatic nerves in the tail of the catfish *Ameiurus* were stimulated over periods of from 5 to 20 minutes at frequencies of 15 to 25 per second, with pulses lasting for 4 to 8 msec., and at approximately 8 volts, the band of innervated melanophores concerned blanched. This confirmed earlier work. When similar nerves were stimulated electrically over periods of 5 to 10 minutes, at frequencies of only 1 to 2 per second, with long pulses, 0·3 to 0·5 seconds, at 6 to 8 volts, the melanophores concerned darkened. Hence these workers concluded that dispersing as well as concentrating chromatic nerves in *Ameiurus* are open to electrical stimulation, a conclusion which brings this whole subject into line with what would be expected from general nerve physiology. Very probably concentrating nerve-fibres, like dispersing ones, may be stimulated by cutting, but this aspect of the subject has thus far not been investigated. These recent results in the electrical stimulation of chromatic nerves in teleosts support the conclusion that the melanophores in these fishes are dineuronic.

What is known of the histology of the chromatic nerves and their terminations in teleosts accords well with this view of double innervation. The work on this subject carried out by Eberth (1893), by Ballowitz (1893 a, b), and by Eberth and Bunge (1895) shows that each colour-cell in the teleosts studied by these workers normally receives branches from several nerve-fibres (Fig. 63), not simply from one, and that in consequence a histological basis is afforded for a dineuronic interpretation of these elements.

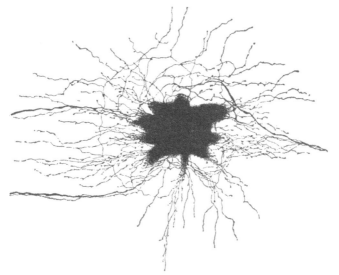

Fig. 63. Innervation of a chromatophore from a perch. Ballowitz, 1893 b, Pl. 38.

Before bringing this section to a close, an interesting peculiarity of the chromatic nerve-fibres in teleosts, particularly of the dispersing fibres for their melanophores, should be mentioned. If caudal bands in these fishes are produced through local activation by the cutting of dispersing nerve-fibres, these fibres might be expected to show central or antidromic activities as well as peripheral ones (Parker, 1936 c). A search for evidence of such activities showed that the region immediately proximal to the initiating cut is an unfavourable one for the appearance of antidromic responses in consequence of the continued local activity of the concentrating fibres. When, however, experimental steps are taken to prevent the interference of the concentrating fibres, antidromic responses in the form of the expansion of melanophores proximal to the cut can be readily demonstrated not only in *Fundulus* (Parker, 1937 h), but also in *Ameiurus* (Parker, 1937 f).

Such responses have also been identified by Tomita (1938 b) in the angelfish *Pterophyllum*. Although this type of response was sought for in the young of *Mustelus*, it was not found (Parker, 1938 b). Thus, at least in certain teleosts, dispersing melanophore nerve-fibres, like ordinary motor fibres, exhibit under appropriate conditions antidromic activities.

E. INTERMEDINE

The fact that teleosts possess both concentrating and dispersing chromatic nerve-fibres does not preclude the occurrence in them of other means of colour control such as hormones. Chief among these may well be intermedine. The term intermedine is here used to designate the chromatophore-activating principle contained in the intermediate lobe of the vertebrate pituitary gland. The effect of this principle on teleost melanophores was variously described by the early workers. Spaeth (1918), Wyman (1924 a) and Gilson (1926 b) stated that the pituitary extract produced a contraction of the melanophores in *Fundulus*. Abolin (1925 a, 1925 b), who worked on *Phoxinus*, was unable to confirm this statement and reported that in the minnow both melanophores and xanthophores expanded when subjected to this agent. Abolin also noted an expansion of chromatophores when the pituitary extract was injected into *Nemachilus*, *Esox*, *Carassius*, and *Leuciscus*. Przibram (1932 a), who worked upon amphibians, attempted to explain away these differences on the basis of dosage. He declared that in frogs a low dosage resulted in a dispersion of the pigment in melanophores and a high one in a concentration of colour in these cells. This view was not substantiated by Abramowitz (1937 c). It appears that attention must be given to the diversity of pituitary extracts, the production of which is far from uniform, and to possible differences in the chromatophores themselves even in the same fishes. Hewer (1926) found that the melanophores of *Phoxinus* contracted to an extract from the posterior lobe of the pituitary gland of the cod, and Meyer (1931) noted the same kind of reaction to pituitrin and hypophysin when used on *Gobius* and *Pleuronectes*. Matthews (1933) made the interesting observation that a pituitary extract from *Fundulus* would contract the melanophores in an isolated scale from this fish though this agent had no effect upon the colour of *Fundulus* when injected into the fish itself. Subsequent work showed that expansion of melanophores by pituitary extracts was the rule in *Ameiurus* (Odiorne, 1933 b; Parker, 1934 e;

Abramowitz, 1936 c; Veil, 1937, 1938 a; Osborn, 1938 a), in *Phoxinus*, *Gasterosteus*, *Rhodeus* (Osterhage, 1932) and in *Carassius* (Verne and Vilter, 1935). Xanthophores and erythrophores were also found to expand to such extracts. This was discovered to be true for *Gobius* and *Pleuronectes* by Meyer (1931) and for *Phoxinus* by Giersberg (1932), by Peczenik (1935), and by Lewis, Lee and Astwood (1937). The responses of *Fundulus* to pituitary extract and particularly to intermedine are somewhat peculiar. Desmond (1924) noted that the loss of the hypophysis was without effect on the melanophore reactions of young *Fundulus*, although a transplant of the pituitary gland from this fish into hypophysectomized axolotls or tadpoles darkened these animals. Matthews (1933) also discovered that a *Fundulus* without a pituitary gland could change dark or pale quite as a normal one did. Lee (1942 a) has very recently shown that pituitary secretion is not necessary for the colour changes in the toadfish *Opsanus*. Matthews, as already noted, also observed that the melanophores in isolated scales from *Fundulus* contract when the scales are immersed in an aqueous extract of the pituitary gland from this fish. Kleinholz (1935) made the interesting discovery that when a *Fundulus* with a blanched caudal band was injected with pituitary extract from another *Fundulus* the band darkened, showing that the melanophores of *Fundulus* when released from nervous control are open to the action of intermedine. Abramowitz (1937 c) was led to study from a quantitative standpoint the melanophore pituitary hormone in *Fundulus*. He found that the pituitary gland of a *Fundulus* 10 cm. long, when extracted with distilled water, yielded about four frog-units of intermedine. Such a gland on chemical treatment could be brought to yield some hundred such units. About a fifth to a tenth of this amount will darken a pale caudal band in *Fundulus*. Blood from a single dark *Fundulus* when extracted and chemically treated also yielded enough intermedine to darken a pale band. Thus the blood of this fish carries in itself enough potential intermedine to be physiologically significant in colour changes. Whether in the normal colour responses of *Fundulus* it is thus significant, as believed by Abramowitz, is still to be shown.

From these rather confused records covering the relations of pituitary extracts to chromatophores and particularly to melanophores, two conclusions seem to emerge: first, that the pituitary extracts used by many of the workers, particularly the earlier ones, were probably far from pure and doubtless contained accidental inclusions of no small significance in colour changes; and, secondly, the black pigment cells of teleosts, though included under the general name melanophore as though they formed a physiologically homogeneous group, are probably far from uniform in their reactions. The irresponsiveness of *Fundulus* melanophores, as compared with the reactiveness of those in *Ameiurus*, to what appears to be a reasonably pure form of inter-medine is indicative of a decided physiological difference between the dark pigment cells in these two fishes.

To the early workers on chromatophoral pituitary extract, it was soon evident that this activator could be obtained from the pituitary complex of any vertebrate from fishes to mammals (Hogben, 1924). This lack of specificity has been many times recorded by later investigators. Even extracts from what is regarded in tunicates as the homologue of the vertebrate pituitary complex has yielded not only an oxytocic principle (Butcher 1929, 1930) but also a melanophore-expanding one as tested on frogs (Bacq and Florkin, 1935a, 1935b). As already noticed, the same general non-specificity has been recorded again and again for the eye-stalk extracts from crustaceans (Koller, 1928; Perkins and Kropp, 1932a, 1932b, 1933; Abramowitz, 1936b, 1936e, 1937d; Hanstrom, 1936, 1937a, 1937b; Abramowitz and Abramowitz, 1938). Not only is there lack of specificity within phyla, but this also occurs between phyla. Thus Koller and Meyer (1930) and later Meyer (1931) showed that the crustacean eye-stalk extract would activate the chromatophores of flatfishes, an observation that was soon extended to other chromatic vertebrates (Perkins and Kropp, 1932a, 1932b, 1933; Kropp and Perkins, 1933b; Abramowitz, 1936e). Conversely, vertebrate intermedine was shown to act upon crustacean chromatophores much as eye-stalk extract does (Abramo-witz, 1936b, 1937d; Abramowitz and Abramowitz, 1938). In con-sequence of these reciprocal interphylar activities, a certain similarity, perhaps chemical in nature, may be assumed to exist between vertebrate intermedine and the crustacean eye-stalk principle, and the chromatophoral reactions of animals even as distantly related as those in these two groups, may be more intimate than their systematic positions would lead one to expect.

A method for making strong extracts of intermedine has recently been published by Landgrebe, Reid and Waring (1943). However, very little is known chemically about this substance. It is evidently one of the components of what Hogben and his associates have called the B-substance. It is soluble in water and is carried in the blood and lymph from the pituitary gland, where it is formed, to the chromato-phores, where it acts. The blood of a dark fish when injected into a pale one will ordinarily induce the formation of a temporary dark spot, but such blood will not call forth a colour change in a dark fish. Intermedine is a typical hydrohumour (Parker, 1935 d) and must play an important part in the normal darkening of many teleosts. It is a reasonably stable material, for it can be identified in the blood of hypophysectomized fishes some 3 days after the removal of their pituitary glands (Osborn, 1938 b). Van Dyke (1943) describes it as "the melanosome-dispersing or chromatophorotropic hormone of the pars intermedia. Probably this hormone is a polypeptid. So far as we know the survival of cells secreting intermedine in the mammalian pituitary represents an atavic heirloom since intermedine is principally of interest as a hormone dispersing pigment granules in cells of certain fishes, amphibia and reptiles." This is probably an inadequate statement, for the fact that intermedine is generously produced in almost all chordates from the tunicates to man is indicative of other functions than the purely chromatic one known in the cold-blooded vertebrates. What these functions may be is yet to be discovered. That intermedine has served to ameliorate the human ailment piebald skin or vitiligo is indicative of functional significance even in the human being. It is probably far from an atavic heirloom.

It must be evident from what has been stated in this section on intermedine and in the preceding one on the innervation of teleost chromatophores that some of these fishes, for instance the catfish *Ameiurus*, possess two means for darkening: dispersing nerve-fibres and the pituitary gland. It may well be asked how these two mechanisms with the same function are related. When the pituitary gland is removed from a catfish, it has left only one means of darkening, the dispersing nerve-fibres. Such fishes when put in a black, illuminated environment will darken only to about half the extent of a normal fish. In a fully pale catfish, the black pigment massed in the macro-melanophores have a diameter of about 45 microns; in fully dark fishes these coloured areas have diameters of some 145 microns. In the hypophysectomized fishes, the areas of melanin have diameters of approximately 100 microns. As already stated, the melanophores in this condition appear to be about half expanded. This is the maximum effect in this fish of activation by dispersing nerves. If into such a catfish thus brought to an intermediate tint some pituitary extract is injected, the fish will darken gradually to its full deep coloration; that is, the pigmented areas of its melanophores will change in diameter from about 100 to 145 microns. The remarkable limitation to darkening by nerves, first noticed by Abramowitz (1936 c) and by Veil (1937), brings out one striking difference between dispersing nerves and pituitary extracts as colour-cell activators.

Another difference between these two means of dispersing melanophore pigment in the catfish is seen in the rapidity with which the two agents act. If a pale catfish with a blanched caudal band is allowed to darken naturally in a black, illuminated aquarium, it will be seen in an hour or so to be wholly dark except the caudal band. Somewhat after an hour, this band will begin to darken till in a brief time the band is as dark as the rest of the tail. As the caudal band is denervated and does not darken in the beginning, the initial darkening of the rest of the animal must be due to the dispersing nerves. Thus the deep tint must be initiated by the nerves. The darkening of the caudal band, which follows in an hour or so the general darkening of the tail, cannot be due to nerves, for the band is denervated. It must result from pituitary intermedine and from now on the fish must continue to increase its deep tint by the combined action of both dispersing nerves and intermedine. If the analysis of this general operation is correct, it follows that of the two darkening agents in *Ameiurus* the colour change must start through the action of the dispersing nerve-fibres, to be followed an hour or so later by the intermedine from the pituitary gland. Which of these two agents maintains the dark condition of the fish, which under appropriate conditions may last indefinitely, cannot at present be stated. Very likely both are involved. What, however, is certain is that in the darkening of *Ameiurus*, nerves are effective before the pituitary secretion comes into play, but are less complete in their action than is the pituitary agent (Parker, 1941 c). How these several relations in the colour changes of the catfish are harmonized cannot at present be stated.

F. ADRENALINE

The preciseness and uniformity of the responses of teleost chromatophores to adrenaline is in strong contrast with the uncertainty of their reactions to pituitary extracts. In 1916, Spaeth pointed out that the melanophores in the scales of *Fundulus* would contract to a solution of adrenaline (epinephrine) one part in a million. Subsequently, Barbour and Spaeth (1917) determined the limit of effectiveness of this solution to be one part in fifty millions. They also pointed out that after a scale had been treated with ergotoxine, its melanophores would expand to adrenaline. The contracting action of this agent on melanophores was also observed on trout (Gianferrari, 1922), on

Fundulus (Wyman; 1924 *a*, Abramowitz, 1936 *d*; Bogdanovitch, 1937 *a*, 1938), on *Phoxinus* (Abolin, 1925 *a*, *b*; Giersberg, 1930; D. C. Smith, 1931 *a*; Osterhage, 1932), on *Carassius* (Fukui, 1927; Beauvallet, 1934, 1938; Verne and Vilter, 1935), on flatfishes (Hewer, 1927; Meyer, 1931), and on *Ameiurus* (Bacq, 1933 *a*). As a blanching agent, adrenaline appears to be remarkably uniform and consistent in the colour responses that it calls forth in teleosts. Bogdanovitch (1937 *b*) pointed out that this reagent was subject to destruction by the tissues of the scales in which the melanophores were naturally imbedded. Abolin added the interesting fact that in *Phoxinus*, though the melanophores were contracted by adrenaline, the erythrophores failed to react in any way to this agent.

Gilson (1922), who confirmed Spaeth's original observation that adrenaline contracted the integumentary melanophores in *Fundulus*, showed that this agent caused a dispersion of melanin in the retinal pigment cells of this fish. Thus adrenaline when injected into a given fish would cause a contraction of its integumentary pigment and an expansion of its retinal melanin. These observations by Gilson confirmed what Fujita (1912) and Bigney (1919) had recorded and were in opposition to the statements of Klett (1908) on this subject. Parker (1934 *b*) pointed out that in Spaeth's original test of the action of adrenaline on the melanophores in the scales of *Fundulus*, the scales as prepared by Spaeth carried with them live nerve endings as well as melanophores and that, therefore, it was impossible to decide which of these parts, the nerve or the colour-cell, was acted on by the reagent. Parker, therefore, prepared caudal bands in the tail of *Fundulus*, and after the nerves had fully degenerated and the fishes were darkened by being kept on a black background, he injected adrenaline into them. The innervated melanophores in general as well as the denervated ones in the caudal band contracted, showing that this agent could act directly on such colour-cells. The same was found to be true of the melanophores of *Ameiurus*. Parker, therefore, sided with Lieben (1906) and Wyman (1924 *a*) in the belief that melanophores may be stimulated directly by adrenaline. and opposed Fuchs (1914), Spaeth and Barbour (1917), Abolin (1925 *a*) and Giersberg (1930), who held that this agent excites the melanophores only indirectly through their nerves.

Veil (1938 *b*) found that the melanophores in the scales of the carp could be made to contract to adrenaline reduced to millionths in dilution. When this extreme efficiency of the adrenaline was set off against that of intermedine, it was estimated that the mildly active intermedine solution must be increased about five thousand fold to counteract the adrenaline. Veil (1936 b) had previously been led to conclude, from a study of the melanophores in the scales of teleosts, that the concentrating melanophore nerves in these fishes produced a substance at their terminals which was adrenaline-like in character.

As early as 1911 von Frisch noticed that dark denervated areas in *Phoxinus* would blanch quickly, following excitation of the fish. This excitement pallor, as it has since been called, has commonly been looked upon as due to the sudden discharge into the blood in the fish of an amount of adrenaline sufficient to blanch the creature. Bray (1918) recorded that when the catfish *Ameiurus* was raised to a high pitch of nervous excitement, it became blanched and remained so for a considerable time. Abramowitz (1936 c) also often noticed that catfishes paled when they were disturbed by attempts to capture them. Dark caudal bands in these animals would blanch under such circumstances and would remain in this condition several hours after the fish had again darkened. Similar effects were evoked by the electrical stimulation of the roof of the mouth, of the medulla oblongata, or the anterior end of the spinal cord. Some evidence was found by Abramowitz to the effect that the blood of such stimulated animals carried a substance that would blanch other dark individuals. Although this evidence is not finally conclusive, it points to the possibility of adrenaline or some like substance as a means of teleost blanching (Osborn, 1939 b; Waring, 1940; Healey, 1940).

Adrenaline is a secondary alcohol which was synthesized over two decades ago. Its molecules are relatively small. It may well represent the W-substance of Hogben and be derived from such sources as the pars tuberalis of the pituitary complex (Hogben and Slome, 1931, 1936; Hogben, 1936). It can act both as a hydrohumour and a lipohumour (Parker, 1940 c). Whether Cannon's sympathine is to be placed among the chromatic neurohumours of vertebrates, as Vilter (1937 d) and Chang and Lu (1939) have done with it, is an open question. Adrenaline as a neurohumour is derived from two general sources, the medulla of the adrenal gland with other less defined chromaffine masses and the concentrating chromatic nerve-fibres. It is not impossible that the substance here described as adrenaline from the chromatic fibres is sympathine, but this question must be left for further investigation.

G. ACETYLCHOLINE

Acetylcholine, because of its well-known importance as an activator of both neurones and effectors, has long been suspected of being an agent in the excitation of melanophores. However, with this agent, as with intermedine, the tests of the earlier workers gave little uniformity. Barbour and Spaeth (1917) observed no response to it from the melanophores of *Fundulus*, even when it was used one part in a thousand. Negative results were also obtained by Wunder (1931), who employed it on *Rhodeus*, and by Smith and Smith (1934), who tested it on the melanophores of *Scorpaena*. Parker (1931 b) suspected it of producing a slight contraction in the melanophores of *Fundulus*, and Beauvallet (1938) substantiated this view on the goldfish. Bogdanovitch (1938) declared in favour of contraction in *Fundulus*. Parker (1934g), who modified his technique by the preliminary use of physostigmin, found in a second set of tests on *Fundulus* evidence of a slight expansion rather than a contraction. Chin (1939), who worked on the paradise fish, *Macropodus*, obtained a dark area on injecting acetylcholine into this creature and believed that he could identify this substance in extracts of the caudal fin of *Macropodus*. Chang, Hsieh and Lu (1939) accepted these results and presented evidence to show that this agent is to be found in the chromatic nerve-fibres of the snakefish *Ophiocephalus*.

It now appears after extended trials with this activator that it disperses melanophore pigment and thereby darkens fishes (Mendes, 1942). Acetylcholine when injected into a fish is rapidly destroyed by the cholinesterase of the animal's tissues. To prevent this destruction, it is necessary to prepare a chromatic vertebrate for the reception of acetylcholine by a preliminary injection of eserine. In the common catfish, *Ameiurus*, one part of eserine by weight may be injected for every 200,000 parts by weight of fish. In a series of catfishes thus prepared, each individual received 0·2 c.c. of solutions of acetylcholine of different concentrations from 1 part of the reagent in 100 parts of Ringer's solution to 1 part in 100,000,000,000 parts of solvent. The weakest of these mixtures had no effect on the colour of the fishes. All others darkened them and the strongest two eventually brought on death to them. The satisfactory range of the solutions for physiological tests was from 1 part in 10,000 to 1 part in 1,000,000,000. The darkening that resulted from these various concentrations usually required from 20 minutes to half an hour to reach its maximum (Parker, 1940c).

Acetylcholine is a choline derivative the functional efficiency of which is enormously increased by acetylation. Its molecules are relatively small and readily inactivated by the cholinesterase of animal tissues. It is soluble both in water and in oils and can therefore act either as a hydrohumour or as a lipohumour. When dissolved in the lipoids of the tissues, it is protected for a considerable period from destruction by cholinesterase (Parker, 1940c).

H. Extracts, Drugs, and Chemicals

In addition to intermedine, adrenaline, and acetylcholine, many other like agents are capable of exciting teleost chromatophores. Wyman (1924a) reported the contraction of the melanophores in *Fundulus* to extracts made from the pineal, thymus, and thyroid glands, and Osterhage (1932) observed a similar response of the melanophores in *Phoxinus*, *Gasterosteus*, and *Rhodeus* to ovoglandol, testiopton, and testiglandol. The dark cells in *Cyprinodon* and *Gambusia* are said by Ciabatti (1929) to contract to ephedrine, ephetonine and sympatol as they do to epinephrine.

Veratrine, which may contract melanophores at first, has been reported by Lowe (1917) and by Wyman (1924a, b) as an expanding agent. Atropine has been universally described as a means of dispersing pigment not only for melanophores (Barbour and Spaeth, 1917; Wyman, 1924a, b; Abolin, 1926; Osterhage, 1932; Beauvallet, 1934; Bogdanovitch, 1937a) but also for erythrophores (Smith and Smith, 1934). This agent is stated by Beauvallet (1938) to counteract the concentrating effect of adrenaline. Pilocarpine acts also as a dispersing agent for melanophores (Barbour and Spaeth, 1917; Kahn, 1922; Abolin, 1926; Bogdanovitch, 1937a) and for erythrophores (Smith and Smith, 1934). Beauvallet (1934) has recorded it as a concentrating agent for the melanophores of *Carassius*. Phenylethylamine, *p*-oxyphenylethylamine and indoethylamine are stated by Barbour and Spaeth (1917) to contract the melanophores of *Fundulus*. These melanophores also contract to ergotoxine after which adrenaline is said to have a contracting effect upon them (Spaeth and Barbour, 1917). Picrotoxine will induce contraction in trout melanophores (Lowe, 1917) and in the dark cells of *Phoxinus* (Abolin, 1926), whereupon the latter expand. The xanthophores and erythrophores in this fish are said to be uninfluenced by picrotoxine.

Curare induces a mixed melanophore response in trout (Lowe, 1917), but expansion in *Fundulus* (Wyman, 1924a) and in *Phoxinus*

(Abolin, 1926). Nicotine is stated by Lowe (1917) to bring about chromatophore contraction, a result at variance with that of the majority of later workers all of whom declare for expansion (Wyman, 1924a, b; Abolin, 1926; Smith and Smith, 1934; Verne and Vilter, 1935). Quinine, morphine, strychnine, and cocaine are claimed by Lowe (1917) to be concentrating agents for the melanophores of the trout and by Wyman (1924a, b) and Abolin (1926) to be dispersing agents for those of *Fundulus* and *Phoxinus*. Cocaine concentrates the erythrophores of Scorpaena (Smith and Smith, 1934).

Santonin and caffeine expand the melanophores of *Fundulus*, according to Wyman (1924a, b), and physostigmine has the same effect upon the melanophores of *Phoxinus* (Abolin, 1926) and the erythrophores of *Scorpaena* (Smith and Smith, 1934). According to Bogdanovitch (1937a), this agent induces the melanophores of *Fundulus* to contract. Tyrosine and yohimbine bring about expansion ·in the dark cells of *Phoxinus*, *Gasterosteus* and *Rhodeus* (Osterhage, 1932; Astwood and Geschickter, 1936), as well as of other fishes (Beauvallet and Veil, 1936), but guanidine induces contraction in these cells in *Carassius* (Beauvallet and Veil, 1932). Ergotamine is claimed by Bacq (1933a) to expand the innervated melanophores in *Ameiurus* and to contract the denervated ones, but Parker (1941d) has shown that it acts only on innervated dark colour-cells by inducing pigment dispersion and has no effect on denervated cells, a conclusion substantiated by Mendes (1942). In *Carassius*, ergotamine is said to reverse the action of adrenaline (Verne and Vilter, 1935).

Ethyl alcohol is an irregularly expanding agent for the melanophores of *Fundulus* (Wyman, 1924a, b) and a more uniform one for those of Phoxinus (Abolin, 1926). The erythrophores in *Scorpaena* respond to it by contraction (Smith and Smith, 1934). Methyl, ethyl, and propyl alcohols are inactive in very dilute solutions (Lowe, 1917). Ether, chloretone, novocaine, and stovaine all call forth expansion of melanophores in *Fundulus* (Wyman, 1924a, b), probably through their action on central nervous parts. Chloroform is said by Wyman (1924a) to expand the melanophores of *Fundulus*, and by Beauvallet and Veil (1932) to contract those of *Carassius*. Deuterium oxide is reported by Bogdanovitch (1937b) to be a contracting agent for the melanophores of *Fundulus*. Whether or not this response is associated with the absence of oxygen, an element necessary for the expansion of melanophores (Lowe, 1917), cannot be stated. It is difficult to point to important general conclusions that can be drawn from the body of detail contained in this section. Many of the apparently isolated responses herein recorded may be of great significance when they are viewed in connection with the problem of the chemical nature of those little known neurohumours by which teleost chromatophores are normally incited to action.

I. TELEOST NEUROHUMOURS

Teleosts such as *Ameiurus* darken in consequence of the expanding action on their melanophores of dispersing autonomic nerve-fibres and of intermedine. They blanch because of the concentrating action on these colour-cells of concentrating nerve-fibres and possibly of some adrenaline-like hormone. Of these two sets of opposing agents, the conditions which activate one probably inhibit the other. Within one set, as for instance that concerned with the dispersion of pigment, the nerves probably act first and their activity is then supplemented by that of the associated neurohumour, in this instance intermedine.. Of these two, intermedine is the more potent, as has been pointed out by Veil (1937) and by Osborn (1938 b). Osborn measured the diameters of melanophores in their fully contracted and fully expanded states in *Ameiurus*, and found that at full contraction the mass of pigment in these cells had an average diameter of 45 to 50 microns and at full expansion one of about 130 microns. After hypophysectomy, the cells under the action of the expanding nerve-fibres reached diameters of only 70 to 75 microns, showing that nervous expansion was much less effective than that due to intermedine.

Other hormones than intermedine and adrenaline-like materials may be concerned with these colour changes. Thus flavine, recently found in the skins of fishes by Fontaine and Busnel (1938), has been suggested as a possible chemical mediator for melanophores. But these matters are purely speculative. In *Fundulus*, the chromatophore responses appear to be almost strictly nervous, for the part played by intermedine must be at best very subordinate. In fact, though the evidence is very fragmentary, it appears that the control of the teleost chromatophores is chiefly nervous.

The nature of this nervous control, however, is not so remote from that of humoral control as was at first supposed. This idea was already intimated as early as 1930 by Giersberg who, in discussing the condition in *Phoxinus*, remarked "Wenn auch wohl für die Melano-phoren die 'nervöse' Regulation allgemein gelten dürfte, ist diese bei den gelben und roten Farbzellen nicht durchgängig der Fall. Die Ellritze gibt uns im Gegensatz zu anderen Fischarten ein Beispiel des hormonalen Farbwechsels der farbigen Chromatophoren. Nervöse und hormonale Farbwechselregulationen kommen also beide bei den Fischen vor, damit schwindet aber ein wesentlicher Unterschied, den man zwischen Fischen und Amphibien z. B. anzunehmen geneigt war. Die Erscheinungen des raschen Farbwechsels dürfte wenigstens bei den Wirbeltieren keine prinzipiellen Gegensätze aufweisen, sondern in allen drei Klassen auf der doppelten Reaktionsbasis der nervösen und hormonalen Farbwechselregulationen sich aufbauen.'' Much the same idea was independently espoused by Parker in a lecture delivered in Cambridge, England, in 1930 and published in 1932. In this publication, Parker put forward a theory of melanophore control which, according to Mills (1932b, d), provided a simple explanation for the behaviour of both innervated and denervated cells. He suggested that the normal nervous control of chromatophores might be induced by hormones secreted from the terminals of chromatophore nerves in such close proximity to colour-cells as to cause them to react. A gradual spread of these substances into denervated areas would explain the response of such cells to changes in the shade of the background. It may be supposed further that the stimulation of melanophore-concentrating nerves would excite the production of one substance, whereas the stimulation of the nerves for dispersion would produce another antagonizing material causing the opposite effect. Thus the two sets of nerve terminals would be the sources of different hormones capable of exciting in one case dispersion and in the other concentration of melanophore pigment.

The blanching of caudal bands in *Fundulus* was studied by Mills to ascertain whether there was reason to assume that such substances were present. A careful inspection of a blanching caudal band in *Fundulus* showed that the responses of the melanophores in such a band following a change in the shade of the fish as a whole does not take place simultaneously. Such responses start among the melanophores on the periphery of the denervated band and spread gradually to those at its axis. This indicates that the secretions affecting the melanophores are not produced at a distant source and carried in the blood stream to the band as a whole, but that they are formed in the neighbouring tissues and reach the denervated cells by gradual lateral spread. Reactions of denervated melanophores when the surrounding innervated cells responded to artificial stimulation also occurred in tails isolated from the body by a cut posterior to the anal fin. Hence such changes could not be due to any secretions from glands anterior to the cut and distributed by the blood stream. This type of blanching in the caudal bands of *Fundulus* has been followed by Parker (1935*c*) photographically. Fifteen minutes after such a band has been formed, the melanophores across its whole width are very uniformly expanded (Fig. 64). Nine and a quarter hours after the

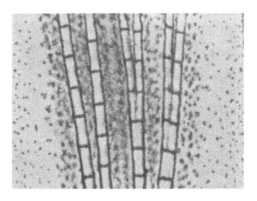

Fig. 64. Photograph of a caudal band in a living *Fundulus* 15 minutes after it had first appeared. All the melanophores in the band show fully dispersed pigment. Parker, 1935*c*, Pl. 3.

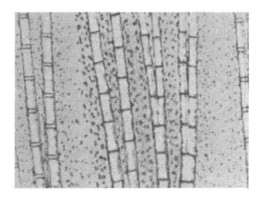

Fig. 65. Photograph of the same spot in the caudal band of the living *Fundulus* as that shown in Fig. 64, 9¼ hours after the band had first appeared. The peripheral melanophores are contracted as compared with the axial ones. Parker, 1935*c*, Pl. 3.

first formation of the band, the peripheral melanophores are somewhat contracted, but the axial ones remain almost fully expanded (Fig. 65). Fifty-four hours after the initial cut, all melanophores in the band, axial as well as peripheral, are very fully contracted (Fig. 66). This kind of peripheral blanching in caudal bands has also been demonstrated very fully by Matsushita (1938) in *Parasilurus*. Thus the steps by which a dark caudal band in a pale fish fades favour the view that this process is brought about through a lateral invasion of the band by a substance from the adjacent pale fields.

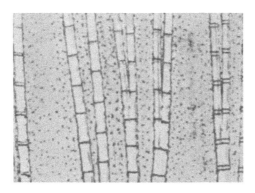

Fig. 66. Photograph of the same spot in the caudal band of the living *Fundulus* shown in Fig. 64, 54 hours after the band had first appeared. All the melanophores, axial as well as peripheral, are about fully contracted. Parker, 1935 c, Pl. 3.

Other evidence bearing on this problem is to be seen in the time taken for caudal bands of different widths to blanch. According to Parker (1934 a), a band about 1 mm. wide in the tail of *Fundulus* will blanch in from 22 to 96 hours with an average on the basis of twenty-five cases of 29·7 hours. A band 2 mm. wide requires from 43 to 143 hours to blanch with an average based on fifteen cases of 78·3 hours. Thus the broader the band, the longer the period for blanching, a condition suggestive of lateral diffusion of a concentrating humour as the occasion of the change. Tests of a like kind but on *Ameiurus* were made by Abramowitz (1936 c), who showed that caudal bands, one ray in width, blanched in 2 days, bands of two rays in 5 days, and of three rays in 10 days, thus confirming in essentials Parker's results on *Fundulus*. In *Macropodus*, Dalton and Goodrich (1937) showed that at a temperature of 20° C. bands blanched laterally in 18 hours, but at 29° C. in only eight hours, a result confirmed by a number of other tests. These results, to the effect that bands blanch from the edges more rapidly the higher the temperature, support the view that this type of blanching is dependent upon lateral diffusion.

Still another test as to the source of the blanching material was made by Parker (1934 d) on *Ameiurus*. If a faded caudal band in this catfish is flanked over half its length by two newly excited half-bands (Fig. 67), that part of the original faded band which lies between these two new half-bands will darken slowly whereas the part of the original band not flanked by the dark half-bands will remain pale. This experiment shows quite clearly how adjacent dark areas may bring about a dispersion of pigment in a pale area, a change easily understood on the assumption of an invading dispersing neurohumour.

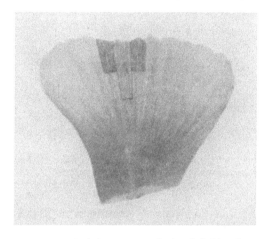

Fig. 67. Drawing of the tail of *Ameiurus* in which a faded band is partly flanked by two newly formed, dark half-bands. The faded band has darkened between the two flanking half-bands, but has remained pale in its proximal portion where it is not flanked by the dark half-bands. Parker, 1934d, Pl. 2.

The experiment just described has been reported by Dalton and Goodrich (1937) on *Macropodus* and by Matsushita (1938) on *Parasilurus* with confirmatory results. From these several types of tests, it was concluded by Parker (1936 a) that in teleosts the terminals of the dispersing as well as the concentrating nerve-fibres secrete special neurohumours which in one instance excite the melanophores to expand and in the other to contract. Thus the nervous control of teleost melanophores is as truly neurohumoral as is the hormonal control, in that both activities are dependent upon secreted activating materials. This unification is regarded by Parker as a significant step in the understanding of chromatophoral activation (1936c).

It has been repeatedly pointed out by Parker (1932b, 1932c, 1933a, 1935g, 1936c) that there is much evidence in favour of the view that the neurohumours produced by the melanophore nerves are not soluble in water and are therefore not transported by the blood and lymph. They do move, however, from place to place, and Parker has suggested that this is accomplished through the lipoid components of the cells; in other words, that these neurohumours are soluble in oily or fatty materials through which they may spread. On this assumption, attempts were made to extract such substances from the skins of teleosts in appropriate conditions. To this end, the dark phase of *Ameiurus* is favourable. Extracts of the skins and fins of dark catfishes were made by means of olive oil, and the resulting residue was emulsified and injected subcutaneously into pale catfishes. This operation was followed in a little less than an hour by the formation of dark splotches on these fishes (Fig. 68). Such splotches, which were

FIG. 68 A catfish, *Ameiurus*, into which an injection of olive oil extract of the dark fins and skins of five other catfishes had been made anteriorly from a point below the adipose fin. The resulting dark area is superficial to the region where the injected fluid escaped from the needle point Parker, 1935 e, Pl. 1.

caused by the dispersion of pigment in the melanophores of the region concerned, disappeared spontaneously after a few days. When they were first formed, they could be temporarily obliterated by an injection of adrenaline. Thus there is good reason to believe that in *Ameiurus* there is a dispersing chromatic neurohumour that can be extracted by oil and can darken pale individuals. That there is also a concentrating neurohumour of a similar kind in *Ameiurus* is probable but not proved (Parker, 1935 e, 1940 c), as will be shown in the latter part of this section.

The chromatophoral activation thus far discussed pertains almost exclusively to the melanophores in teleosts. These fishes, however, possess in many instances other kinds of chromatophores, such as xanthophores, erythrophores, leucophores and the like. How are these chromatophores controlled? Are they dependent upon the same neurohumours as the melanophores, and do they in consequence act in unison with these black cells, or have they independent systems of neurohumours of their own? These questions are mainly for future settlement in that the evidence now at hand is relatively scanty.

Of the teleost chromatophores other than melanophores, more is known of the xanthophores than of any other group of colour-cells. Xanthophores have been surmised to react to environmental changes in precisely the opposite way to that in which melanophores do. Such, however, is not the case, as the studies of Connolly (1925) and especially of Fries (1931) have shown. Thus, according to these two workers, both xanthophores and melanophores are contracted in a *Fundulus* which has been kept some time on an illuminated white background, and both are expanded in one that has been on an illuminated black background. On blue, however, the melanophores are mainly expanded and the xanthophores contracted, and on yellow the xanthophores are expanded and the melanophores contracted. For tests of this kind, *Fundulus majalis*, the mayfish, is much more favourable than is *F. heteroclitus*, the species studied by Connolly and by Fries. Consequently, Abramowitz (1936 d) carried out an extended series of tests on the more advantageous form. Caudal bands were cut in a large number of mayfishes, which were then put some over white backgrounds and others over black ones. After 8 days, pale fishes on white backgrounds had lost the yellow colour in their caudal bands and those on black were uniformly dark. On inspection under the microscope, both kinds of chromatophores in the bands on the pale fish were found to be contracted and both in those on the dark fishes fully expanded. The pale fishes were now distributed among blue, yellow, and black dishes and the dark fishes among white, yellow, and blue dishes. Two days later, the bands were again examined microscopically. The fishes on blue were then divided among black, yellow, and white vessels; those on yellow among white, black, and blue; those on black among white, yellow, and blue, etc. Thus fishes adapted to one background were transferred to other backgrounds in all possible combinations. The conditions of the melanophores and the xanthophores, both innervated and denervated, were examined before and after these changes, two to three days being allowed between observations. The results are shown in Table 8. They indicate in a most unequivocal manner the independence in action of the two sets of chromatophores, melanophores and xantho-phores, and the consequent separateness of their neurohumours.

A similar conclusion has been recently reached by Fries for these colour-cells in several European fishes (1942 c) as well as for the leucophores in the American *Fundulus* (1942 d). Whether other possible sets of chromatophores in teleosts possess independent systems of neurohumours is quite unknown, though the work of Dalton and

TABLE 8. Responses of innervated and denervated caudal chromatophores to various backgrounds—*Fundulus majalis* (Abramowitz, 1936 d, p. 376). Symbols: Mel., melanophores; Xan., xanthophores; E, fully expanded; mE, mainly expanded; C, fully contracted; mC, mainly contracted

Animals adapted to the following backgrounds	Conditions of the chromatophores				Animals changed to the following backgrounds	Conditions of the chromatophores			
	Inner-vated		Dener-vated			Inner-vated		Dener-vated	
	Mel.	Xan.	Mel.	Xan.		Mel.	Xan.	Mel.	Xan.
1 White	C	C	C	C	Yellow	C	mE	mC	E
2 White	C	C	C	C	Black	E	E	E	E
3 White	C	C	C	C	Blue	mE	C	mE	C
4 Black	E	E	E	E	Yellow	mC	E	mC	E
5 Black	E	E	E	E	Blue	mE	C	mE	C
6 Black	E	E	E	E	White	C	C	C	C
7 Blue	mE	C	mE	C	Black	E	E	E	E
8 Blue	E	C	E	C	White	C	C	C	C
9 Blue	mE	C	mE	C	Yellow	C	mE	mC	E
10 Yellow	C	mE	C	E	White	C	C	C	C
11 Yellow	mC	E	mC	E	Blue	mE	C	mE	C
12 Yellow	C	E	C	E	Black	E	E	E	E

Goodrich (1937) indicates this for some of the colour-cells in *Macropodus*. So far as neurohumours are concerned, the situation in teleosts is one that gives promise of increasing numbers of activators rather than the reverse. If the condition thus far worked out for melanophores and xanthophores can be taken as a sample, the number of chromatic neurohumours in teleosts will be multiplied several times when the whole field of these activators in teleosts has been exhaustively studied.

The nature of the concentrating and the dispersing neurohumours in teleosts has only very recently attracted attention. It has already been stated that both the concentrating and the dispersing nerve-fibres in these fishes belong to the autonomic system. This system also supplies the vertebrate heart with two sets of nerve components, the vagus fibres which slow the heart-beat and the sympathetic branches which hasten it. The first of these influences the heart muscle through acetylcholine and the second through adrenaline. Is there any ground for supposing that these two neurohumours are associated in like opposition one with the dispersion and the other with the concentration of teleost pigment? In attacking this question, it may be asked, Do these neurohumours exist under appropriate circumstances in the skins of chromatic vertebrates, for instance, teleosts? A reliable specific test for acetylcholine is found in the response given by leech muscle to this substance. Appropriately prepared extracts of dark skin from the catfish *Ameiurus* were therefore applied to leech muscle against standard solutions of acetylcholine. By this means, it was found that on the average 0·078 gamma of acetylcholine was contained in each gram of wet, dark, catfish skin or roughly one part of acetylcholine in 13,000,000 parts of skin (Parker, 1940c).

While this work was in progress, Chang, Hsieh and Lu (1939) published a determination for acetylcholine by the same method applied to the skin of the Chinese snakefish, *Ophiocephalus*, in which the amount of acetylcholine was shown to be 0·077 gamma per gramme of skin. Although the close agreement of these two determinations must be in a measure accidental, they are confirmatory one of the other and show at least the order of magnitude of the amount of acetylcholine in dark fish skin. They furthermore indicate how enormously efficient this neurohumour is. It is not without interest to note that the amount of acetylcholine found in the skin of the catfish would result from the injection into a 50 g. fish of 0·2 c.c. of a mixture of 1 part of acetylcholine in 52,000 parts of solvent, a concentration within the limits of the physiologically effective mixtures already used; namely 1 part in 10,000 and 1 part in 1,000,000,000 (Parker, 1941a).

Although acetylcholine has thus been demonstrated in the dark skins of snakefishes and of catfishes in amounts that will darken them, it has not been shown definitely to be a product of the dispersing integumentary nerve-fibres. Nor has it been shown that this neurohumour is absent from pale skin as might be expected. This step, though attempted, has not been carried out with success for the reason that the removal of a pale skin involves the cutting of all its integumentary nerves, which renders the skin dark before it can be used in a test and thus precludes the possibility of a final determination (Parker, 1940c). These two important deficiencies must be made good by further research. However, so far as such preliminary tests go, they show that acetylcholine is present in the skin of dark fishes and in amounts that are known to be effective in dispersing melanophore pigment. This substance, therefore, may well be the neurohumour liberated by dispersing nerve-fibres.

.A search for adrenaline in the skin of pale catfishes was made by the specific use of an organic indicator, the Straub frog heart. Freshly removed catfish skins, which had of course momentarily darkened, were mixed with horse serum to remove such acetylcholine as had been formed, and ground to a pulp with sand. The pulp was then diluted, dialysed, and the dialysate was divided into two parts. One of these was kept as it was. The other was boiled to destroy the adrenaline and the two samples were then tested on the prepared frog's heart. To the unboiled sample, the heart responded by a typical lengthening of its stroke. To the boiled one, there was only a very slight response. The strong response was believed to be due to adrenaline; the slight one to some inorganic heart activators that had passed the dialysing membrane. As a specific test for adrenaline, a frog's heart previously treated with ergotamine was used. Under these conditions, the heart responded positively. It was, therefore, concluded that the pale skin of the catfish contained adrenaline, though to what amount could not be stated for the method used was not quantitative. With these tests for adrenaline as with those for acetylcholine the evidence is insufficient, but that adrenaline is present in the pale skin seems certain, though that this neurohumour is a product of the concentrating nerve-fibres has not been proved (Parker, 1940c).

From these studies it may be concluded that the teleost *Ameiurus* darkens through the action of the pituitary neurohumour, intermedine, on its melanophores and probably through that of the dispersing neurohumour, acetylcholine, from the appropriate autonomic nerve-fibres. It blanches in all probability through the action of the neurohumour, adrenaline, from concentrating autonomic fibres. Acetylcholine is very probably the agent to which is to be attributed the lateral darkening of blanched caudal bands. It is, in other words, a good example of what has been called a lipohumour, and doubtless spreads through tissues by way of the contained lipoids. When it is injected into the watery fluids of such a fish as *Ameiurus*, it is so quickly destroyed by the contained cholinesterase that it disappears before it can produce a noticeable effect upon the melanophores of this fish. In using this agent experimentally, it is necessary as a preliminary step to inject the fish with eserine so as to remove the cholinesterase and thus prepare the test animal for the satisfactory reception of the acetylcholine. Under these circumstances, the fish when injected with acetylcholine will darken. Acetylcholine is readily soluble in olive oil. If a 0·5 per cent solution of it in this oil is agitated with a small amount of Ringer's solution and as such is injected under the skin of a pale catfish, in about two hours dark spots will begin to appear immediately over the region of injection. These spots, as can be seen under a microscope, are due to the dispersion of pigment in the melanophores. They will remain clearly visible for some four or five hours, after which they will disappear. They are indistinguishable in all respects from the spots obtained by injecting an oil extract of dark catfish skin into a pale fish (Parker, 1934e), and are a perfect imitation of such spots. The life of these spots, far beyond that of other types of darkening produced by acetylcholine, suggests that oil may serve as a protective reservoir for such a delicate agent as that here used, and that from such centres the active agent may make its way to the neighbouring melanophores and affect them characteristically before it is destroyed. In a similar way, the activity of adrenaline may be much prolonged by an olive-oil vehicle. The protective action of oils and fats, of lipoids in general, is probably a very significant feature not only among the lipohumours of the chromatic system, but among many other similar activating agents in animals from their egg stages to their adult conditions (Parker, 1940c).

Such are some of the more important means of neurohumoral action in teleosts. They depend upon the three neurohumours, intermedine, acetylcholine, and adrenaline. Additional chromatic neurohumours may well be expected, but whatever these may prove to be it is fair to conclude that the three already mentioned will remain the chief ones and that the others will be at best subsidiary agents.

J. ORGANIZATION OF CHROMATIC SYSTEMS

A survey of the organization of chromatic systems in cold-blooded vertebrates, particularly in teleosts, may well begin with the eye. This organ has been known from the time of Lister (1858) and even earlier to be the one primarily concerned with the initiation of colour changes. The eye in fact is the receptor *par excellence* for all such changes in chromatic animals. It is also generally recognized that the effectors by which colour changes are brought about, the various types of melanophores, xanthophores, erythrophores, and the like, act with great independence of one another. It is conceivable that this independence comes about either by specific adjustments in the vertebrate central nervous apparatus or by specializations within the most peripheral receptors, the retinas. Hence arises the question, Are the retinas of chromatic vertebrates differentiated in such a way as to determine at the outset the particular chromatic response to a given situation or is this determination relegated to a deeper part of the nervous mechanism in these animals? During the last decade this problem has excited the attention of a number of investigators.

The problem of the differentiation of the vertebrate retina for colour changes has been worked upon experimentally by Sumner (1933 a), who studied the colour responses of *Fundulus parvipennis* after the eyes had been covered more or less with blinders. The blinders used in this work consisted of celloidin caps or false corneas so made as to fit over the eyes of the fish and painted in part or whole with opaque india ink. Such a false cornea could be slipped over an eyeball and would remain in place for some hours or even a day or more. By means of these corneas, light could be excluded from one or other half or the whole of the eye, and under such circumstances the colour changes of the fish could be tested. When both eyes were covered completely with the opaque blinders, an intermediate tint was induced in the fish such as appears after the animal has been kept in full darkness. When the lower half of the field of vision was darkened, the entire dorsal surface of the fish became dark even on a white background. When the upper half was darkened, no visible change was noted in a fish on a white background, but a fish so treated was commonly paler than a control on a dark-grey background. Sumner concluded that the shade which a fish assumes on a given background is determined by the relative intensity of illumination in the upper and in the lower portions of the visual field, the upper acting positively and the lower negatively.

In their study of colour changes in the South African toad, *Xenopus*, Hogben and Slome (1936) were led to conclude that the changes in tint were incited through two neurohumours, a dispersing agent, B-substance, from the intermediate lobe of the pituitary complex and a concentrating one, W-substance, from the pars tuberalis of the same organ. How are these two agents controlled from the eyes of this animal? The eyes of *Xenopus* are on the top of its head and are directed upward. Hence these toads must be approached from above and tests were made by surrounding them overhead with white or black walls illuminated from below. When only the side walls of the container were white, it was assumed that the retina of the eye was excited peripherally; when only the top of the container was white, it was believed that the floor of the retina was excited. When both sides and top of the container were white the whole retina was taken to be activated and when both were black the retina was supposed to be inactive. The results of such tests are summarized in Table 9.

From the records given in Table 9 as well as from others in which tests with monochromatic light were used, Hogben and Slome concluded that the white and black background responses of *Xenopus* depend upon distinct localized retinal areas for each type of activation.

TABLE 9. Melanophore indices, where 5 stands for full dispersion of pigment and 1 for full concentration, as a result of excitation by reflected light from a white surface directed on to the floor of the retina of *Xenopus* or on to the periphery of such a retina (Hogben and Slome, 1936, p. 164)

Floor of retina	Periphery of retina	Melanophore index
Excited	Inactive	4·5
Excited	Excited	1·8
Inactive	Inactive	3·0
Inactive	Excited	1·3

An important body of evidence on the differentiation of the vertebrate retina in connection with chromatic reactions comes from the work of Butcher and his associate, Adelmann. Butcher and Adelmann (1937) tested dorsal and ventral parts of the retinas in *Fundulus heteroclitus* as chromatic activators by using incomplete blinders attached to the eyes much as Sumner (1933 a) had done, and by rotating the eyes as practised by Vilter (1937 a, c). When, by means of a half-shield, light was limited to the dorsal part of the retina, the fish was found to blanch. When light was admitted to the ventral half, but excluded from the dorsal portion, the fish darkened. These results were confirmed on fishes whose eyes had been rotated through 180° and then appropriately illuminated. Thus Butcher and Adelmann were led to conclude that the dorsal and the ventral portions of the retinas in *Fundulus heteroclitus* are physiologically different. It is to be remarked, however, that, though their general conclusion that the two parts of the retina in *Fundulus* are functionally unlike agrees with that of Sumner, the particular observations upon which this conclusion is based are almost the reverse of those given by Sumner. Sumner states that on darkening the ventral half-retina (dorsal half illuminated), the fish darkens, while Butcher and Adelmann declare that on illuminating a dorsal half-retina (ventral half dark), the fish blanches.

The histological structure of the retina in *Fundulus heteroclitus* in relation to its conditions of illumination has been studied by Butcher

(1937 a, b, 1938 a, b), who has shown that the dorsal 70 per cent of this organ contains rods, and single and double cones, the ventral 30 per cent rods and only double cones. The ventral region also contains a crescentic area with more rods and double cones than any other part of the eye. These structural differences characterize the two parts of the retina in *Fundulus*. When light from a yellow background impinges on the dorsal part of this retina, the xanthophores of the fish expand, but when it falls on the ventral part, this reaction does not occur. As the dorsal part of the retina is the only area of this organ in which single cones are present, Butcher has been led to suspect that these elements are the receptors for yellow light. Butcher has tested further the differentiation of the regions of the retina in *Fundulus* by restraining the fish in a tube which could then be inverted or righted in a given environment. He has also attempted to destroy the dorsal or ventral halves of the retina by cutting through the coats of the eyeball either dorsal or ventral to the optic nerve. In this way the part of the retina opposite that which was cut into was left intact. Tests with fishes under these various conditions have confirmed Butcher in the opinion that the two portions of the retina, dorsal and ventral, are physiologically different and that the dorsal region is concerned with adaptation to light backgrounds and the ventral with the darkening of the fish.

In a subsequent communication (1939), Butcher discussed the significance of the ratio of direct light to reflected light in its bearing on the colour changes of *Fundulus*. This fish was tested on grey backgrounds of different depths of tint illuminated from an overhead source of direct light. When the reflected light was twenty or more fiftieths of the direct light, the fishes were always pale. When it was ten-fiftieths, only about half of the fishes were pale and at three-fiftieths all fishes were grey or intermediate. On increasing the illumination from below to more than three and eliminating that from above, the fishes immediately paled. On restoring the illumination from above so that the relation of direct to reflected light was fifty to ten, the fishes quickly assumed an intermediate shade. Butcher concluded from these observations that the shade assumed by the fishes depends upon the ratio between the direct and reflected light because illumination of one region of the eye inhibits melanophoric response caused by illumination of the other regions of the eye.

Danielson (1939, 1941) studied contrasts in the visual fields of two fishes, *Nocomis* and *Semotilus*, by illumination from below and above the containing aquaria. By changing the ratio of the illumination of the dorsal and the ventral retinas, Danielson confirmed the growing belief that when this ratio is large, as is the case of a fish on a black background, pigment dispersion follows, and when the ratio is small, as happens on a white background, concentration of pigment is the result. So far as differentiation of the retinas is concerned, Danielson concludes that "although all parts of the eye are not equivalent the state of the melanophores appears to be determined not by the stimulation of particular regions of the retina but by the degree of contrast in the visual field as a whole".

Hogben and Landgrebe (1940), in a general investigation of the colour responses of the stickleback, *Gasterosteus*, took up the question of specialization in its retinas. By appropriate illumination, these investigators were led to conclude that the black background response was initiated by the activities of the floor of the retina, or what in this survey has been called the ventral retina, and that the white background response started from the centre of the retina. Thus they supported the idea of retinal differentiation and confirmed fairly closely the results of other investigators. Sperry's work (1943) on the effects of rotation of the eyes of the newt *Triturus*, though largely concerned with motor responses, contains a brief statement on the animal's chromatics to the effect that after the rotation of the eyes through 180°, "the animal's colour tends to match the overhead lighting rather than that of the bottom", and Sperry concludes by favouring the view of a "dorsal-ventral differentiation of retinal function".

These several contributions to the question of retinal specialization for chromatic responses have been written with so little textual relation to one another that comparisons are scarcely possible. Enough, however, has been accumulated by workers in this field to point with considerable probability to the conclusion that the retinas in chromatic animals possess a certain degree of differentiation associated with the chromatophoral activities of the animals concerned. Apparently excessive illumination of the ventral retina with little or no light on the dorsal area is the condition that induces dispersion of melanophore pigment, and illumination of both halves of the retina, but especially the dorsal one, results in the concentration of melanophore pigment.

From the eyes as initiating receptors, systems of control in the chromatic organization of the lower vertebrates must spread by various courses to the colour effectors. These systems afford a relatively novel topic for discussion. Animal colour changes as described in most modern surveys are stated to be accomplished through two agencies, nerves and hormones. This situation has been fully recognized by Vilter, who in a series of papers covering the last decade has developed a plan of chromatic organization for vertebrates that is based on an antagonism between the dark colour phase and the pale one as seen on the surfaces of such animals. These two phases, according to Vilter, are polarized dorso-ventrally; the dark phase is at its maximum on the dorsum of the creature, whence it becomes gradually reduced as it passes ventrally, whereas the pale phase is at its maximum ventrally to become attenuated step by step as the dorsum is approached. A fish in which both phases are moderately accentuated would then be dark on the back and pale below. The dark phase, according to Vilter's view, is excited by the pituitary chromatic hormone, the pale one by the sympathetic nerves. Thus the pituitary hormone is especially active when the creature is on a black background, thus inducing its melanophores to disperse their pigment and itself to darken. When the sympathetic nerves come into full play, as happens when the animal is in a white environment, the melanophore pigment concentrates and the creature blanches. Thus these two systems work as antagonists one against the other (Vilter, 1939 b, 1941 c). In this respect, Vilter's conception agrees, at least in its simplicity, with Hogben's view that vertebrate colour changes are dependent upon two factors, the W-substance and the B-substance.

Vilter (1941 a) further believes that the retinas in chromatic vertebrates are polarized and that their ventral portions are associated particularly with the ventrally centralized skin changes and their dorsal parts with those dorsally centralized. It has. been further pointed out by Vilter (1941 c) that the trophic activities of melanophores may be differently affected by these two systems in that the pituitary secretion not only induces the dispersion of melanophore pigment but also promotes assimilation in the dark colour-cells, a process which is retarded by the sympathetic nerves. In the eel, Vilter (1941 b) discovered that the xanthophores as well as the melanophores were polarized dorso-ventrally. These yellow cells, however, do not work in harmony with the melanophores. On the whole, the xanthophores are more responsive to the pituitary secretion, the melanophores to sympathetic nerve influences. Vilter (1940) gives an interesting account of the African siluroid *Synodontis*, which has assumed the habit of swimming on its back with the result that it exhibits, as compared with other fishes, an inverted colour scheme, pale on the dorsum which is directed downward and dark on the belly which is turned upward. Vilter shows how his polarity theory may be applied to such reversed cases. Vilter attained to his conception of colour polarity by studies chiefly on the Mexican salamander, axolotl. He then extended his work to fishes, and was soon led to the belief that his system of polarized organization applied to chromatic vertebrates in general and thus had a wider scope than was at first suspected. It will be necessary to consider his views again in the chapter dealing with amphibian colour changes.

Vilter's theory is not entirely free from weakness. It implies the invariable presence of two opposing agents, the darkening pituitary secretion and the blanching sympathetic nerves. But at no place in Vilter's extended exposition does he discuss what is now generally believed to be true of chromatic vertebrates, that in many of these animals there are more than two agents concerned with colour change. In a number of teleosts, as the preceding account shows, there are commonly three chromatic agents, one humoral and two nervous, with incidentally a number of other subordinate activators. This multiplicity of factors finds no place in Vilter's system. Furthermore, Vilter's system implies a distribution of melanophores in chromatic vertebrates and a type of response in the two assumed activators that agree only very poorly with what occurs in nature. According to Vilter, melanophores are scattered over the bodies of fishes and other chromatic vertebrates with more or less uniformity. When one of these animals assumes the pale phase, this state is brought on by the action of the sympathetic nerves which is most pronounced ventrally and which diminishes dorsally. The blanching thus produced is at its maximum on the underside of the creature where the nerves are believed to be most effective, and dies out as the upper regions are approached. On the other hand, the dark phase, due to pituitary secretion, begins on the back of the animal and spreads ventrally, to be extinguished on the creature's lower surface. This general plan of organization is in conflict with what is actually present in most chromatic fishes. These animals are pale below and dark above, not simply because of the concentration of melanophore pigment in the ventral colour-cells and its dispersion in the dorsal ones, but because in most vertebrates there are few or no melanophores ventrally and a great abundance of them dorsally. This difference in the distribution of the colour-cells is of prime importance. It is, of course, greatly emphasized by the shift of the melanophore pigment, the effectiveness of which, however, is due mainly to the differential distribution of the colour-cells. Though it is conceivable that a blanching agent nervous in character can be developed in a localized manner so that it is at its maximum ventrally and its minimum dorsally, it is very difficult to apply this conception to a secretion freely carried in the blood. A blood-borne agent such as the pituitary secretion cannot be so localized in its action as to produce a maximum effect dorsally and a minimum one ventrally. Such a condition would require a vasomotor mechanism of such complexity that its realization is very improbable. Hence, both in the matter of melanophore distribution and the local activation of these colour-cells, particularly in relation to hormones, Vilter's theory meets serious obstacles, and when these are added to the multiplicity of chromatic activators present in fishes but not called for in Vilter's view, his conception, because of its lack of uniformity to the facts of colour changes in fishes, seems not to apply to these animals. As already stated, its importance in the amphibians will be taken up later.

As a matter of fact the organization of the chromatic systems in vertebrates is not based on such a simple scheme as that suggested by Vilter. It is quite clearly complicated. As the present account shows, vertebrate chromatophores and particularly their melanophores are controlled in the main by either nerves, concentrating or dispersing, or by certain glands of internal secretion, chiefly the pituitary complex. The special combinations of these elements which characterize the chromatic organization of some of the better known teleosts are shown roughly in Table 10 where the determination of the presence (P) or the absence (A) of a given element is credited to the observation of an author as designated by his name and year of publication. A question mark after the sign for presence or absence of an element implies that the evidence upon which the determination is based is not so conclusive as might be wished. A blank indicates that the presence or absence of a given element has not yet been determined by test.

The combinations of elements by which the melanophores of several fishes recorded in Table 10 are controlled are somewhat diverse. In certain examples, as for instance *Phoxinus* and *Anguilla*, all four elements seemed to be involved; in *Fundulus* there appear to be at most three; and in *Pterophyllum* perhaps only two. The differences shown in Table 10 are based upon the conditions presented by only one set of chromatophores, the melanophores. Were it possible to tabulate in a similar way the other types of colour-cells, xanthophores, erythrophores, etc., the individual differences among the fishes would be still more pronounced and the complications greatly increased.

This intricacy in chromatic organization also appears when the outlines of the nervous connections that may be assumed to control the melanophore system in the fishes named in Table 10 are compared (Parker, 1943 b). As a typical example of such an outline, that of the eel (*Anguilla*) may well be taken. The consideration of such an outline naturally begins with the eye, where in the eel in all probability, as in many other chromatic vertebrates, the ventral retina is especially concerned with darkening and the dorsal one with blanching. As is

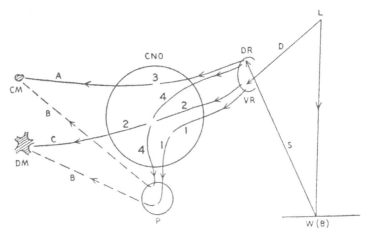

Fig. 69. Diagram of the melanophore system in the eel (*Anguilla*). The outer field is represented at the right, the central nervous connections in the centre, and the melanophores at the left. The reflex arcs are numbered and the special parts of the diagram are lettered: 1, retino-pituitary arc; 2, retino-cholinergic arc; 3, retino-adrenergic arc; 4, retino-tuberal arc. Abbreviations: *A*, adrenergic nerve-fibres; *B*, blood and lymph; (*B*), black background; *C*, cholinergic nerve-fibres; *CM*, melanophore with concentrated pigment; *CNO*, central nervous organs; *D*, beam of direct light; *DM*, melanophore with dispersed pigment; *DR*, dorsal retina; *L*, source of light; *P*, pituitary gland; *S*, beam of scattered light; *VR*, ventral retina; *W*, white background. Parker, 1943 b, 220.

shown in the outline (Fig. 69), when an eel is on a white background and both it and the background are illuminated from above, the eye of the fish receives light in part directly from the overhead source (*L*) and in part from the light-scattering, white background on which the fish rests (*W*). The direct light (*D*) falls upon the ventral retina (*VR*) and the scattered light (*S*) from the white background upon the dorsal retina (*DR*). Under these conditions, the eel, as is well known, becomes pale. When, on the other hand, the eel is on a black background (*B*) and illuminated from above, the light that enters its eye is almost exclusively the direct light from the source overhead, for the light which falls from that source on the black background is there mostly absorbed and thus fails in the main to reach the eye. Hence, under these circumstances the ventral retina is the one illuminated, the dorsal one receiving little or no light. Under these conditions, fishes regularly become dark. It is obvious, however, that this brief

TABLE 10. Elements concerned with particular colour changes in a number of teleosts, as indicated by the names of the authors who recorded the determining responses and dates of their publications (Parker, 1943 b, p. 219). Abbreviations: A, element absent; P, element present; blank, condition untested; 1, unpublished determination by recutting of nerves

	Activating elements			
			Nerves	
		Concentrating		Dispersing
Phoxinus	P	Healey, 1940	P	Healey, 1940
Ameiurus	P	Parker, 1935 d	P	Parker, 1935 d
Parasilurus	P	Matsushita, 1938	P	Matsushita, 1938
Anguilla	P	Waring, 1940	P	Parker, 1
Fundulus	P	Mills, 1932	P	Mills, 1932
Gasterosteus	P	Hogben et al. 1940	P	Hogben et al. 1940
Ophiocephalus	P	Chang et al. 1939	P	Chang et al. 1939
Pterophyllum	P	Tomita, 1940	P	Tomita, 1940
Macropodus	P?	Chang et al. 1939	P	Kamada, 1937
Tautogolabrus	P	Hunter et al. 1941	P	Hunter et al. 1941
Paralichthys	P?	Osborn, 1939 b	P	Osborn, 1939 b
	Pituitary neurohumours			
		Dispersing		Concentrating
Phoxinus	P?	Healey, 1940	P	Healey, 1940
Ameiurus	P	Parker, 1935 d	A	Parker, 1934 e
Parasilurus	P	Enami, 1939		
Anguilla	P	Waring, 1940	P	Waring, 1940
Fundulus	A?	Matthews, 1933		
Gasterosteus	P	Hogben et al. 1940		
Ophiocephalus	P	Chang et al. 1939		
Pterophyllum				
Macropodus	P?	Chang et al. 1939		
Tautogolabrus	P?	Hunter et al. 1941		
Paralichthys	A	Osborn, 1939 b	P?	Osborn, 1939 b

and somewhat superficial analysis scarcely justifies the conclusion that the dorsal retina is concerned exclusively with the pale tint of the fish and the ventral one with the dark tint. More likely both retinas are more or less involved in each colour phase as has been emphasized especially by Sumner who, in his discussion of this and other closely allied problems, has pointed out the importance of the ratio of the illumination on the two parts of the retina. This relation certainly has much to do in determining the final tint of the creature. Yet for the sake of simplicity in discussing the outlines it may be assumed, as in fact Butcher (1939) and others have done, that the dorsal retina is mainly a blanching receptor and the ventral a darkening one.

If, in attempting to gain some insight into the complicated chromatic organization of the eel, the reflex tracts concerned with its melanophore colour changes are enumerated, they will be found to be at least four. The first of these is shown in the outline (Fig. 69, 1) as extending from the ventral retina through the brain to that part of the pituitary gland (P) concerned with the production of the neurohumour intermedine. This neurohumour here produced is then carried by blood and lymph (B) to the melanophores which are by this means brought to disperse their pigment (DM) and darken the animal. Such an arc stretching from the eye through the pituitary gland to the colour-cells may be called the retino-pituitary arc. Described as a reflex mechanism, this arc is composed of two arms, one nervous from eye to gland, and the other humoral from gland to colour-cell. Hence it is an arc of what may be called a neurohumoral type (Parker, 1940d). The second arc in *Anguilla* (Fig. 69, 2) also begins in the ventral retina, extends through the central nervous organs, and as cholinergic fibres (C) makes its way with other chromatic elements over autonomic branches to the melanophores. The terminals of these fibres are believed to liberate acetylcholine, which induces a dispersion of melanophore pigment and a darkening of the fish. This arc has been called a retino-cholinergic one (Parker, 1941e) and is a typical nervous reflex arc. The third arc in the eel (Fig. 69, 3) extends from the dorsal retina through the central nervous organs and as adrenergic fibres (A) continues over the autonomic tracts to the melanophores. Its terminals are believed to produce adrenaline, whereby the melanophores are excited to concentrate their pigment (CM) and thus to cause blanching. This arc, a purely nervous one, has been designated the retino-adrenergic arc. The fourth and last arc in this fish (Fig. 69, 4) extends from the dorsal retina through the brain to the tuberal part of the pituitary complex where it may excite the production of the W-substance described by Hogben (Waring, 1940). This neurohumour is carried by the blood and lymph (B) to the melanophores, the pigment of which is thereby induced to concentrate and blanch the eel. This arc may be termed the retino-tuberal arc and, like the first one described, is of neurohumoral type. The four arcs thus outlined give some idea of the complications in the organization of the chromatic elements in the melanophore system of the eel, a type of organization which may well characterize many teleosts.

As a second example of chromatic organization that in the catfish (*Ameiurus*) may be described. In this fish the colour system is still more complex than that in the eel. The first, second, and third arcs described for the eel occur also in the catfish (Fig. 70). The fourth arc appears to be absent. Illuminations of the skin (*I*) of a catfish will cause darkening, as can be shown in blinded fishes. This response is

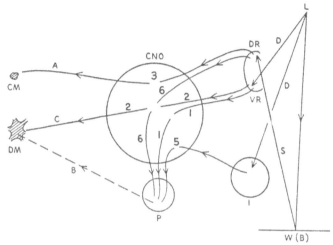

Fig. 70. Diagram of the melanophore system in the catfish (*Ameiurus*). The arrangement in the diagram is the same as in Fig. 69. In addition to arcs 1, 2, and 3, all of which are excitatory, there is in *Ameiurus* arc 5, the dermo-intermediate excitatory, and arc 6, retino-intermediate inhibitory. The lettering is the same as in Fig. 69 with the addition of *I*, integument. Parker, 1943*b*, 222.

not due apparently to a direct stimulation of the melanophores, but is a true reflex (Fig. 70, 5) which originates in the photoreceptors of the integument, *I* (Van Heusen, 1917), passes through the central nervous organs, *CNO* (Wykes, 1938), and thus reaches the pituitary gland, *P* (Parker, 1940*c*). The intermedine from this gland is then carried in the blood (*B*) to the melanophores (*DM*). This fifth type of reflex is a dispersing or darkening one and may be termed a dermo-intermediate excitation arc. It is apparently active whenever the skin of the catfish is illuminated, and its stimulating effect on the intermedine of the pituitary gland must be inhibited when, for instance, the catfish being on an illuminated white background (*W*) blanches. Such an inhibition, as has been suggested by Chang and Lu (1939) for the snakefish, appears to depend in the catfish upon a sixth arc (Fig. 70, 6) extending from the dorsal retina through the brain to the pituitary gland. The arc is probably purely nervous and may be designated a retino-intermediate inhibitory arc. Thus in several respects the chromatic organization in *Ameiurus* is more complex than that in *Anguilla*.

A less complicated type of chromatic organization than that of the eel is to be met with in the killifish, *Fundulus*. In this fish the pituitary portion of the melanophore system, though present (Kleinholz, 1935), is functionally ineffective, and the normal colour changes in *Fundulus* are dependent upon adrenergic and cholinergic nerves activated from the eye. The two tracts thus indicated correspond to the second and third arcs described for the eel, the retino-cholinergic and the retino-adrenergic. The reduction of the chromatic organization in *Fundulus* to an almost purely nervous one is associated in all probability with pituitary degeneration.

A simpler type of chromatic organization than that seen in *Fundulus* has thus far not been identified in teleosts. Such a type, however, appears to be present in other chromatic vertebrates. One of these is *Mustelus*. In this fish two of the four chromatic arcs described for the eel are present, the retino-intermediate and the retino-adrenergic. Of these the retino-intermediate is the more efficient, commonly overriding its opponent in its effects upon the coloration of this fish (Parker, 1937a). Some workers have called in question the presence of the retino-adrenergic arc, but on what appears to be insufficient ground.

The most simplified type of colour organization among chromatic vertebrates is that in which there is only one reflex arc. This probably occurs in some elasmobranchs (Parker, 1942) and frogs (Parker and Scatterty, 1937; Chang and Lu, 1939) and certainly in the lizard *Anolis* (Kleinholz, 1936b). It is an interesting fact that in all the examples of this type mentioned the kind of reflex arc present is the retino-intermediate (number 1 in the diagram for the eel). This arc with its neurohumour intermedine appears to be characteristic of all chromatic vertebrates, for the two exceptional cases that might be quoted, the chameleon and the killifish, are both probably incorrectly regarded as completely devoid of intermedine (Kleinholz, 1935; Parker, 1938a). Not only is the belief in the presence in all chromatic vertebrates of this reflex arc with its neurohumour intermedine very probable, but it is also likely that this part of the vertebrate chromatic system is, from an evolutionary standpoint, the original part around which the other chromatic elements subsequently developed (Waring, 1942).

As the present discussion shows, simplicity and uniformity in colour organization are not to be found among chromatic animals in the way in which the older workers looked for them. The purely nervous interpretation of colour changes has long since ceased to be tenable. Nor has the view that there are two essentially different principles of chromatophore activation, one nervous and the other hormonal (Hogben, 1924; Waring, 1942), fared much better. The conclusion that appears to have been reached is that colour effectors are controlled by numerous substances, the chromatic neurohumours, some of which are from glands of internal secretion and others from nerve terminals. Occasionally, as in the case of adrenaline, the same neurohumour may come in one instance from a gland and in another from nerve terminals. Thus a uniform and simple solution of the problem of chromatic organization in animals is not to be expected.

The preceding account, intricate as it is, gives no true idea of the complexity of animal chromatics, for as here given it is based almost exclusively on a study of melanophores, and most animals have in addition to these colour-cells xanthophores, erythrophores, leucophores, not to mention the numerous types of chromatosomes and other complex colour groupings. The control of each of these additional colour effectors calls for quite as complex a neurohumoral organization as the melanophores do. When this aggregate of elements and the multiplicity of their possible combinations is kept in mind, some conception of the enormous intricacy of the chromatic system in animals can be envisaged. And yet with all this complexity the number of activating neurohumours may not be over great. The nature and interrelation of these agents afford the modern student of colour changes an inviting field for adventure and research.

K. THE BLINDED STATE

Very few teleosts are exceptions to the rule that enucleation is followed by an early darkening of the skin (Parker, Brown and Odiorne, 1935). Pouchet (1877, 1880) discovered the remarkable fact that in the trout the removal of one eye, which in most fishes has no observable effect on the colour changes of the animal, results in a contralateral darkening in this form, a condition naturally associated with the optic chiasma. Pouchet's discovery was confirmed by von Frisch (1911), who attributed to abnormal conditions the inability of Steinach (1891) to reach similar results. The general slowness of colour reaction on the part of certain flatfishes is probably accountable for some records in which these forms were not reported as darkening by blinding (Bauer, 1910; Buytendijk, 1911; Polimanti, 1912). What appears to be a real exception to the rule that enucleation is followed by darkening is seen in *Phoxinus*. On the loss of its eyes, this fish is stated by von Frisch (1911 c) to be first dark and then variable. Abolin (1925 b) declared that after the loss of its eyes *Phoxinus* exhibited moderately expanded chromatophores. When it is recalled that after the fish has been blinded the chromatophores of *Phoxinus* will still respond with more or less appropriateness to environmental changes, its exceptional position must be evident. Its powers of chromatophoral response under such conditions have led to tests to locate photoreceptors other than the lateral eyes as explanatory of this state, a search that has pointed to the pineal organ rather than to the skin as the part concerned (von Frisch, 1912 a; Scharrer, 1928). Thus *Phoxinus* is peculiar in several respects and from this standpoint is to be set off against other fishes, the loss of whose eyes seems to be associated with a moderate or full darkening of the skin.

Critical studies on the chromatic effects of blinding have been made only in recent years and on very few animals. Eyeless catfishes, *Ameiurus*, have been reported by all recent investigators as very dark (Van Heusen, 1917; Bray, 1918; Pearson, 1930; Odiorne, 1937; Wykes, 1938; Osborn, 1938 b). Contrary to the statement of Parker and Lanchner (1922), a blinded *Fundulus* likewise takes on a dark tint (Wyman, 1924 a; Sumner, 1933 a; Wykes, 1938). This change also characterizes *Lepadogaster* and certain other shallow-water teleosts (Wykes, 1937). The details of these changes have been most fully worked out by Wykes (1938) on *Ameiurus*. This catfish when blinded resembles an eyeless minnow in that it can still change its tint at least to changes in illumination. Like *Phoxinus*, it becomes dark in bright illumination and pale in darkness. The degrees of these changes were estimated by Wykes, who measured the diameters of the melanophores of this fish under the two conditions. In bright light these colour-cells had an average diameter of 114·6 microns and in darkness 49·8 microns. In other words, they were almost fully expanded in the light and much contracted in darkness. In fishes prepared for this kind of test, the nerves to one pelvic fin were cut by Wykes, the other pelvic fin being left as a control. When such fishes were subjected to bright light and to darkness, the melanophores in the normal fin changed as had been previously recorded, but those in the denervated fin, which measured 127·6 microns in the light, changed only to 119·4 microns in darkness. In other words, they really remained expanded. This loss of the power to contract on loss of innervation led Wykes to conclude that contraction was dependent upon a spinal reflex, a conclusion supported by further observations on the responses of fragments of living skin from this fish and on the responses of the melanophores to point illumination. Wykes finally concluded that the changes in tint shown by blinded *Ameiurus* under differences of illumination were not due to the direct stimulation of the melanophores by light, but to their responses as part of a concentrating nerve reflex. What photoreceptor was involved in this reflex was not ascertained. From this standpoint, the colour responses of eyeless teleosts to illumination call for no other effector mechanism than that already described for the fish with vision. Parker (1939*b*, 1939*c*) has pointed out that catfishes with only one eye first darken and then become variable, changing either to a pale or a dark phase. This variability characterizes the whole fish, which exhibits no evidence of bilaterality as is seen in the trout. Osborn (1939*c*) has observed that flatfishes with only one eye adapt to changes of background more slowly than normal individuals but are otherwise uninfluenced.

L. FLATFISHES

Probably no animal, not even the far-famed chameleon, exhibits such remarkable colour changes as the flatfish. The colour responses of such fishes to environmental conditions, both natural and experimental, have been very fully studied by Sumner (1910, 1911) and by Mast (1916), both of whom have pointed out the remarkable capacity of these forms to adapt themselves not only in colour but also in pattern to their surroundings. In these respects, *Paralichthys albiguttus* is probably the most remarkable of all flatfishes (Mast, 1916). When an active individual of this species is placed successively on white and black chequerboard backgrounds with squares of different sizes, the fish, though resting on backgrounds all of which have equal amounts of black and white, does not respond in the same way to all, but is finely mottled on the fine chequerboard and coarsely mottled on the coarse one (Fig. 71). Thus such a fish may be said to respond to pattern as well as to the simple illumination of the background, a subject discussed by Hewer (1931). Responses to differences in pattern do not seem to have been observed in animals other than flatfishes; hence, the exceptional nature of this group of teleosts.

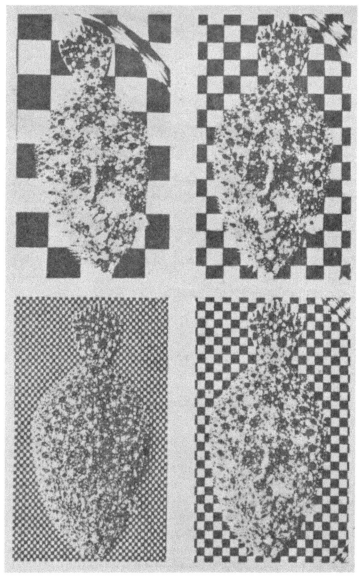

Fig. 71. A flatfish, *Paralichtys albiguttus*, on chequerboard patterns of different sizes. All figures are from the same fish. The sides of the chequerboard squares were 2 mm., 5 mm., 10 mm. and 20 mm. Mast, 1916, Pl. 21.

The chromatophore system in *Paralichthys* has been worked out in detail by Kuntz (1917), who has recorded in this fish the presence of melanophores, xanthophores, erythrophores, and guanophores to which Hewer (1927) had added iridophores. Schafer (1921) in his study of *Pleuronectes* has shown that whenever nerves are cut in this fish dark dermal bands result, and when nerves are stimulated electrically the innervated areas blanch. Thus concentrating chromatophoral nerves appear to be present and, if the interpretation of dark bands brought forward in this survey be accepted, also dispersing chromatophoral nerves. Meyer (1930, 1931) discovered that the injection of 0·3 c.c. of serum from a flounder that had been 8 days on a black background into a pale flounder induced in the recipient the formation of local dark spots. These spots appeared within 5 minutes after the injection had been made and reached their maximum in half an hour. In a similar way, serum from a pale flounder when injected into a dark one produced pale spots. These observations lead to the conclusion that in *Pleuronectes* the chromatophores are under the control of at least two sets of humours as well as under that of nerves.

The colour changes of *Pseudopleuronectes* have been discussed recently by Osborn (1939c). This fish shows appropriate responses to a white or a black background and takes on an intermediate tint in complete darkness. Denervated areas darken at once and then gradually fade to the given body colour. Excitement causes temporary pallor in dark fishes and darkening in pale ones. Pale spots on dark fishes and dark spots on pale ones disappear when the fishes are excited. The injection of adrenaline into *Pseudopleuronectes* calls forth conditions like those due to excitement. Osborn concludes that the melanophores of *Pseudopleuronectes* are not only under nerve control, but also under that of an adrenaline-like humour. Extract from the pituitary gland of *Pseudopleuronectes* capable of darkening three frogs for 12 hours failed to darken pale fishes and did not overcome the excitement pallor of denervated areas in dark fishes. Hence *Pseudopleuronectes* appears to be less influenced by its own intermedine than is even *Fundulus*. These results have been substantiated by Osborn (1939b) in a still more recent paper on the colour changes in *Paralichthys* and *Lophopsetta*. It is clear from these accounts that the chromatophoral systems of flatfishes, like their colour reactions, are of a most complicated, perhaps the most complicated, type.

Flatfishes as a rule are pigmented on the side of the trunk that is away from the 'substrate on which they rest and they are without pigment on the side next this substrate. Some species are pigmented normally on the right side, others on the left. From time to time reversed specimens are found. Thus a specimen belonging to a species normally pigmented on the left side may have pigment on the right side, the left side being unpigmented. In other instances, the pigmented and unpigmented areas may not be limited to sides but may spread partly over each side. Numerous such ambicolorate flatfishes, as they are termed, have been described (Hussakof, 1914; Banta, 1921; Gudger, 1934, 1935a, b, 1936; Gudger and Firth, 1935, 1936, 1937; Taki, 1938) without, however, throwing much light on the origin of this condition. Hussakof has suggested that ambicoloration may result from life on a rough bottom where the fish is

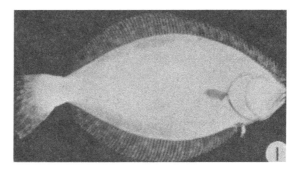

Fig. 72. Undersurface of a freshly caught summer-flounder *Paralichthys*. No melanophores are present on this surface. Osborn, 1941b, 347, Pl. 1.

unable to keep its applied side always in deep shadow but must expose it more or less to horizontal light. However, it takes only a superficial inspection of the instances already recorded to show that such an explanation cannot apply. Recently Gudger (1941) has described a flounder fully coloured on both sides. As Gudger and Firth remark (1937), no clear understanding is at hand for any or all of these anomalies. This presumably is to be sought for in the developmental aspect of these fishes, where factors determining the distribution of their chromatophores may throw light on such strange conditions.

An experimental approach to the subject of flatfish coloration was initiated some years ago by Cunningham (1921), whose results, based on the illumination of these fishes from below, were extremely meagre and indecisive. The subject has been revived recently by Osborn (1939a, 1940a, 1941a, b), who in a series of extended tests on adult flatfishes has attained a truly remarkable outcome. Osborn's work was done on the summer flounder, *Paralichthys*, which is normally quite pale on its under side (Fig. 72). When such flounders are kept in an aquarium with black sides and a black ceiling and with a frosted glass bottom through which light is admitted to the interior of the container, the fishes show on their unpigmented sides the first active melanophores in about 7 days. In 30 days, such melanophores have become abundant and in some 7 weeks they may be present in excess. This production of active melanophores, for such colour-cells can be induced to concentrate and disperse their pigment, takes place regularly when the sides and top of the aquarium illuminated from below are black; it does not occur when the sides and top are white

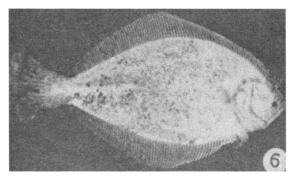

Fig. 73. Undersurface of a summer-flounder *Paralichthys*, blinded immediately after capture and continuously illuminated from below. Although this fish was illuminated only 18 days its melanination on the under side, originally white, is very extensive. Osborn, 1941 *b*, 347, Pl. 1.

and the fishes illuminated from below. Blinded flounders, which are of course dark-coloured in consequence of enucleation, when illuminated underneath develop melanophores on their pale surfaces irrespective of the colour of the walls and top of the aquarium, white or black (Fig. 73). Evidently the eyes of the fish are not essential to the formation of melanophores on its pale side. This formation to be successful requires two conditions: first, the fish must be in its dark phase either through having been retained in a black-walled aquarium or from blinding; and, secondly, it must be illuminated from below. This remarkable development of melanophores in an unusual region is in full accord with a principle now generally recognized that the growth of new dark colour-cells is favoured by those conditions which in any given chromatic animal bring about the dispersion of its melanophore pigment.

Osborn's main conclusions, as already briefly stated, have been confirmed by his latest studies (1941 a) on the regeneration of pale and of dark scales on the undersides of flatfishes illuminated from below. On the whole, these results show most conclusively that the

Fig. 74. Ventral surface of a catfish blinded and ventrally illuminated (direct light) for 79 days. Note the excessive ventral pigmentation. Osborn, 1941 c, 357, Pl. 2.

Fig. 75. Ventral surface of a catfish blinded and kept as a control with stock fishes in an unlighted tank of neutral shade 76 days. The fish has darkened as a result of blinding but excessive ventral pigmentation has not developed. Osborn, 1941 c, 357, Pl. 2.

pale sides of these fishes are by no means beyond chromatic resuscitation and may be brought by experimental means to reassume a pigmentation ancestrally normal to them.

From Osborn's discovery of the revival of pigment-cells on the pale sides of flatfishes, this investigator was led to inquire into the possibility of a similar revival of such colour-cells on the ventral, pale aspects of teleosts normally coloured. As a subject for such inquiry, Osborn chose the catfish *Ameiurus*. These fishes, as is well known, are deeply pigmented except for a restricted ventral area in what might be called the chest region which is very pure white. Catfishes, as already stated, are characterized by a well developed colour change ranging from a very pale greenish yellow to coal-black. These changes are chiefly on the dorsum, the ventral aspects of this fish being almost pure white (Fig. 75). In Osborn's experiments, blinded catfishes were subjected to direct ventral illumination, as in the case of the flatfishes, for over two months. The fishes under such circumstances developed a very abundant growth of melanophores on what had previously been a white ventral area (Fig. 74). For this change as for the similar one in flatfishes the eyes were not necessary, but it was essential that the fish should be kept in the dark phase and continuously illuminated. Catfishes blinded and kept in darkness as controls failed to darken ventrally (Fig. 75). By tests on hypophysectomized fishes, Osborn was able to show that intermedine was essential to the growth of new melanophores. Thus this investigator (1940 b, 1941 c, d) demonstrated that an ordinary teleost such as *Ameiurus* could be brought to develop melanophores on areas of skin normally free from these dark cells and a new light was thrown on the histogenesis of this usual but remarkable type of chromatophore. Teleosts, and particularly flatfishes, have thus served to open up a most inviting field for chromatic investigation.

Chapter VI

COLOUR CHANGES IN VERTEBRATES (continued)

5. *COLOUR CHANGES IN AMPHIBIANS*

ALTHOUGH the colour changes in the common frogs, toads and other amphibians are fairly conspicuous, they do not seem to have attracted the attention of the very early naturalists. The first record of such changes appears to come from Vallisnieri who, in an appendix to his natural history of the chameleon (1715), published a brief description of certain colour changes in the frog. Such changes were subsequently very admirably portrayed by Roesel von Rosenhof in his *Historia naturalis ranarum nostratium*, a work issued in Nuremberg in 1758. These early publications were followed by a succession of later ones in which the control of the colour changes in amphibians was ascribed to the nervous system. Thus, in 1858, Lister observed that frogs on the loss of their eyes ceased to respond by colour changes to differences in their environment. Lister was, therefore, led to conclude that the cerebrospinal axis was, to use his own expression, "chiefly, if not exclusively, concerned in regulating the functions of the pigment cells". This nervous interpretation of chromatophore activity has proved to be acceptable with certain investigators even down to the present time (Perotti, 1928).

During the last few decades, however, an entirely novel means of changing the colours of amphibians has been gradually brought forward. In 1898, Corona and Moroni recorded that when an extract of the adrenal gland was injected into a frog the animal quickly blanched, thus suggesting a humoral means of colour change. This observation was confirmed in 1906 through results independently arrived at by Lieben. In 1914, Eycleshymer noted that on decapitating the young larvae of *Necturus* all quickly become pale in consequence of the contraction of their melanophores and remained so indefinitely, thus indicating that some organ in the head of these animals had to do with their colour changes. But the real humoral interpretation of these changes in amphibians came from work done directly on their pituitary glands. When from an embryo frog the tissue destined to form the pituitary gland is removed, the resulting tadpole, as was found by a number of experimental embryologists (P. E. Smith, 1916*a*, *b*; Allen, 1916; Atwell, 1919), is consistently colourless. It was recognized that this colourless state was due to the contracted condition of the animal's melanophores rather than to any real lack of pigment. Such observations led Hogben and his associates to undertake studies on the relation of the pituitary gland to the colour changes in amphibians.

These studies began to appear in 1922 when Huxley and Hogben showed that the feeding of pituitary substance to pale axolotls resulted in a marked temporary expansion of their melanophores. At this time Krogh (1922) also pointed out that the loss of the pituitary gland by adult frogs will cause them to blanch. In the same year and the year following, Hogben and Winton published four important papers on amphibian coloration (1922 a, b, c, 1923), in the course of

Fig. 76. Two frogs of which the right-hand individual had been injected six hours previously with pituitary extract from a foetal ox and the left one kept pale as a control. Hogben, 1924, 54.

which they declared that with the loss of the pituitary gland frogs became permanently pale even though they were retained on an illuminated black background. On injecting into such animals an extract of this gland from any of a long series of vertebrates, fishes to mammals, the frogs could be made to darken temporarily (Fig. 76). The pituitary gland of a single frog was found by these workers to contain enough of the active principle to darken more than fifty frogs. Hogben and Winton also demonstrated that the cutting and stimulation of nerves in the frog resulted in no obvious change in the animal's coloration. From such observations and many others of a more detailed kind these investigators were led to conclude that the colour changes in amphibians were scarcely if at all under direct nervous influence, and that the darkening of these animals was due to the action of a pituitary secretion carried from the gland by the blood to the melanophores. The concentration of pigment in these colour-cells and the consequent assumption of the pale state by the frog was believed by Hogben and Winton to be due to the gradual disappearance from the animal's blood of the original pituitary secretion. Thus, very strong evidence was advanced by these workers for a purely or almost purely humoral interpretation of amphibian colour change. This is in strong contrast with the explanation of such changes through the direct action of nerves on the chromatophores as maintained by the older workers in this field.

A. COLOUR CHANGES

The colour changes in amphibians are apparently never of the spectacular kind such as are shown by flatfishes and by chameleons. They are nevertheless easily discernible and follow the usual rule of pale coloration on an illuminated white background and dark coloration on an illuminated black one. This is strikingly true, for example, of both the dark and pale breeds of axolotls (Vilter, 1931 a). In *Necturus*, the adult pigmentation is so dense over the back of the animal that colour changes are scarcely visible there and may be seen to advantage only on the legs and the venter where pigment is less abundant (Dawson, 1920). According to a multitude of workers, pale and dark phases characterize most anurans, both European (*Rana esculenta, R. temporaria* and *R. fusca*) and American (*R. pipiens, R. clamitans*). These amphibians are commonly most pale on a white background, less so on a grey one and darkest on black (Przibram, 1932 b). The common European treefrog, *Hyla arborea*, ranges from pale green to dark green but may assume in addition an ashen grey or a lemon yellow tint (Hargitt, 1912; Berggrün, 1914; Schmidt, 1920 e; Przibram, 1932 b). The common European toad, *Bufo viridis*, according to Goubeaud (1931), changes not only from pale to dark and the reverse but also assumes greenish and brownish tones.

In all these colour changes, the pattern for each species remains definitely fixed, becoming merely paler or darker with the changes of tint (Fuchs, 1914). This conclusion has recently been reaffirmed by Parker (1930) in a study of the American treefrog, *Hyla versicolor*. The change of pattern in amphibians as tested by means of skin transplantation has led to no very uniform results. In some instances, the specificity of the skin appears to have been retained (Lindeman, 1928; Herrick, 1932; Vogel, 1940), and in others not (Vilter, 1936). Twitty (1935) has stated that in transplanting very early embryonic melanophore tissue in amphibians the pattern finally produced by the descendant colour-cells is specific for the donor. Such evidence favours the view that a pattern once established in amphibians is likely to exhibit great permanency.

In the past, much emphasis has been placed upon the relative importance of temperature and humidity as well as of light in connection with amphibian colour changes. Hooker (1912) pointed out that in a warm, dry situation *Rana fusca* was pale and in a cold, damp one it was dark. The same conditions were reported by Hogben and Winton (1923) for *Rana temporaria* and *R. esculenta*. Hewer's less consistent observations (1923) on *R. temporaria* confirm, so far as they go, the statements of Hogben and Winton. In *Xenopus*, which is a predominantly aquatic amphibian, the colour changes, as might be expected, are little influenced by humidity and temperature, but are primarily due to the effects of light (Slome and Hogben, 1928; Hogben and Slome, 1931).

The time required for the completion of colour changes in amphibians seems to be always relatively long. None of the colour changes in these animals appears to be accomplished in seconds or minutes, but all require hours or even days. Thus, according to Dawson (1920), a deeply tinted *Necturus* when placed in darkness will blanch in some three hours and on being returned to light will darken in about the same time. The South African toad, *Xenopus*, studied by Slome and Hogben (1929), will darken in part of a day but will require from one to several days to blanch. This amphibian was found to change from pale to the intermediate tint assumed in darkness in 180 hours and from dark to the same intermediate tone in 70 hours. The reverse changes from intermediate to pale required 100 hours and to dark 24 hours. Pierce (1942) observed that *Rana clamitans* darkened on a black background in 4·5 to 5 hours, and blanched on a white one in 10 to 12 hours. Stoppani (1942 b) records that *Bufo arenarum* changes in the light from a full dark green to a full pale green in about two hours and makes the reverse change in darkness in about one hour. Notwithstanding the relatively short intervals given by Stoppani for the toad, all amphibian chromatic reaction times are long in comparison with those of most other animals and favour the opinion that amphibian colour changes are under a control pronouncedly humoral.

A daily or nyctohemeral rhythm has been pointed out in the colour changes of *Xenopus* by Hogben and Slome (1931). Minkiewicz (1933) has noted a similar activity for other amphibians, especially frogs. The age of individual amphibians apparently has much to do with their colour responses. Thus, young axolotls are said by Babák (1910b) to respond more quickly than old ones to environmental colour changes. According to Fischel (1920), *Salamandra maculosa*, which as a young animal changes its colour easily, loses this capacity completely in its old age, an observation confirmed by Himmer (1923).

When a pale frog darkens as a result of an injection of obstetrical pituitrin, the electrical potential of its skin is said by Ford and Steggerda (1940) to increase 70 per cent. When a dark frog is blanched by an injection of adrenaline, the skin potential is stated to decrease 62 per cent. Ultraviolet rays have been observed by Torraca (1914) as darkening agents for the skin of *Triton*.

B. Chromatophores

The majority of amphibians possess three classes of chromatophores, melanophores, xanthophores, or, as they are often designated, lipophores, and guanophores. These three kinds have been identified in *Hyla* by Berggrün (1914), in Salamandra and *Triton* by Schmidt (1920b), and by Himmer (1923), in the tadpoles of *Discoglossus* and of a number of other anurans by Prenant (1920), in the common frogs by Millot (1929) and by Hadjioloff (1929a, b), and in *Pelobates* by Juszcyk (1937). Hadjioloff's statement that, in addition to the three classes of chromatophores mentioned, frogs possess cyanophores or blue cells and phyophores or brown ones has not been confirmed by Millot (1929). Collins and Adolph (1926) mention the occurrence of melanophores in *Diemyctylus* and indicate the possible presence of xantholeucophores and erythroleucophores in this newt without however definitely deciding this matter. Slome and Hogben (1928) note melanophores and xanthophores in *Xenopus* but make no mention of guanophores which may well be absent from this South African toad. Elias, Cohen and Lieberman (1942) state that *Rana pipiens*, in addition to the three kinds of chromatophores usual with amphibians, possesses also erythrophores. Schmidt (1920b) has described from the tail region of tritons and axolotls remarkable colour-cells which have the form of double attached stars, one unit of which is nucleated while the other is purely cytoplasmic. These unusual cells may be either melanophores or xanthophores (Fig. 77). Double xanthophores had already been noticed by Pernitzsch (1913) in young axolotls.

Melanophores have long been recognized as the chief elements in the colour changes of amphibians (Schmidt, 1920c). They occur in both the epidermis (Fig. 78) and the derma of almost all species of amphibians that have been closely examined: tadpoles of *Rana pipiens* (Hooker, 1914a), of *Rana clamitans* (Herrick, 1932, 1933),

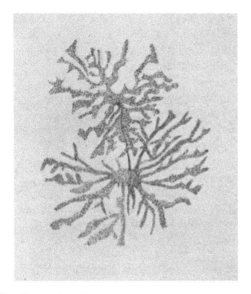

Fɪɢ 77. Double-star lipophore from the tail of a salamander larva 3·5 cm. long. The cell processes are not wholly filled out. Schmidt, 1920b, 234.

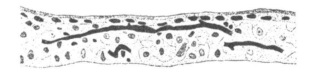

Fɪɢ 78. Section of the epidermis of a frog tadpole, showing the position of the epidermal melanophores which are represented by fragments only. Herrick, 1933, 306.

Necturus (Dawson, 1920), *Salamandra maculosa* (Fischel, 1920; Himmer, 1923), axolotl (Schnakenbeck, 1921), *Bufo viridis* (Goubeaud, 1931), and others. The origin of the epidermal melanophores is by no means settled. They may be cells that have migrated from the derma into the epidermis, as has been claimed by Vilter (1931c,1932a, 1935), or they may be true epidermal cells that have taken up pigment from the derma (Murisier, 1913). Hooker (1914a) found that plasma cultures of the embryonic epidermis from *Rana pipiens* contained pigmented cells, a condition suggestive of the local origin of epidermal melanophores. An invasion of the epidermis of axolotls by dermal melanophores is denied by Schnakenbeck (1921).

Amphibian melanophores are usually richly branched (Fig. 79), but they may be only sparsely provided with processes, as in the epidermal melanophores described by Herrick (1933) for the tadpoles of various frogs (Fig. 80). A system of branching processes from the subepidermal melanophores in the larvae of *Alytes* and of a number of other anurans and known as the network of Asvadourova (1913) has been discussed by Borrel (1913 *a, b*, 1914 *a, b*) and by Prenant (1920, 1923), but without elucidating its significance. This problem has been taken up by Bytinski-Salz and Elias (1938), particularly by Elias. According to this investigator, the melanophores in amphibians fall into four groups definable by their positions in the skin: epidermal, those within the epidermis; adepidermal, in the derma next the epidermis; intracutaneous, in the derma not far from the epidermis; and subcutaneous, under the derma. No single amphibian

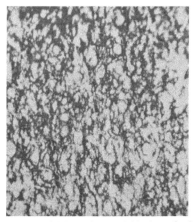

Fig. 79. Melanophores richly branched and with fully dispersed pigment from the web of the foot of a normally dark frog, *Rana pipiens*. Parker, Brown and Odiorne, 1935, 442.

Fig. 80. Epidermal melanophores from a frog tadpole in living, uninjured tissue. Herrick, 1933, 306.

possesses all four of these melanophore types, which occur throughout the group in a way which indicates a certain phylogenetic sequence. Urodeles, both larval and adult, are characterized by epidermal and adepidermal melanophores. Of the three discoglossids studied by Elias, *Discoglossus*, *Bombina*, and *Alytes*, all have adepidermal and subcutaneous melanophores, and *Bombina* and *Alytes* have in addition epidermal colour-cells. In the frogs and toads, no adepidermal cells at all are present, but the other three types are well represented. Adepidermal melanophores are evidently of a primitive type, for they occur in the urodeles and the discoglossids, primitive anurans, but not in the higher frogs and toads. In this sense, they are indicative of phylogeny within the group of amphibians. The adepidermal cells in the discoglossids are incapable of concentrating or dispersing their pigment. They have thus ceased to be concerned with chromatic changes, which are carried out by one or more of the other types of melanophores. These motionless melanophores in the discoglossids are ordinarily arranged in meshes, thus forming the so-called network of Asvadourova, the function of which has been quite mysterious. It is Elias' opinion that this network is sustentative in character and serves as a support for the region of the skin in which it occurs. If such be the case, the adepidermal melanophores must be regarded as chromatic cells which have changed their function from that of colour agents to that of supporting structures. Such an hypothesis of course awaits confirmation.

Fischel (1920), Perotti (1928) and Herrick (1933) favour the view that the processes of amphibian melanophores anastomose with each other, an opinion that is denied for the melanophores in the discoglossids by Bytinski-Salz and Elias (1938). The last-named authors as well as Schnakenbeck (1921) have recorded amphibian melanophores in process of mitotic division and have shown that by this means they may increase in number even after they have developed an abundance of melanin (Fig. 81).

The dermal melanophores in amphibians are generally admitted to be much more responsive in colour changes than the epidermal ones (Schnakenbeck, 1922); in fact the epidermal melanophores in certain species of batrachians have been declared by some investigators to be entirely inactive (Hooker, 1914*d*, *Rana pipiens*; Dawson, 1920, *Necturus*). The inactivity of these epidermal cells may be due to their relative inaccessibility to humoral agents as compared with the freedom with which such agents may come in contact with dermal colour-cells. Although epidermal melanophores may be in certain instances entirely inactive, amphibian melanophores as a group exceed all other chromatophores in these animals in effectiveness. They are always present in larger numbers in all amphibians normally exposed to light. In the cave-inhabiting and subterranean forms *Proteus* (Kammerer, 1912), *Typhlotriton*, and *Typhlomolge*, they are largely absent, and these forms in consequence are almost if not quite colourless. Albinotic tadpoles of *Bombinator* have been described by Ippisch (1928) and colourless tadpoles and adults of the common frog *Rana temporaria* by Eales (1933). Juszcyk (1937) has studied the skin

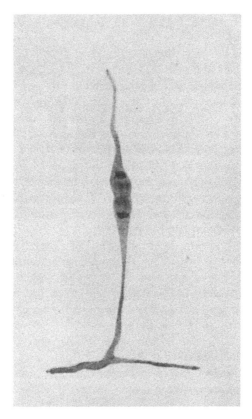

Fig. 81. A paraepidermal melanophore in the telophase stage of cell division from *Discoglossus pictus*. Bytinski-Salz and Elias, 1938, 20.

of albinotic *Pelobates fuscus* and has noted that, though these animals show a complete lack of melanophores, they still possess the usual complement of lipophores, guanophores, and allophores.

Lipophores ranging from yellow, xanthophores, through orange to red, erythrophores, have been identified in the integument of urodeles as well as in that of anurans. Schmidt (1918c, 1920a, b) has described them in the frog, axolotl, *Triton*, and *Salamandra*, and Himmer (1923) in *Salamandra*. In *S. maculosa* they are said by

Himmer to be quite inactive. In *Necturus*, according to Dawson (1920), they remain expanded under all conditions. Goubeaud (1931), who has studied them in *Bufo viridis*, states that in this toad they can expand and contract slightly but that they are by no means as active as are the melanophores. They are abundant in the coloured inter-spaces of the skin of *Bufo* rather than in the spots and are concerned with the yellow and the green coloration of this amphibian. In most species they are located in the derma, but in the embryos and in the young of *Salamandra* they are said to be found in the epidermis (Himmer, 1923). Elias (1941a), who studied the development of xanthophores in various amphibians, distinguishes four types of these colour-cells depending upon the state of the yellow mass, xanthoma, within the cell: crystalline xanthoma (*Disco-glossus*), liquid xanthoma (*Hyla, Rana esculenta*), half-diffused xanthoma (*Bufo*), and direct yellowing of stromata (*Bombina, Bufo, Rana latastei*).

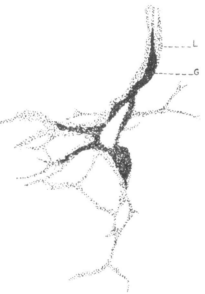

Guanophores, containing granular, often crystalline, re-flecting particles, and called at times leucophores, have been seen in *Salamandra maculosa* by Himmer (1923). They have also been identified in the skin of *Bufo viridis* by Goubeaud (1931), where, in conjunction with melanophores, they produce a blue tint and with lipophores a yellow one. Allophores, con-taining a relatively insoluble red pigment and first described by Schmidt, have been identified by Ballowitz (1929) in the skins of a number of toads and frogs.

Fig. 82. Xantholeucosome from the tail of a frog tadpole: *G*, leucophore; *L*, lipo-phore; both chromatophores expanded. Schmidt, 1918a, 497.

C. CHROMATOSOMES

In amphibians, as in teleost fishes, chromatophores of different kinds may unite to form chromatosomes. The usual type of batrachian chromatosome consists of a xanthophore (lipophore) united with a leucophore (guanophore), and commonly designated a xantholeuco-some. This type of chromatosome has been extensively studied by Schmidt (1918a, 1919b, 1920e, 1921). In the tail of the frog tadpole richly branched xantholeucosomes are to be seen. In these combinations both xanthophore and leucophore may be expanded (Fig. 82)

Fig. 83. Xantholeucosomes from the tail of a frog tadpole: G, leucophore; L, lipophore; both chromatophores contracted. Schmidt, 1918a, 499.

or both may be contracted (Fig. 83) or one may be expanded and the other contracted (Fig. 84). In *Rana esculenta*, the xantholeucosomes form a well defined layer immediately under the epidermis and peripheral to a deep layer of dermal melanophores (Schmidt, 1921). This condition is also present in *Rana fusca* and in *Bufo viridis*. It is especially well seen in *Hyla arborea*, where the two component chromatophores are more exactly fitted to one another than in the other amphibians mentioned. In each xantholeucosome in *Hyla* a lens-shaped xanthophore fits into a cup-shaped leucophore, the combination resting in such a position that the xanthophore is in contact with the inner face of the integumentary epidermis and the leucophore away from that face. Immediately behind the leuco-

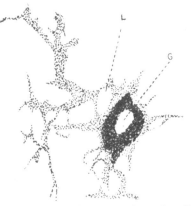

Fig. 84. Xantholeucosome from the tail of a tadpole: L, lipophore expanded; G, leucophore contracted. Schmidt, 1918a, 498.

phore is a melanophore which in this instance often appears to be as much a part of the combination as either of the other two cells (Fig. 75). A succession of such xantholeucosomes constitutes a definite layer in the peripheral derma of the common treefrog.

The chief colour phases of *Hyla arborea* include a pale green, a

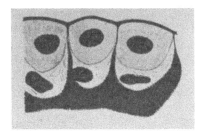

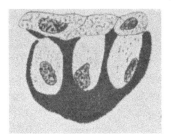

Fig. 85. Section of the pigment layer from the derma of *Hyla arborea* in the pale green phase; in the xantholeucosomes the xanthophores are above and the leucophores below; the melanophores are below the xantholeucosomes. Schmidt, 1920*e*, Taf. 21.

Fig. 86. Section of the pigment layer from the derma of *Hyla arborea* in the dark green phase; the xanthophores form a layer above the leucophores which are almost surrounded by the melanophores. Schmidt, 1920*e*, Taf. 21.

Fig. 87. Section of the pigment layer from the derma of *Hyla arborea* in the lemon-yellow phase; the xanthophores are superficial but some have also pressed in between the leucophores. The melanophores are all deep-seated. Schmidt, 1920*e*, Taf. 21.

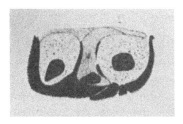

Fig. 88. Section of the pigment layer from the derma of *Hyla arborea* in the grey phase; the upper faces only of the leucophores are free of melanin; the xanthophores are wedged in between the leucophores. Schmidt, 1920*e*, Taf. 21.

dark green, a lemon yellow, and a grey. The conditions of the xantholeucosomes and their associated melanophores by which these colour phases are produced have been worked out by Schmidt (1920e). In the pale green phase only the deep face of the xantholeucosome is covered by the melanophore, the two components of the combined cells being exposed in an equal and unobstructed way (Fig. 85). In the dark green phase, the sides as well as the deep face of the xantholeucosome are fully shrouded by the melanophore (Fig. 86). In the lemon-yellow phase, the melanophores are fully withdrawn and the xanthophores very completely expanded (Fig. 87). And finally in the grey phase, the leucophore is completely surrounded by the melanophore and the xanthophore is more or less withdrawn (Fig. 88). Thus the chief colour phases of *Hyla* depend to some extent upon the movements of the leucophores, but in the main are due to a considerable adjustment in the melanophores. In this instance, as in the simpler colour changes of amphibians, the chief part is played by melanophores, the other chromatophores being either purely passive or at most only slightly active.

D. INNERVATION

That the eye is the all-important organ in the frog for the excitation of colour changes was the opinion of many of the older workers (Lister, 1858), and, notwithstanding the adverse opinion held by some (Biedermann, 1892), this still remains the commonly accepted opinion (Przibram, 1932b). But the view that from the eye to the chromatophores a reflex arc stretches in part through the central nervous organs and in part over the peripheral nerves as hypothesized by Lister is by no means so certain. In a long series of papers from the middle of the last century to the present time the results of cutting and stimulating central nervous organs and peripheral nerves in amphibians have left this question still in uncertainty. As a result of this work, however, the belief has gained ground that when chromatophoral nerves in frogs are cut the melanophores of the denervated areas disperse their pigment slightly and the regions darken somewhat, and that when these nerves are stimulated electrically or otherwise the innervated areas blanch noticeably. These responses, even as described by their most ardent advocates, are admittedly slight, and often difficult to observe. They never approximate the changes that can be called forth by nerve cutting and nerve stimulation in fishes and in some lizards, such as chameleons for instance. It is not surprising, therefore, that many students of this subject have expressed doubt as to the validity of the evidence put forward by others in favour of amphibian chromatophoral nerves. This question was subjected to a very complete experimental investigation by Hogben and Winton (1922 c), who used common frogs for their research. When the right lumbosacral plexus of a frog was severed and the frog was kept under observation for several days, no differences in coloration or in the state of the melanophores of the web of the foot were at any time observable. Branches of the dorsal cutaneous nerves were cut and during two days no change in the tint of the denervated skin could be seen. The sciatic nerves in each leg of a frog were severed near the hip and exposed to the knee. The nerve of one side was then stimulated electrically over a period of 15 minutes, that of the other side being kept quiet as a control. At the end of half an hour, the melanophores in both webs had remained equally expanded. Other modifications of these tests were carried out but in all instances, according to these workers, without the appearance of any special change of coloration in the frogs tested. In consequence of this and further work, Hogben and Winton (1923) concluded that if a direct chromatophoral innervation exists in frogs, it is not significant for the normal colour changes in these animals. This conclusion was subsequently reaffirmed by Hogben (1924). At about the same time, Kahn (1922) tested the control of colour changes in frogs by operating in such a way as to eliminate the sympathetic chain. As a result of such operations, he could observe no associated colour changes in the animal, but on other grounds, chiefly pharmaceutical, he declared in favour of chromatophoral innervation, probably both sympathetic and parasympathetic in character.

Notwithstanding the strong evidence against the nervous control of chromatophores put forward by Hogben and Winton, there has continued to appear from time to time observations that were difficult to harmonize with this view. As early as 1915 Königs had noted the contraction of frog melanophores on the electric stimulation of the nerves of the hind leg, even when this operation was carried out in a way to exclude vasomotor effects. Elze (1923) also observed a darkening of the legs and flanks of frogs from which the sympathetic chains had been removed, an observation confirmed in general by Speranskaja-Stepanova (1930) and Hafter (1932). Kropp (1927), on the basis of carefully conducted experiments on *Rana pipiens* and *R. clamitans*, was led to declare in favour of chromatophoral innervation in frogs in that he found that on severing a nerve in the leg of this animal the melanophores of the web of the given leg usually expanded. He noted further that on stimulating spinal nerves and sympathetic roots in the frog's sciatic plexus the skin of the leg supplied by these trunks blanched. Although such responses were not considerable, they were certain. These observations, so far as they touch on the darkening of denervated areas of a frog's skin, have been confirmed by Thörner (1929) and by Karásek (1933), both of whom have shown further that the darkening is independent of such vasomotor changes as accompany the operation. Moreover, very recently Stoppani (1941 c, 1942 b) has reported, in confirmation of what has already been done, that when the sciatic nerve of the South American toad *Bufa arenarum* was cut the corresponding leg darkened slightly and when such a nerve was stimulated electrically the leg blanched somewhat.

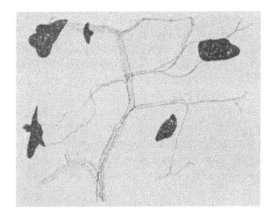

Fig. 89. Nerve fibres and melanophores in *Salamandra maculosa*. Perotti, 1928, 197.

In the year following Kropp's work, Perotti (1928) published an extended histological study by means of the Golgi method of the innervation of the chromatophores in *Rana esculenta*, *Triton cristatus* and *Salamandra maculosa*. According to Perotti, in preparations of amphibian skin not too heavily impregnated by the Golgi deposit, nerve-fibres may be seen in varying numbers to approach and surround melanophores and in some instances to terminate within them. The relation of the fibres to the melanophores is so characteristic and so intimate as to leave no doubt that these nervous elements are true chromatophoral nerve-fibres (Figs. 89 and 90). No fibres could be traced to the guanophores which were supposed by Perotti to be indirectly activated. Histological results somewhat similar to those obtained by Perotti, though much less complete, have been reported by Tusques (1939) on frog tadpoles. Thus the belief in a certain degree of chromatophore innervation in amphibians has been revived

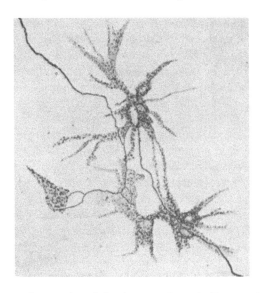

Fig. 90. Innervation of the chromatophores in *Rana esculenta*.
Perotti, 1928, 198.

by the work of these most recent investigators. The validity of their assumptions has been accepted by Vilter (1933 to 1941) who, in a series of recent contributions already discussed, has advanced a general theory of colour control for amphibians and other vertebrates in which sympathetic innervation is set off as an antagonist of the pituitary secretion.

E. INTERMEDINE

In the study of the relations of the pituitary gland to amphibian colour changes, hypophysectomy, early employed by Krogh (1922), by Hogben and Winton (1923), and described in detail by Hogben (1923), has been practised by a host of investigators. The outcome of their work has been most uniform, for wherever hypophysectomy has been applied to amphibians the animals after the operation have quickly blanched and have remained continuously pale (Stoppani, 1942*b*).

Hogben and Winton hypophysectomized dark frogs and noted that six hours after the operation the animals exhibited marked general pallor. This change took place in a damp, cool situation and on a black background, conditions under which normal frogs used as controls darken in four hours. Tests of this kind were carried out on frogs by very many later workers and always with confirmatory results. Beams and King (1938) attacked this problem in *Rana pipiens* by means quite different from simple hypophysectomy. They centrifuged the eggs of this species in the early gastrula stage and found that they could thus prevent the formation of the hypophysis in some of the tadpoles. In those tadpoles where the hypophysis failed to appear, the animals were always albinotic in consequence of the contraction of the melanophores that were present and of the relatively small number of such cells. Thus in an indirect way the pallor of this amphibian was shown to be dependent upon the absence of the pituitary gland.

Not only did frogs show blanching on the loss of the pituitary glands but, as the following enumeration shows, many other amphibians exhibited the same: *Hyla* (Houssay and Ungar, 1924 *a*, *b*; Przibram, 1932 *a*; Geiringer, 1935; Burch, 1938), *Bufo* (Giusti and Houssay, 1924; Houssay and Ungar, 1924 *b*; Stoppani, 1942 *b*), *Leptodactylus* (Houssay and Ungar, 1924 *a*; Stoppani, 1942 *b*), *Axolotl* (Blacher, 1927; Woronzowa, 1928 *b*), *Amblystoma*, *Pleurodeles*, *Molge* (Blacher, 1927), *Xenopus* (Hogben and Slome, 1931), *Salamandra* (Geiringer, 1937), *Hynobius* (Sasaki and Nakamura, 1937), and others.

Some ambiguity exists in the statements of various writers as to which lobe of the pituitary complex elaborates the material that induces melanophore expansion. Some authors locate the source of this material in what they call the posterior lobe (Blacher, 1927; McLean, 1928; Hogben and Slome, 1931; Geiringer, 1935) and others in what they believe to be the intermediate lobe (Swingle, 1921 *a*; Giusti and Houssay, 1924; Burch, 1938). The separation of the lobes in this gland is by no means easily accomplished and it is entirely possible, as has been intimated by Hogben and Winton (1922 *a*), that the tissue of one lobe infiltrates more or less that of the others, thus making the separation really impossible. Under such circumstances to name one lobe or the other as the source of the activating substance may have very little meaning. Further confusion has been added through the use of commercial pituitary extracts, the sources of which cannot always be relied upon with certainty. Thus "pituitrin", "infundin", and the like, usually assumed to be extracts of the posterior lobe, almost invariably contain more or less substance from the intermediate lobe. Under such circumstances, melanophore responses that have been ascribed to a hormone from the posterior lobe may in reality have been excited by one from the intermediate lobe. The more conclusive pieces of work on this question point very definitely to the intermediate lobe of the pituitary complex as the source of the colour hormone. Thus Swingle's discovery (1921 a, b), that when an intermediate lobe is implanted into a hypophysectomized and consequently pale tadpole the animal regularly darkens, is strong evidence in favour of, this lobe as the source of the activating material. In very recent work Burch (1938) has shown that, by an appropriate operation on the embryos of Hyla, the development of the intermediate lobe in the tadpoles of this amphibian can be completely suppressed, whereas the rest of the pituitary gland continues to develop. The resulting tadpoles thus deficient in intermediate lobes are reported as albinos. Here again the hormone controlling melanophore expansion appears to depend upon the intermediate lobe. Evidence of the kind here reported convinced Jores and his associates, (Jores, 1932, 1938; Jores and Lenssen, 1933; Jores and Will, 1934) in common with many other workers, that the hormone concerned with colour changes is a product of the intermediate lobe. To this substance the name intermedine has been given by Zondek and Krohn (1932a). Jores, however, has carried the analysis a step further by attempting to show that the secretion of the intermediate lobe contains two distinct hormones. One of these, which he calls the melanophore hormone (Hogben and Winton, 1922a), is believed to bring about the expansion of frog melanophores and the other, for which he reserves the name intermedine (Zondek and Krohn, 1932a), is supposed to induce a like response on the erythrophores of the minnow. This distinction is based on certain chemical differences. Thus when the extract of the intermediate lobe is boiled with alkali the effect on the melanophores is increased while that on the erythrophores is decreased as compared with the efficiency of the untreated extract. This and other pieces of evidence have led Jores to the assumption already stated. Jores' work, however, has been repeated by Böttger (1937a, b, c), who has been unable to confirm it. Böttger has therefore returned to the position held by the majority of workers in this field; namely, that a single hormone from the intermediate lobe, which may well be called intermedine, is the means of dispersing the pigment in both melanophores and erythrophores. This whole subject, however, is far from settled.

The early investigators in this field of research not only demon-strated the blanching of amphibians on the loss of their pituitary glands, but they also showed that when these glands or extracts from them were introduced into blanched and often hypophysectomized individuals, these animals temporarily recovered their dark colour and retained it even under conditions in which they as normal amphibians would have become pale. Hogben and Winton (1923) carried out tests to this end on *Rana temporaria* and *R. esculenta* and used as an expanding agent commercial infundin. To this agent hypophys-ectomized and consequently pale frogs showed no change when strengths of one part in a million to five parts in one hundred thousand were injected, but to injections of one part in ten thousand or stronger the frogs darkened. Eventually these frogs blanched again, thus showing the temporary nature of this effect. The darkening of pale frogs has been accomplished by Houssay and Ungar (1924 b), by Omura (1930), by Przibram (1932 a), by Koller and Rodewald (1933), by Rodewald (1935), by Zondek (1935 a), by Kleinholz (1935), and by Parker and Scatterty (1937), of pale toads by Giusti and Houssay (1921), by Houssay and Giusti (1929), and by Aubrun (1935 a, b), of *Xenopus* by Hogben and Slome (1931, 1936), and of *Necturus*, *Amphiuma*, *Axolotl*, and *Triturus* by Osborn (1936). Houssay and Ungar (1924 b) darkened frogs by the use of intermedine from the pituitary glands of twenty-two species of vertebrates, from the rays to man.

Franz and Gray (1941) have called attention recently to a significant type of response in frog melanophores to obstetrical pituitrin (Parke-Davis). This extract of the pituitary gland, in a series of sixteen dilutions from 1 : 3 to 1 : 8191, was injected into normal and hypo-physectomized frogs. In the normal frogs the melanophore responses increased as the dilutions decreased in strength to 1 : 24 and thereafter decreased. At 1 : 1023, the colour response was about the same as at 1 : 3 beyond which it diminished till it was hardly appreciable at all at the weakest dilution used. When the several dilutions were injected into hypophysectomized frogs, there was a steady falling off of melanophore response following the order of the dilutions used from the strongest to the weakest. These results make it evident that the pituitary gland itself must have an effect on frog melanophores in consequence of injected pituitrin.

The intermedine system in amphibians is assumed to operate in the following manner. When the eyes of the animal are illuminated by light from a light-absorbing environment, the nerve impulses that are thus produced pass from the retinas of the animal over its optic tracts and through the deeper parts of its brain to the intermediate lobe of the pituitary gland. Here intermedine is freed and is carried in the blood and lymph to the melanophores of the skin which are thereby induced to expand. The correctness of the various steps in this assumed course of connections has been in part established. It has long been known that cutting the optic nerves of amphibians is as effective a method of checking colour changes in these animals as is the removal of the eyes. Furthermore, Geiringer (1937) has cut the nervous connections between the optic centres and the pituitary gland in the neighbourhood of the root of the hypophysis in *Hyla* and has shown that though vision is unimpaired, the colour changes so far as intermedine is concerned cease. This loss of response does not follow other operative procedure on the midbrain so long as the optic-pituitary tracts are left undisturbed.

Vunder (1931), who was aware that a single pituitary gland would darken a pale hypophysectomized axolotl, found that the same darkening would result from pricking with a needle the normal pituitary gland of this amphibian. He, therefore, regarded the darkening caused by a transplanted gland as due in part to trauma. This view was adopted finally by Etkin (1935), who with Rosenberg (1938) demonstrated that in normal life during the pale phase of an animal the pituitary gland must be inhibited from discharging its intermedine. This inhibition is believed by Etkin and Rosenberg to be carried out over the infundibular tracts.

Shen (1937 b, 1939 a, b), who has studied with much fullness the action of drugs on the melanophores of the frog, *Rana temporaria*, has pointed out that so far as the pituitary complex is concerned there are two classes of reagents. The first of these includes piperidino-methyl-3-benzodioxane (F. 933), diethyl-amino-methyl-3-benzodi-oxane (F. 883), chloraloxane, nicotine, and yohimbine, all of which darken frogs only when the pituitary gland is present, and the second includes ether, chloroform, amyl nitrite and strychnine, which darken hypophysectomized frogs. It is probable that those in the first class act indirectly on the frog melanophores by exciting the discharge of intermedine from the pituitary complex, while those in the second act directly on these colour-cells.

Numerous incidental observations scattered through the literature of this subject bear witness to the first importance of intermedine as a melanophore activator. Thus the blood from a dark frog or toad will darken a pale one by inducing melanophore expansion (Stoppani,

1941 b), but where the recipient is kept in darkness the intermedine in its blood is rendered ineffective probably by the action of some substance non-pituitary in origin (Rodewald, 1935; Jores and Hoeltje, 1936). The expansion of frog melanophores is a convenient and delicate biological test for the presence of intermedine in mammalian blood (Belehradek, 1937). The importance of acidity in tests of this kind has been pointed out by McLean (1928), who states that on the acid side of neutrality the melanophores of *Rana pipiens* tend to expand, on the alkaline side to contract. Przibram's contention (1932 a), based on a study of *Rana fusca, R. esculenta*, and *Hyla arborea*, that a weak dose of infundin brings about melanophore expansion and a stronger one contraction has not been substantiated (Abramowitz, 1937 c). The effect of intermedine on amphibian melanophores may depend in part upon the threshold of the skin for this substance (Woronzowa, 1928 a, b). Thus the melanophores in the pale race of axolotls is believed by Woronzowa to have higher thresholds for intermedine than those in the dark race, a condition which this author believes to rest on a genetic difference in these races. From such condensed statements, the supreme importance of intermedine as a darkening agent in amphibians must be evident. Intermedine is the same as the B-substance of Hogben and Slome's terminology (1929, 1931, 1936).

F. ADRENALINE

Adrenaline is probably the most efficient of the known agents for the concentration of pigment in amphibian melanophores. A solution of one part in 200,000 will contract the melanophore pigment in *Necturus* in about three and a half hours, after which it will remain contracted for a day (Dawson, 1920). Much the same is true of the axolotl (Huxley and Hogben, 1922), of common frogs, toads and treefrogs (Hogben and Winton, 1922 c; Przibram, 1932 a; Geiringer, 1937; Parker and Scatterty, 1937; Stoppani, 1942 b). If 1 c.c. of 1 : 10,000 adrenaline is injected into the dorsal lymph space of a frog, *Rana pipiens*, the animal will begin to blanch within 10 minutes and will shortly thereafter reach full paleness. In the course of a day or so the frog will return to its former dark state. A microscopic examination of the webs of such blanched frogs will show the melanophore pigment in them to be nearly if not fully contracted, a condition to be expected from the extreme paleness of the animals. Pierce (1942) had investigated the blanching by adrenaline of the common frog *Rana clamitans* in relation to the weight of the individuals tested. To a uniform dose of adrenaline, frogs weighing 10 g. remained pale some 12 hours, those weighing 70 g. only 4 hours. Stoppani (1941 b, 1942 b), after the experimental removal of the adrenal glands from *Bufo arenarum*, has been led to conclude that the adrenaline from these glands is the chief factor in causing the blanching of this toad. On rare occasions frogs when roughly handled have been seen quickly to turn pale, a step that is reminiscent of what has been called "excitement pallor" in other chromatic animals and has been attributed to an outpouring of adrenaline by the creature concerned. Such an assumption, however, is by no means certain (Parker and Scatterty, 1937).

G. Extracts, Drugs, and Chemicals

To test the specificity of the pituitary substance that darkens frogs, Hogben and Winton (1922 b) injected into pale specimens of *Rana temporaria* solutions of a large number of drugs and the like in order to discover the extent to which the melanophores would be expanded. According to these workers, the frogs remained pale on receiving injections of pilocarpine, atropine, strychnine, curare, histamine, ergotoxine, tyramine, veratrine, digitalis, barium chloride, sodium nitrite, and caffeine. To only two solutions, those of apocodeine and of nicotine, did the frogs darken and to these reagents the darkening was only slight. From such determinations, it would seem probable that the darkening substance contained in the pituitary secretion was highly specific. Hogben and Winton (1922 c) subsequently extended these tests by using dark as well as pale frogs. Frogs darkened by having been kept in an appropriate environment or by the injection of pituitary extract became pale when injected with tyramine, ergotoxine, caffeine, and cocaine, but remained unchanged when they were injected with histamine, barium chloride, sodium nitrite, digitalis, veratrine, pilocarpine, atropine, or Ringer's solution. Pale frogs did not change colour when they received injections of atropine, pilocarpine, histamine, ergotoxine, tyramine, veratrine, caffeine, barium chloride, sodium nitrite, strychnine, and digitalis. As in the previous paper by these authors, nicotine and apocodeine are said to have darkened frogs. In a later paper, Hogben and Winton (1923) reinvestigated the action of pilocarpine in view of Kahn's statement (1922) that this agent induced melanophore expansion. They were unable to confirm Kahn's findings.

The effects of a number of pharmacological agents on the colour of the South American frog *Leptodactylus* have been briefly described by Houssay and Ungar (1925 a). According to these authors many more agents induce darkening than pallor in this frog. The following darken: veratrine, brucine, thebaine, strychnine, picrotoxine, silver nitrate, papaverine, codeine, morphine, caffeine, barium chloride, theocine, nicotine, acetylcholine (in strong doses only), insulin, atropine, curare, various alcohols, ether, chloroform, chloral hydrate, chloralose, chloretone, serpent venom, and lactic acid. All agents tested were believed to act in an indirect way except the pituitary extract, caffeine, and barium chloride. The few blanching agents that were found for *Leptodactylus* were ergotoxine, adrenaline (which after the frog had been injected with ergotoxine would darken it), potassium chloride, and ergamine. In the action of these various drugs on this frog no evidence was found for distinguishing sympathico- from parasympathico-mimetic conditions. Blanching was induced by both ergotoxine (sympathetic paralyser), and adrenaline (sympathetic activator), and darkening by atropine (parasympathetic paralyser) and by acetylcholine, nicotine, and pilocarpine (all parasympathetic activators). The following inactive agents were recorded: uric acid, urea, narcotine, anatine, chlorhydrate, peptone, guanidine, santonine, theobromine and many organic salts. Many of these determinations have been very recently confirmed not only on *Leptodactylus* but also on *Bufo* (Stoppani, 1942 a, c).

Teague and Noojin (1938), in a recent contribution to this subject, have submitted the following lists of agents with the responses to them that they have observed in *Rana pipiens*. The following materials, said to darken frogs, have been found by these workers to have little effect upon normal frogs and none at all upon hypophysectomized ones: strychnine, alcohol, picrotoxine, silver nitrate, copper sulphate, barium chloride, atropine, eserine, choline, acetylcholine, digitalis, curare, morphine, paraldehyde, chloral hydrate, chloroform, ether, amyl nitrite, nitroglycerine, apocodeine, cantharides, weak hydrochloric acid, sodium hydroxide, coramine, F. 933 (piperidinomethyl-3-benzodioxane), insulin, extracts of testis, of liver, brain, thyroid and muscle. The following substances darkened normal frogs but had no action on hypophysectomized individuals: veratrine, brucine, pilocarpine, chloralose, metrazol, and yohimbine. When the results of these three sets of investigators (Hogben and Winton, Houssay and Ungar, and Teague and Noojin) are compared, it is evident that the discrepancies in their work are so considerable that a critical reinvestigation of the whole field will be necessary before even probable generalizations can be reached. One conclusion, however, appears to be fairly supported and that is that the majority of the drugs tested appear to act on the melanophores indirectly rather than directly. But even this is by no means certain.

Both Uyeno (1922) and Hewer (1923) state that oxygen induces a concentration of melanophore pigment. Carbon dioxide, though said to be without effect on the dark colour-cells by Hewer (1923), is claimed by Uyeno (1922) and by Houssay and Ungar (1925 b) to be a darkening agent whose action, however, may depend upon acidity (McLean, 1928).

Kropp (1929) has pointed out that when an aqueous extract of the eyes of dark-adapted tadpoles of *Rana clamitans* is made in the absence of light and the solution thus obtained is injected into other tadpoles, an expansion of melanophores is brought about. This condition appeared in from 5 to 30 minutes after the injection and lasted some 2 to 3 hours. Extracts from the eyes of light-adapted tadpoles yielded no clear-cut results. Kropp was unable to obtain similar reactions from adult frogs, but in this respect Gray and Little (1939) were more successful. These workers found that when extracts from the eyes of pale frogs were injected into dark-adapted and into blinded frogs, in both instances a concentration of melanophore pigment took place. When extracts from the eyes of dark frogs were injected into pale ones, dispersion of pigment occurred but less regularly. That these results were not due to neurohumours in the residual blood of the extracted eyes is probable from the fact that extracts from large amount of muscle had on injection no chromatic effect. Extracts of eyes when injected into hypophysectomized frogs induced no colour changes. Hence the effective agent in the eye extracts did not act directly on the melanophores, but probably influenced them through the pituitary gland. Extracts of central nervous organs appeared also to be able to play a significant part in these colour responses. These chromatic changes from eye-ball extracts and from central nervous parts recall vividly the conditions already described in crustaceans.

H. THE BLINDED STATE

That the ordinary, adaptive colour changes of frogs (Fig. 91) and other chromatic amphibians cease with the loss of the eyes is a fact recorded by almost all workers in this field from the earliest (Lister, 1858) to the latest (Stoppani, 1942 b). Such blinded animals, however, have

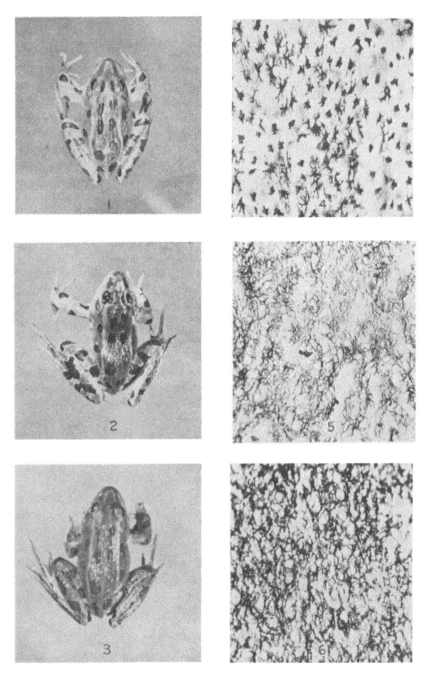

Fig 91 Dorsal views of the same individual frog, *Rana pipiens*, in three states, with microscopic views of the conditions of the melanophores in the web of the foot for each state, 1, normal frog in pale state, 2, eyeless frog, somewhat dark, 3, normal frog in dark state, 4, 5 and 6 are microphotographs of the melanophores in the web of the foot corresponding to the colour phases shown in 1, 2, and 3 respectively Parker, Brown and Odiorne, 1935, 442

not entirely lost their power to respond by changes of tint to changes in the illumination of the environment. Babák (1910*b*, 1912) showed that after the larvae of *Amblystoma* had been deprived of sight they still were able to respond to illumination differences. They became slightly dark in bright light and somewhat pale in darkness. These changes were not pronounced but they were unquestionable. Fischel (1920), in his study of *Triton alpestris*, stated that though this was an amphibian with characteristic colour changes, these ceased when the animal was blinded and that the eyeless individuals were dark in bright illumination. *Salamandra maculosa* when blinded lost its ability to change in colour with changes in the surroundings, but was pale in darkness and was very dark in the light. *Anuran* larvae responded in much the same way, but were slower and less certain in these respects than were young urodeles. Slome and Hogben (1928, 1929; Hogben and Slome, 1931), who experimented with blinded *Xenopus*, noted that this amphibian on the loss of its eyes ceased to respond to a white or a black background, and assumed an intermediate tint which, however, became slightly darker in light than in darkness. The same condition has been recorded by Parker, Brown and Odiorne (1935) for *Rana pipiens*, and a similar state has been observed by Stoppani (1942*b*) in *Bufo arenarum*. Thus a general rule appears to hold that amphibians when blinded become somewhat dark in tint, lose most of their former capacity to respond to the illumination of their surroundings, but still respond slightly to it by being a little darker in bright light than in darkness.

The responses of these blinded amphibians are obvious illustrations of those colour states that were described in the introductory portion of chapter IV of this volume under the headings of primary and secondary phases of colour change. In the present instances, the given animal by the loss of its eyes is made to pass from a secondary to a primary condition. As a matter of fact this whole subject of the alteration in the type of chromatic response was first noticed among amphibians and described for these animals by Babák, Laurens and others. Babák noted these colour alterations, not only in young amblystomas, but also in the larvae of *Rana fusca* and of *R. esculenta*, though they were not so striking in these animals as in the young of *Amblystoma*. It is doubtful whether they occur at all in *Bombinator* or in *Hyla*. Laurens (1917) reported that in *Amblystoma tigrinum* the seeing as well as the eyeless animals showed melanophore expansion in the light and contraction in darkness, a retention of what may have been an earlier ontogenetic phase.

I. Persistent Environmental Colour Effects

That the larvae of *Salamandra maculosa* when reared on a light-tinted background become gradually pale and on a black background become gradually dark has been shown to be true not only by older workers but by those of most recent times (Geiringer, 1935). This type of change was confirmed by Kammerer (1913), who proceeded then to test these changes for possible inheritance. After having bred salamanders continuously over a series of generations that extended through more than a decade, Kammerer was led to conclude that the continual crossing of pale individuals with each other and their retention on a light-tinted background led to the establishment of a permanently pale race and the corresponding treatment of dark individuals on a black background to a dark race. Thus the inheritance of an acquired colour pattern appeared to be established and the interest of biologists was greatly aroused in this example as supporting evidence in favour of the Lamarckian principle in evolution.

Herbst (1919) undertook a repetition of Kammerer's experiments with the result that he was unable to confirm the first step; namely, that young salamanders became pale on a light-tinted background and dark on a black one. He therefore denied without going further the validity of Kammerer's contention and regarded Kammerer's results as probably the outcome of unconscious selection. Von Frisch (1920), who had also been testing this question, surveyed the field and pointed out certain peculiarities in Herbst's technique which in his opinion had prevented Herbst from attaining results which von Frisch himself had been able to reach. Von Frisch maintained that it was possible to darken or blanch young salamanders by keeping them continuously on an appropriate background, a fact well confirmed since the time of von Frisch's work. Von Frisch, however, found no ground for supposing that such changes were heritable. This investigator showed further that the colour changes under consideration, though characteristic in general of the life of the salamander before metamorphosis, extended somewhat into adult life, to disappear completely, however, before full maturity was reached. This position was in the main supported by the subsequent work of Herbst (1924, 1927) and of Herbst and Ascher (1927), in which the non-inheritance of such changes was demonstrated from breeding tests, a conclusion maintained by Herbst notwithstanding statements to the contrary by Przibram and Dembowski (1922) and by MacBride (1925).

It now seems quite clear that the final tints of salamanders can be profoundly and permanently influenced by the environment as has been shown by Vilter (1931 a), who has worked on this question with axolotls. These salamanders reared for 17 months increased from 13 to 17 cm. in length; those on a black background became very dark and those on a white one pale olive green. When the melanin was extracted by the method of Piettre from representative individuals of these two groups, it was estimated to be about four times as great in amount in the dark specimens as it was in the pale ones. The formation of melanin appears to be excited by black surroundings and to be inhibited by white ones. There is, however, no conclusive evidence as yet that such changes are inherited and intensified as was originally maintained by Kammerer.

J. Amphibian Neurohumours and Colour Systems

The great majority of modern investigators of amphibian colour changes accept without reservation the view that the dark phase in these animals is excited by the pituitary secretion intermedine. Since the time of Hogben and Winton (1922 a), evidence in favour of this opinion has been steadily growing. The pale phase in the colour changes of these animals has been less satisfactorily explained. In discussing the general question, Hogben and Winton (1923) declared that the experimental evidence at their command "points to the probability that the colour response in the frog is determined by a balance between adrenal (medullary) and post-pituitary secretion" From this standpoint the frog would be a bihumoral amphibian (Parker and Scatterty, 1937) in that its pale phase would depend upon one neurohumour, adrenaline, and its dark phase upon another, intermedine. Subsequently Hogben was led to conclude that only one neurohumour, intermedine, was necessary for these colour changes. In his volume on the *Pigmentary Effector System*, Hogben states that "it is at least justifiable to conclude...that pituitary secretion is the only significant factor which we need to postulate to account for the regulation of colour response in adult frogs". Under such circumstances the pale phase of the frog is presumably the result of the disappearance from the frog's blood of the intermedine which had darkened the animal. From these quotations the uncertainty of the situation must be evident.

In a later investigation, Hogben and his associate, Slome (1929, 1931, 1936), advanced evidence to show that in addition to intermedine a second and heretofore unknown neurohumour, a concentrating one, was present in amphibians. This evidence was derived from a study of *Rana fuscigula*, but especially from work on *Xenopus laevis*. When *Xenopus* is made to change from a pale tint or from a dark one to the intermediate tint produced by keeping the animal in the dark or by blinding it, and when it is made to change from this intermediate tint to a fully blanched or to a fully darkened one, it does so at different rates, being quicker in its changes toward the pale state than toward the dark one. Furthermore, when it makes full changes from one extreme in tint to the other, it accomplishes the change to the dark

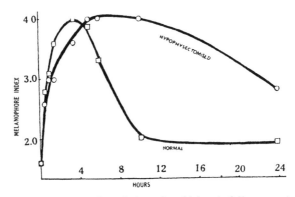

Fig. 92. Plottings of melanophore indexes in which 1 is full contraction and 5 is full expansion against time in hours of normal and of hypophysectomized toads, *Xenopus*, into which equivalent amounts of pituitary extract had been injected. The final blanching of the toads is much more rapid and considerable in the normal individuals than in the hypophysectomized ones. Hogben and Slome, 1931, 34.

state more quickly than to the pale one. These temporal differences, according to Hogben and Slome, are inconsistent with the view that only one humoral agent is herein concerned and call in their opinion for at least two such agents. They have also shown that when injections containing intermedine are given to pale normal or to hypophysectomized toads the subsequent responses of these two sets of animals are different. Both classes of toads are at first profoundly darkened, after which the normal individuals become rather quickly and pronouncedly pale, while the hypophysectomized ones blanch slowly and less completely (Fig. 92). This retardation and incompleteness in the blanching of the second set is interpreted by Hogben and Slome as due to the absence from their blood of a neurohumour that would concentrate pigment, the W-substance. Since the W-substance was lacking in the hypophysectomized individuals, Hogben and Slome were led to look for its place of origin somewhere in the pituitary complex. This complex is differently constituted in *Rana* and in *Xenopus*. In *Rana* the pars tuberalis of this complex, the part under special consideration, is represented by two small plaques which are fixed to the tuber cinereum some distance from the anterior lobe. In *Xenopus* the pars tuberalis is a single forward lip continuous with the pars anterior and so located that the removal of the pars anterior or of this and the posterior lobe involves the removal of the pars tuberalis. In experimental procedure these anatomical relations are of first importance. The removal of the anterior lobe from *Rana* (without the removal of the pars tuberalis) did not affect the colour changes in this animal. In *Xenopus*, the removal of the anterior lobe with the pars tuberalis completely abolished the white background response (the toad remained maximally dark). In *Rana*, after the removal of the pars tuberalis the frog was less responsive to intermedine than was the normal pale frog. In *Xenopus*, the removal of the pars anterior, the pars posterior, and the pars tuberalis left the toad more responsive to intermedine than was the normal pale toad. The pallor of *Rana* after the removal of the pars anterior and the pars posterior, but without the removal of the pars tuberalis, was more intense than that of a normal animal on a white background. The removal from *Xenopus* of the entire pituitary gland (including the pars tuberalis) did not produce in this toad a pallor as intense as that due to a white background. Such observations led Hogben and Slome to conclude that in these amphibians the pars tuberalis produced the W-substance by which the animals are blanched and they declared in favour of the view that these forms were bihumoral in that they controlled their melanophores in part through W-substance from the pars tuberalis and in part through B-substance (intermedine) from the intermediate lobe. A renewed study of the structure of the anuran pars tuberalis has been recently undertaken by Atwell (1937).

This system of colour change was tested on *Rana pipiens* by Parker and Scatterty (1937). This frog on the loss of its whole pituitary gland will begin to blanch in half an hour or less and will continue to become lighter until, after four or five hours, it reaches full paleness. On injecting into the subdermal space of such a pale frog commercial pituitrin, aqueous extract of frog pituitary gland, or intermedine, the frog will start to darken in a quarter of an hour and will become fully dark in two to three hours. The blood from a dark frog when injected into a pale one will darken the recipient. Hence it was believed by

Parker and Scatterty that *Rana pipiens* darkens in consequence of the action of intermedine on its melanophores. The pale phase of this frog can be excited by white surroundings, by adrenaline, and by hypophysectomy (Table 11).

TABLE 11. Means of blanching, numbers of frogs used in each set of tests, and average states of the melanophores expressed on a scale in which full concentration of pigment is represented by 1 and full dispersion by 5 (Parker and Scatterty, 1937, p. 304)

Means of blanching	White surroundings	Adrenaline	Hypophys- ectomy
Numbers of frogs	32	31	18
Melanophore states	1·82 ± 0·39	1·31 ± 0·11	1·27 ± 0·10

As Table 11 shows, *Rana pipiens* does not blanch as fully on a white background as it does from an injection of adrenaline or from hypophysectomy. Adrenaline and hypophysectomy reduce it almost to full paleness. The blood of a normally pale frog (white background) when injected into a dark one will not cause the dark frog to blanch to any degree. Presumably, therefore, the blood of a normally blanched frog contains no agent concerned with the concentration of its melanophore pigment. When a decerebrate dark frog is fully irrigated with Ringer's solution, it becomes as pale as one blanched with adrenaline or by hypophysectomy. It is, therefore, believed by Parker and Scatterty that *Rana pipiens* blanches not in consequence of a special concentrating neurohumour in its blood but because of the loss of intermedine from this fluid; in other words, that this frog is unihumoral and that intermedine is the only neurohumour concerned with its colour changes. This view has been criticized by Soderwall and Steggerda (1937; Steggerda and Soderwall, 1939), who claim that when the pars tuberalis of *Rana pipiens* is destroyed by local cauterization, the frog is then unable to blanch completely. These authors, however, have not discussed the possibility of a slight excitation of the intermediate lobe by the cautery whereby a mild darkening might ensue.

The various investigations on amphibian colour changes surveyed in this section show that whatever views may be held about the nervous control of the melanophores in these animals their colour-cells are certainly excited in the main by hormones, either one, intermedine (the B-substance of Hogben and Slome), in such unihumoral species as *Rana pipiens*, or two, intermedine and the secretion of the pars tuberalis, in such bihumoral species as *Xenopus laevis*. Such a grouping is of course very conventional, for it must be evident that further discovery may at any moment convert an assumed unihumoral amphibian into a bihumoral or even a multihumoral one.

In the discussion of amphibian neurohumours thus far disclosed, no suggestion of a special chromatic system among these agents has been brought forward. Such a system, however, has been proposed by Vilter, whose polarization theory as a matter of fact was first conceived of by this investigator in his early studies of the chromatic changes in the axolotl. This theory has already been discussed in its application to fishes, particularly to teleosts, but its bearing on amphibians must here be briefly considered. When, according to Vilter, the spinal cord of an axolotl is cut, the dark spots in this salamander's skin posterior to the incision lose their distinctness and take on the paler tint of the rest of its body. This change can also be demonstrated in segments of the axolotl's body by cutting one or more spinal nerves. The pale phase of the axolotl in consequence of these results is assumed by Vilter to be under autonomic nerve control. The dark phase is believed by him in common with most other workers to be induced by the pituitary neurohumour. That the nerve control and the pituitary control are oppositely polarized on the body of the axolotl is shown, Vilter believes, by the following experiment. If from the flank of an axolotl, a dorso-ventral band of skin extending from the dorsal dark region of this amphibian to its ventral pale region is cut off and transplanted on to the flank of another axolotl so that the dark dorsal end of the band is placed ventrally and the pale ventral end dorsally, the band on healing will gradually change its colour to agree with that of its new immediate surroundings. This change, according to Vilter, cannot result from a generally dispersed neurohumour such as intermedine but must depend upon the "function du terrain" which can be nothing other than the concentrating autonomic nerve terminals in the flank of the animal's trunk. These terminals are believed to give out a substance, a neurohumour, which is present and active ventrally and there blanches the band but which is not present in sufficient amounts dorsally to counteract the pituitary secretion. The concentration of this secretion dorsally is such as to darken the skin of that region. Hence, the dispersing neurohumour from the pituitary gland acts as an antagonist to the concentrating one from the nerve terminals and the transposed band of skin under the conditions of the test takes on a coloration normal to its position. According to Vilter, opposed chromatic polarization is thereby demonstrated in this amphibian.

The scheme of colour organization thus outlined for amphibians by Vilter is in essence the same as that which he also advanced for fishes. In the amphibians, however, it calls for a degree of nervous participation that is scarcely warranted by what is known of the chromatic innervation of these animals. According to Vilter's view, the concentrating chromatic nerves in amphibians must be about as effective in colour changes as is the pituitary gland, for the neurohumours from these two sources are mutually and effectively antagonistic. As the preceding discussion of this subject has shown, however, the concentrating chromatic nerves in amphibians are at best extremely insignificant in the colour changes of these animals, if in fact they play any part at all in these operations. They are in all instances completely overshadowed by such an agent as their assumed antagonist, the pituitary secretion. Hence the actual chromatic mechanism in amphibians rests not on two equal and antagonistic activators, but almost exclusively on one agent, intermedine, as has been assumed by the majority of recent investigators. It therefore seems probable that Vilter's hypothesis of antagonistic, polarized activators finds as little support in the amphibians as it does in the fishes, and the conception, suggestive as it is in certain respects, must be regarded as inapplicable.

The foregoing survey of amphibian colour changes indicates that these changes, complicated though they often may be, are dependent upon relatively few classes of chromatophores of which the melanophores are as a rule the really effective ones. These black cells, by covering or uncovering in various ways xanthophores and leucophores or their combinations, bring about much of the obvious play of colours in batrachians. That these changes are to some degree under the control of nerves seems probable though they are surely in no instance subject to this control to the extent assumed by the majority of the older workers nor by Vilter among the modern students. Amphibian colour changes are induced primarily by a neurohumour from the intermediate lobe of the pituitary gland, intermedine, whose disappearance from the blood of the given amphibian induces blanching. This pale phase, however, may also result from an opposing neurohumour produced by another part of the pituitary gland, the pars tuberalis, or possibly from the adrenal glands. Thus amphibians at present appear to possess a chromatic system based upon either a unihumoral or a bihumoral plan supplemented probably by a relatively weak nervous component. The slowness of amphibian colour response is in agreement with this interpretation of its mechanism.

Chapter VII

COLOUR CHANGES IN VERTEBRATES (concluded)

6. COLOUR CHANGES IN REPTILES

AMONG reptiles the only group which shows important colour changes is that of the lacertilians or lizards. Of these, the chameleon, an African representative, has had a noteworthy history, for it received mention by Aristotle and by Pliny, and since antiquity it has always been recognized as an animal remarkable for its chromatic abilities. Chameleons are not only animals whose capacity for colour change was thus known to the older naturalists, but they are forms in which this capacity reaches a very unusual height. In this respect they resemble the flatfishes among teleosts, and yet these fishes, exceptional as they are, were not noticed by the very early workers nor were they studied until comparatively recent times. Chameleons have even found their way into current literature, witness de Miomandre's *Mon Caméléon* (1938).

That the colour changes in lizards are controlled by the direct action of nerves was the opinion held by all the older investigators. This was the only view entertained by Brücke (1852) in his classical monograph on the African chameleon, and it has been accepted and handed down without dissent until a little over two decades ago. In 1916, Redfield, who worked on the horned toad, *Phrynosoma*, pointed out that the normal control of the pale phase in this lizard depended probably in part on the hormone adrenaline. Thus a hormonal or better a neurohumoral interpretation of the colour changes in reptiles was initiated. That the dark phase in the chameleon was due to a pituitary secretion was denied by Hogben (1924), who injected 0·5 c.c. of 1 per cent "infundin" into chameleons without observing any colour change. He therefore concluded that reptilian melanophores do not respond to pituitary extracts as do those of amphibians. Subsequently in his article 'Pigmentary effector system' (1924), he remarked that it was to be hoped that someone would reinvestigate this subject. In course of time this was done by a number of workers including Hadley (1928–31), Noble and Bradley (1933), Kleinholz (1935–38), Chin, Liu and Li (1938), and Parker (1938a), all of whom have demonstrated the importance of the pituitary secretion for the colour changes in the lacertilians on which they worked. In consequence of these more recent studies, the action of such neurohumours as intermedine as well as that of nerves must be reckoned with in considering lacertilian chromatics.

Of the other groups of common living reptiles, the snakes have long been known to show slight colour changes. Modern workers have not fully studied these reptiles except in the case of the rattlesnake, whose colour changes have been made the subject of special inquiry by Rahn (1940, 1941 b). According to this investigator, the epidermis of the skin of *Crotalus viridis* contains numerous small evenly branched melanophores. The derma of this snake carries a rich supply of yellowish guanophores, among which are numerous large melanophores whose branches reach into the epidermis. Hypophysectomized rattlesnakes become pale (Fig. 93) by the concentration of the pigment in their melanophores. This black pigment can be made to disperse and the snake thereby to darken temporarily by the injection into the animal of intermedine. After the injection of the agent, the snake will show the first signs of darkening in about an hour, the process reaching completion in about 12 hours. A large

Fig. 93. Two normal (dark) and two hypophysectomized (pale) prairie rattlesnakes, *Crotalus viridis*. One normal snake (*x*) has just shed its skin and appears paler than the other dark one which is about ready to cast its skin. Rahn, 1941 b, 233.

dose of intermedine will maintain complete darkening for several days. Rahn assumes that the colour changes in the rattlesnake are controlled by its own intermedine. He gives no account of a possible blanching neurohumour nor of chromatic nerves either dispersing or concentrating. Preliminary tests on other snakes made by Rahn led him to believe that a chromatic organization essentially like that in the rattlesnake was also to be found in the garter-snake, *Thamnophis ordinoides*, the ribbon-snake, *Thamnophis radix*, the bull-snake, *Pituophis Sayi*, and the Florida water snake, *Natrix sipedon pictiventris*.

In addition to snakes, another chromatic reptile has been discovered in the alligator. According to Kleinholz (1941), young alligators possess melanophores in the skin of the throat and of the ventral abdomen. These colour-cells concentrate and disperse their pigment in response to changes in the environment. An injection of adrenaline brings about pigment concentration in them, and an injection of pituitrin, pigment dispersion. There is, therefore, good reason to assume that the melanophores in the alligator are functional. No one apparently has identified active chromatophores in any representative of the remaining group of living reptiles, the turtles.

The rest of this chapter will be devoted to the colour changes in lizards.

A. Colour Changes

The colour changes in lizards range from very simple to most complex. In this respect, these reptiles resemble the teleost fishes rather than for instance cyclostomes, elasmobranchs, or amphibians. The animals

Fig 94. Two specimens of *Phrynosoma blainvillii* showing extremes of paleness and of darkness The specimen on the right had been some days in a white-walled box under bright illumination, that on the left had been for about the same time in a black-walled illuminated box. Parker, 1938 a, Pl. 1.

in these three groups exhibit only relatively slight degrees of colour change. Among the simplest of the lizards in respect to colour changes is *Phrynosoma*. The back of this animal is marked by a series of some five, jagged, dark, transverse bands; similar bands occur over the animal's short tail (Fig. 94). These bands are permanent and are set off against a pale background which extends over the trunk, head, and legs and which is subject to considerable change in tint (Redfield, 1918; Parker, 1938a). In illuminated, white surroundings this integumentary background is very pale and the dark cross bands in consequence stand out in strong contrast. In black surroundings, the integumentary background darkens considerably, but not enough to hide the cross bands which under all circumstances show as clearly observable markings. This blanching and darkening of the integumentary background in *Phrynosoma* is the one conspicuous colour change in this lizard and is dependent upon a concentration and dispersion of pigment in its melanophores. Such changes are best seen locally in the marginal scales of this lizard (Figs. 95, 96). Other lizards that seem to

Fig 95. Marginal dentate scales from a pale *Phrynosoma* (much enlarged). Parker, 1938a, Pl 1.

Fig. 96. Marginal dentate scales from a dark *Phrynosoma* (much enlarged). Parker, 1938a, Pl 1.

have as simple a system of colour change as *Phrynosoma* are *Cosmybotus, Peropus, Hemidactylus* (Ruth and Gibson, 1917), *Ptychozoön* (Boschma, 1925) and *Gecko* (Tokura, 1933). In all these forms a pattern, when it is present, persists and becomes merely more or less obvious as the integumentary background pales or darkens.

A more complicated type of colour change occurs in *Anolis*, particularly in *Anolis carolinensis*, whose chromatic activities in the field have been studied by Strecker (1928) and by Wilson (1939, 1940).

Fig 97 Two specimens of *Anolis carolinensis*, that on the left is a normally dark (brown) individual with a white mid-dorsal stripe extending from the neck to the tail, that on the right is a hypophysectomized individual and is consequently fully pale (green). Kleinholz, 1938a, 481

The colour play in this lizard ranges from dark brown to bright green, the intermediate shades including light brown, yellow, yellowish green, and emerald green (Kleinholz, 1936b, 1938a, 1938b). In some animals a mid-dorsal longitudinal band is present, extending from the cervical region for varying distances toward or even along the tail. In many individuals, this band is completely absent (Fig. 97). When the animal is in the dark phase, a regular darker pattern is discernible against the brown colour of the rest of the body. There is also to be seen in this state a darker quadrangular patch immediately posterior to each orbit and in some individuals a similar area over each scapula.

When a hypophysectomized *Anolis* is stimulated electrically by placing one electrode in the mouth and the other in the cloaca, clusters of dark spots appear in an irregular pattern over the generally green integument (Fig. 98). The possession of this changeable, mottled pattern together with a greater number of colours, green, yellow, and dark brown, show the colour system of *Anolis* to be more complex than that of *Phrynosoma*.

Perhaps the most complicated colour system of any lacertilian is that of the chameleon. *Chamaeleo vulgaris* is said by Gadow, in his account of this lizard in the *Cambridge Natural History*, to be capable of responding in combinations of green, yellow, brown, black, and white, with their various shades in almost endless variety. Keller is quoted by Hogben in his 'Pigmentary effector system' (1924) as ascribing to the chameleon "a fine gradation between orange, through yellow and green to a dark olive, blue-grey or black, while another may turn from a whitish or flesh-coloured tint to red-brown, then lilac-grey and finally neutral grey". Zoond and Eyre (1934), who worked on the dwarf chameleon of South Africa, *Chamaeleo pumilus*, state that this lizard possesses a

Fig. 98. *Anolis carolinensis*, showing the mottled pattern immediately after electrical stimulation for two minutes. Kleinholz, 1938 a, Pl. 3.

constant integumentary pattern consisting of stripes, patches, and even individual tubercles of different colours, within each of which a change from pale to dark or the reverse may take place. The markings of a typical individual are indicated in Fig. 99. Here the components of the pattern are designated, as band (*A*), islands (*B*), margin (*C*), and back (*D*). In extreme pallor the band is pale fawn colour or even white, the islands are pale grey to pale fawn, the margin pale grey or pale blue, and the back yellow-green. In the intermediate state the band ranges through orange to brown, the islands usually assume dark shades of blue or grey, the margin is bright blue, and the back is green. In the fully dark phase the whole skin becomes black. In many chameleons large orange tubercles are arranged in rows on the back. As a result of the variations found in any random group of chameleons in any given situation they may present a heterogeneous appearance which really defies a general description. These statements show clearly how complex the colour system in the chameleon is and how different from that of the simple colour plan of such a form as *Phrynosoma*.

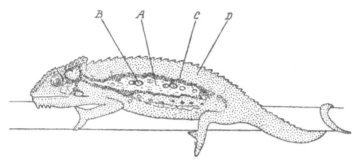

Fig. 99 Typical pattern of the skin of *Chamaeleo pumilus*. *A*, band; *B*, island ; *C*, margin, *D*, back   Zoond and Eyre, 1934, 30.

The periods of time required by lacertilians to change from pale to dark and the reverse are long rather than short. A number of these intervals are recorded in Table 12.

As can be seen by inspecting Table 12, the changes from dark to pale usually require more time than do the reverse ones. Though a few

TABLE 12.   Times for changes in various species of lizards from pale to dark and the reverse as recorded by different authorities

Species of lizard	Authorities	Pale to dark	Dark to pale
Chamaeleo vulgaris	Brucke, 1852	Few min.	½ hr.
C pumilus	Zoond and Eyre, 1934	4 min	Slow
Anolis carolinensis	Carlton, 1903	4 min.	25 min
A carolinensis	Kleinholz, 1935	5–15 min.	20–30 min.
A equestris	Hadley, 1928	1 min	25 min
A porcatus	Hadley, 1928	1 min	15–18 min.
Phrynosoma blainvillii	Parker, 1938	¼ hr.	½ hr.
P douglassii	Pierce, 1941	30 min	50 min.
P modestum	Pierce, 1941	1 hr	2 hr
Cosmybotus platurus	} Ruth and Gibson, 1917	} About ½ hr	} About ½ hr
Peropus mutilatus			
Hemidactylus frenatus			
H luzonensis			
Ptychozoon homalo-cephalum	Boschma, 1925	15 min.	

changes occur in short intervals, the great majority of them are slow, suggestive of a neurohumoral or mixed nervous and neurohumoral activation rather than a purely nervous one. A daily rhythm in the colour changes in lizards is probably not unusual. According to Redfield (1918), it is easily observable in *Phrynosoma cornutum*. At night this lizard is reported as pale. In the early morning it becomes dark but during midday it is again pale. As' evening approaches it redarkens to be followed by the final blanching characteristic of night. Thus the melanophore pigment in *Phrynosoma* is concentrated at night and midday and dispersed in the morning and afternoon. Apparently this sequence of change is also followed by *P. blainvillii* (Parker, 1938a). In *Chamaeleo pumilus* a similar series of alterations in colour has been recorded by Zoond and Eyre (1934), who showed further that this rhythm was disturbed by blinding these lizards, after which they were dark in bright light and pale in dim light or darkness. Rahn and Rosendale (1941) studied the diurnal rhythm in *Anolis*, which is green at night and usually brown in the daytime. This rhythm has been maintained in continuous light or in darkness for as long as 18 days. It is supposed by these authors to depend upon a secretory rhythm of the pituitary gland. Chin, Liu and Li (1938) have emphasized the great importance of moisture as an exciting agent in the colour changes of the gecko. Wilson (1939, 1940) has noted that at temperatures below 13° C. *Anolis* is always brown, and at those above 44° C. it is always green.

B. CHROMATOPHORES

The chromatophores in several lacertilians have been incidentally described by Schmidt (1910, 1911, 1912, 1913a, 1913b) in his accounts of the skin in these animals. In 1917 Schmidt published a most complete monograph of the colour-cells of reptiles. This has been well condensed by Ballowitz (1931) in his survey of vertebrate pigment cells.

According to Schmidt, the lacertilian skin may contain as many as four kinds of chromatophores: melanophores, guanophores, lipophores, and allophores. ' Melanophores with their black-pigment granules have been seen by Schmidt in the embryonic epidermis of *Geckolepis*. The usual location for this type of chromatophore, however, is in the midst of the derma and below all the other types of colour-cells. Dermal melanophores are found in all chromatic lizards. They are large cells densely filled with pigment granules and with lengthy processes that reach outward between the other pigmented bodies to the inner face of the epidermis (Fig. 100). When their pigment is concentrated, it is accumulated in the cell-body situated in the deeper

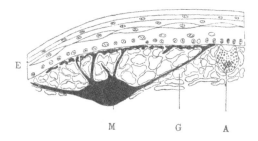

Fig. 100. Chromatophores in the skin of *Uroplatus*: *A*, allophore; *E*, epidermis; *G*, guanophore; *M*, melanophore with dispersed pigment. Schmidt, 1917, 117.

derma (Fig. 101). In the lacertilian colour changes, melanophores are always active and are commonly the exclusively active elements.

The guanophores (leucophores) with their reflecting semicrystalline contents are next to the melanophores in frequency of occurrence in reptiles. They are small in size, but very numerous and form a relatively thick layer immediately peripheral to that of the melanophores (Fig. 101). Lipophores (xanthophores), when they are present,

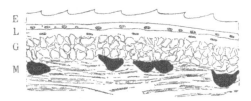

Fig. 101. Chromatophores in the skin of *Calotes*: *E*, epidermis; *G*, guanophores; *L*, lipophores; *M*, melanophores with concentrated pigment. Schmidt, 1917, 117.

constitute the most superficial layer in the pigmented portion of the skin and are located between the guanophores and the inner face of the epidermis. They contain very small spherical droplets of a yellowish substance soluble in alcohol or ether. Notwithstanding their extremely peripheral position, they may be covered externally by the pigmented branches from the melanophores (*Anolis*, von Geldern, 1921). Lastly, the allophores, relatively rare constituents of the reptilian colour complex, are found usually among the guanophores (Figs. 100, 102).

They contain a pigment, yellow, red, or violet in tint, and insoluble in alcohol. The last three constituents of the colour systems in lizards,

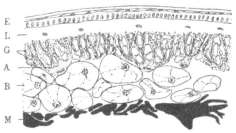

Fig. 102. Chromatophores in the skin of *Phelsuma*: *A*, allophore; *B*, bladder cells; *E*, epidermis; *G*, guanophores; *L*, lipophores; *M*, melanophores. Schmidt, 1917, 117.

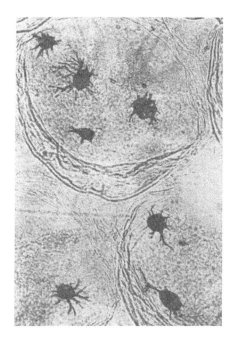

Fig. 103. Skin of *Gecko japonicus* with melanophores, some of which have their pigment much concentrated. Tokura, 1933, Pl. 1.

the guanophores, the lipophores, and the allophores, appear to be quiescent in the colour changes of these animals and are merely exposed to view or covered by the action of the melanophores.

Schmidt has recorded from different lizards five sets of combinations of these four kinds of chromatophores as shown in the following tabulation:

1. melanophores,
2. melanophores + guanophores,
3. melanophores + guanophores + lipophores,
4. melanophores + guanophores + allophores,
5. melanophores + guanophores + lipophores + allophores.

A good example of a purely melanophore system is to be found in *Voeltzkowia*, whose subterranean habits may have induced in its chromatophore system a certain degree of degeneration. The second combination, melanophores and guanophores, is well represented by the geckos (Tokura, 1933). The third combination, which includes lipophores (xanthophores), is seen in lizards that are green when the pigment of their melanophores is concentrated. This combination is exemplified by *Calotes* and by *Anolis* (von Geldern, 1921). The fourth combination in which allophores replace lipophores occurs in *Uroplatus*. Finally, the full array of the four types of chromatophores (Fig. 102) is to be noted in such lizards as *Lacerta*, *Phelsuma*, and *Chamaeleo*, all of which are remarkable for their range of colour change.

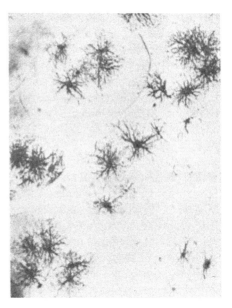

Fig. 104. Skin of *Gecko japonicus* with melanophores, pigment of which is much dispersed. Tokura, 1933, Pl. 1.

Since the publication of Schmidt's monograph (1918b), two noteworthy contributions have appeared dealing with the histology of the lacertilian chromatophores: von Geldern's paper (1921), rich in details and illustrations on the colour system of *Anolis carolinensis*, and Tokura's account (1933) of *Gecko japonicus*. The latter contains good photographs of the melanophores of this gecko with pigment in the concentrated (Fig. 103) and in the dispersed states (Fig. 104).

The reptilian melanophore pigment when fully dispersed fills the distal ends of the cell processes, covers almost completely all the other chromatophores, and renders the lizard very dark if not black in tint (Fig. 100). When it is completely concentrated, the cell processes are transparent and the subjacent chromatophores determine the colour of the animal. If these are guanophores, the tint may be whitish or even somewhat blue; if lipophores, it may be yellow; or with lipophores over guanophores, green. If now the melanophore pigment spreads slightly so that the peripheral processes of the cells directly under the epidermis contain some granules, the colours already assumed will darken and, if this operation is extended, these colours will gradually disappear to give place to a full dark brown or black tint. If allophores are present and exposed, orange, red, or violet may be added to the scheme. It is probable, though not definitely determined, that the lizard skin can produce interference colours, particularly in the direction of blue, which would thus add to its range of colour possibilities (Schmidt, 1918b). By combinations of these various elements the complex chromatic conditions of such lizards as the chameleon are doubtless produced.

C. Innervation

It has been almost universally conceded from the time of Brücke (1852) that the colour changes of lizards are under the direct control of nerves. This general conclusion is based upon experimental work in the cutting and destruction of parts of the spinal cord or of peripheral nerves in a number of reptiles, but especially in the chameleon. On cutting or destroying a part of the spinal cord of this lizard, its skin posterior to the region of operation darkens. When several of its spinal nerves are severed, the denervated area of the skin becomes dark. Moreover, these dark regions fail to respond to environmental differences by colour changes such as are shown by the rest of the animal. These and other more detailed tests carried out not only on the chameleon but on other lizards led investigators to believe that the chromatophores in these animals are under direct nerve control. Of the recent workers who have entertained this view may be mentioned Redfield (1918), Hogben and Mirvish (1928a, 1928b), Zoond and Eyre (1934), Zoond and Bokenham (1935), Sand (1935), Hogben (1936), and Parker (1938a).

Redfield (1918) observed that when a faradic current was led by platinum-point electrodes to the roof of the mouth of a dark *Phryno-*

soma the whole animal blanched in a few minutes, after which in course of time it again darkened. That this blanching was not due to an indirect stimulation of the brain, as intimated by Sand (1935), was shown by the fact that it could be elicited from the cloacal surfaces and the floor of the mouth as well as from the roof of that cavity. Blanching of a similar kind has also been observed by Hogben and Mirvish (1928*a*, 1928*b*) in the South African lizard *Chamaeleo pumilus*.

When the spinal cord of a dark *Phrynosoma* is severed near the middle of the trunk, the portion of the body anterior to the cut becomes slightly paler than that posterior to it, which remains relatively dark (Redfield, 1918). If now in such a preparation the roof of the mouth is stimulated electrically, the whole front portion of the lizard from its extreme anterior end to the cut will blanch very quickly and fully. The animal will then remain in this state for nearly a quarter of an hour, after which the pale portion will slowly darken. The contrast between the pale anterior and the dark posterior portions can be seen clearly on the trunk, but it is most conspicuous on the legs, the anterior pair being very pale and the posterior very dark. The blanching of the posterior portion of a *Phrynosoma* with severed spinal cord on stimulating the cloaca is not always successful, probably because of some peculiarity in the organization of the spinal structure (Parker, 1938*a*). Under similar tests carried out by Hogben and Mirvish (1928*a*, 1928*b*), the South African *Chamaeleo pumilus* shows a beautifully graded and segmented condition of the areas of skin affected by cuts through the cord at different levels (Fig. 105). These peculiarities in colour distribution have naturally been assumed to be associated with concentrating nerve fibres. Yet when a single spinal nerve in the midtrunk of a living *Phrynosoma* is exposed, cut, and even stimulated electrically, no band of paleness develops over the denervated area (Redfield, 1918; Parker, 1938*a*). A similar observation has been made on the chameleon by Zoond and Eyre (1934), who have suggested that in consequence of the overlapping of spinal nerves the cutting of a single nerve is insufficient to denervate a single segment, and consequently the melanophores of such a region remain, even after the severance of a given nerve, still under more or less nervous control. When in *Phrynosoma* three or four spinal nerves one next the other are exposed and simultaneously stimulated, a pale area distal to the region of operation does appear (Parker, 1938*a*). All these observations favour the idea that lizards possess concentrating nerve fibres.

A part of the body of *Phrynosoma* in which colour responses are
well shown is the lateral edge of the trunk between the front and hind
legs. This region in *Phrynosoma blainvillii* is bounded by about thirty

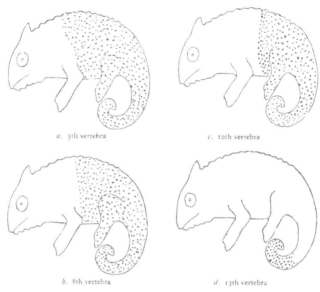

a. 5th vertebra   c. 10th vertebra

b. 8th vertebra   d. 13th vertebra

Fig. 105. Diagrams of *Chamaeleo pumilus* with spinal cord severed at different
levels after which the animals were stimulated electrically in the mouth; the anterior
regions blanched and the posterior ones remained dark (coarsely dotted). Hogben
and Mervish, 1928 b, 299.

conspicuous dentate scales which are strikingly different in their pale
(Fig. 95) and their dark phases (Fig. 96). When a deep cut is made
parallel to the edge of the trunk and for some distance along its length
the nerves going to these scales are severed (Fig. 106). If soon after
this wound has healed the mouth of the lizard (now in a dark condition)
is stimulated electrically, the whole animal within a few minutes will
blanch except over that portion of the lateral edge including its dentate
scales which was peripheral to the original cut. This region will
remain dark. On injecting the animal with adrenaline, however, this
outer dark edge and its dentate scales will also blanch, showing that
it has in no sense lost its capacity to respond (Parker, 1938 a).

Another test to ascertain if there is a nervous control of the pale
phase in *Phrynosoma* was carried out on the legs of this reptile.
Lizards which had been darkened by an injection of pituitrin were
decapitated, and their body-cavities opened so as to expose the

femoral nerves close to the attachment of the hindlegs. Both right and left nerves were then cut and the peripheral end of the right one was stimulated for three minutes with a faradic current. In from five to six minutes later the leg whose nerve had been stimulated had

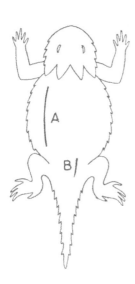

Fig. 106. Outline of a *Phrynosoma* seen from above, showing the positions of cuts· *A*, to denervate the left lateral margin of the lizard with its marginal scales, and, *B*, to expose the lumbosacral plexus. Parker, 1938a, Pl. 2.

Fig 107. *Phrynosoma blainvillii* in dark phase, decapitated, with body cavity opened ventrally and the nerves to both hindlegs cut. The nerve to the right leg was subjected to electrical stimulation for two minutes with the result that that leg soon blanched; the nerve to the left leg was left unstimulated as a control and that leg remained dark. Parker, 1938a, Pl. 1.

blanched while the opposite one had remained dark (Fig. 107). The opposite nerve, the left one, was then cut close to the middle of the length of the femur and the central end was stimulated electrically. No colour change followed, but when the peripheral end was stimulated the leg distal from the knee blanched, showing that this response was induced by that part of the nerve that was distributed to the distal part of the leg.

No evidence of any kind in favour of dispersing nerve fibres was found by Redfield (1918) in *Phrynosoma cornutum* nor by Parker (1938*a*) in *P. blainvillii*. This is in contrast with the chameleon. In this lizard, the severance of a nerve of any considerable size or of a group of nerves was invariably followed by a darkening in the skin of the denervated area (Brücke, 1852; Hogben and Mirvish, 1928*a*, 1928*b*; Zoond and Eyre, 1934; Sand, 1935). From the time of Brücke these darkened areas have been interpreted as due to simple paralysis. Zoond and Eyre (1934) and Sand (1935) have regarded them as the result of paralysis associated with the blocking of central inhibitory influences. Parker, however, has pointed out (1934*f*, 1936*a*) that they may be due to excessive stimulation of dispersing nerve fibres as they appear to be in teleost fishes, but so far as the chameleon is concerned this suggestion is purely hypothetical. Contrary to the opinion held by Bert (1875), there is thus at present no direct evidence of dispersing chromatophoral nerve fibres in reptiles.

The opinion that the melanophores in *Anolis* are controlled by nerves, a view held by almost all the older investigators in this field, has been discussed in three recent papers by Kleinholz (1936*b*, 1938*a*, 1938*b*). Transection of the spinal cord in *Anolis*, according to Kleinholz, does not retard the concentration of melanophore pigment in the posterior half of this lizard. Proof that the green colour, which in *Anolis* corresponds to the pale phase of other lizards, is not effected by local reflex stimulation of the melanophores is given by preparations where, in addition to cutting the spinal cord and the sympathetic chains, the posterior half of the cord was destroyed by pithing. In all anolids thus treated the ability to adapt to different backgrounds seemed in no way impaired; the animal as a whole in both its denervated and innervated portions turned brown on an illuminated black background and green on an illuminated white one. When electrical stimulation was applied to these lizards by way of the cloaca, muscles, or spinal cord, and the animals were kept on a black background, a generalized pallor quickly resulted and this pallor appeared in denervated as well as in normal regions of the skin. These results led Kleinholz to conclude that in *Anolis* neither the dispersion nor the concentration of melanophore pigment is under nerve control and that in this lizard the colour changes are accomplished by endocrines. These matters will be dealt with in a subsequent section.

In conclusion, it may be stated that the recent work on the innervation of lacertilian chromatophores, especially the melanophores, points to decided specific differences. In *Anolis* no nerves whatever appear to be directly involved in the colour changes. In *Phrynosoma* and in *Chamaeleo* there is good reason to assume the presence of concentrating fibres. No dispersing fibres, however, could be identified in *Phrynosoma*, but their presence in *Chamaeleo* is possible.

D. INTERMEDINE

The first direct evidence that intermedine plays a significant part in the colour changes of lizards was presented by Hadley (1931), who observed that when 0·5 c.c. obstetrical pituitrin, even at a dilution of one part in nine of Ringer's solution, was without effect. The next step was taken by Noble and Bradley (1933), who showed that when complete hypophysectomy was carried out on *Hemidactylus* the animal blanched completely and remained so permanently. These workers likewise demonstrated that the part of the pituitary gland concerned with this response was the intermediate lobe, for when this lobe was removed, even though the other parts of the pituitary complex were left intact, the lizard became pale and remained continuously so. If a part of the intermediate lobe was left in the lizard, the animal retained its dark colour. Kleinholz (1935, 1938a, 1938b), who, as already recorded, worked on *Anolis*, found that not only was the loss of the pituitary gland in this form followed by the complete and permanent assumption of the green colour (pale phase), but that the injection of eight units of pituitary extract from *Fundulus* caused the *Anolis* to become brown in a dozen minutes and to remain so for six hours. Parker (1938a) also demonstrated that in *Phrynosoma* hypophysectomy was followed in five hours by the first signs of blanching and seventeen hours by complete paleness. Individual lizards survived in this operation and remained continuously light tinted for from four to five weeks. On injecting such a pale lizard with 0·05 c.c. obstetrical pituitrin, it began to darken in a quarter of an hour, was fully dark in two hours, and was pale again after six hours from the time of injection. An extract in Ringer's solution of a single pituitary gland from *Phrynosoma* when injected into another hypophysectomized and consequently pale horned toad rendered the recipient dark in about five minutes, and kept it so in an illuminated white-walled box for about two and a half days, after which it gradually returned to full paleness. Defibrinated blood from a dark horned toad when injected subcutaneously into a pale individual induced the formation of a dark spot in seven minutes. This spot had completely disappeared two hours later. Blood from a dark horned toad on being injected into a second dark individual had no effect on the colour of the recipient, as was first demonstrated by Redfield (1918). These observations when taken collectively show that, contrary to Hogben's suspicions, reptiles are profoundly influenced by intermedine, which appears to play an important role in their normal darkening. This at least seems to be true of *Anolis*, *Hemidactylus* and *Phrynosoma*. Whether or not it holds for *Chamaeleo* remains to be ascertained.

Other organic agents than intermedine are known to darken lizards. Thus according to Potter and Brown (1941) *Phrynosoma* has been observed to change from the pale to the dark phase on the injection of oestrogens in oil.

E. ADRENALINE

Redfield (1916, 1918) was the first to call attention to the possible role of the adrenal gland in the blanching of lizards. His views, which were based upon a study of *Phrynosoma*, were received with much scepticism by Hogben (1924), Hogben and Mirvish (1928 a, 1928 b), and especially by Zoond and Eyre (1934) and by Sand (1935), all of whom worked upon the chameleon. That injections of moderate concentrations of adrenaline into lizards is followed regularly by blanching was first shown by Redfield (1918) on *Phrynosoma* and has been confirmed on this lizard by Parker (1938 a) and by Pierce (1941 a), on *Anolis* by May (1924), by Hadley (1931), and by Kleinholz (1938 b), on *Chamaeleo* by Hogben and Mirvish (1928 a, 1928 b), and on the gecko by Chin, Liu and Li (1938). Table 13 presents a survey of the concentrations of adrenaline, effective and ineffective, for *Chamaeleo* and *Phrynosoma*.

From Table 13 it will be seen that all effective concentrations of adrenaline act in about ten minutes and that the resulting blanchings last longer, the stronger the concentrations. It is also evident that the chameleon is less responsive to adrenaline than the horned toad, for the chameleon ceased to respond at 1 : 800,000 whereas *Phrynosoma blainvillii* did not fail to react until 1 : 10,000,000 was reached. *P. cornutum*, upon which Redfield (1918) worked, reacted occasionally to 1 : 100,000,000. Redfield also demonstrated that *P. cornutum* blanched to extracts of its own adrenal glands, a statement that was confirmed by Parker (1938 a) on *P. blainvillii*.

Pierce (1941 a) has pointed out the remarkable fact that *Phrynosoma douglassii*, which normally requires from fifty minutes to an hour and a half to turn pale, will blanch from an injection of adrenaline in the very brief period of from four to six minutes. *P. modestum* blanched in even a shorter period—two to three minutes.

When the floor of the mouth of *Phrynosoma* is stimulated electrically, this lizard, as was pointed out by Redfield (1918), will turn pale. If this test is tried on a moderately dark horned toad the nerves to one

TABLE 13. Times of beginning and ending of blanching from injections of various dilutions of adrenaline into *Chamaeleo pumilus* (Hogben and Mirvish, 1928 *a*, 1928 *b*), and into *Phrynosoma blainvillii* (Parker, 1938 *a*)

Dilutions of adrenaline	Times after injection for blanching			
	Chamaeleo (1 c.c.)		*Phrynosoma* (0·5 c.c.)	
	Began	Ended	Began	Ended
1 : 1,000			10 min.	24 hr.
1 : 10,000			10 min.	8 hr.
1 : 25,000	15 min.	10 + hr.		
1 : 50,000	15 min.	10 + hr.		
1 : 100,000	15 min.	10 + hr.	10 min.	2 hr.
1 : 200,000		2 hr.		
1 : 400,000		2 hr.		
1 : 800,000	No reaction			
1 : 1,000,000	No reaction		10 min.	15 min.
1 : 10,000,000			No reaction	

of whose hind legs have been cut, the lizard will blanch generally except for the denervated leg which will remain dark. The general blanching under these circumstances will take place in about a minute and the recovery of the dark phase will require a quarter of an hour or more. Such blanching is conceded to be nervous in character. When a dark horned toad with nerves to one hind leg cut is vigorously molested, the toad will in time blanch and the denervated leg will become pale as the rest of the animal does. This type of blanching depends upon the blood, for if the artery supplied to a given leg is tied off before the lizard is made to blanch, that leg will remain dark. Hence some substance carried in the blood, adrenaline or other similar neurohumour, must be accountable for this kind of blanching. In *Phrynosoma*, blanching of this type begins to appear in from ten to twenty minutes after stimulation and will disappear in some two hours. Thus these two types of blanching, nervous and humoral, differ in their time relations. It is probable that when *Phrynosoma* blanches normally its nerves act first, after which the melanophore contraction thus produced is supported and maintained by the contracting neurohumour carried in the blood (Parker, 1938 *a*). That this neurohumour is adrenaline, at least in *Phrynosoma*, seems highly probable (Redfield, 1918; Parker, 1938*a*), though Kleinholz (1938*b*) has shown that in *Anolis* neither nerves nor adrenal glands are responsible for paleness.

Redfield (1918) first pointed out that when noxious stimuli are applied to *Phrynosoma*, this lizard regularly turns pale. Hogben and Mirvish (1928*a*, 1928*b*) recognized a similar condition in the chameleon and designated this state as "excitement pallor". Brücke (1852) had already noticed that when chameleons were much disturbed their colour patterns became more pronounced. Strecker (1928), and later Hadley (1931), recorded that *Anolis* if unduly excited becomes green (pale). *Phrynosoma* is reported by Parker (1938*a*) as pale on vigorous stimulation. Sand (1935) stimulated pale chameleons by pinching their feet and thereby induced them to hiss and attempt to bite. In from one to two minutes thereafter they became dark. Sand objected, therefore, to the term excitement pallor, since there may be an excitement darkening as well as an excitement pallor. Among lizards, pallor, however, seems to be the usual response to excessive stimulation, particularly of a noxious kind. Whether this response is dependent upon adrenaline (Redfield, 1918) or upon concentrating nerves (Hogben and Mirvish, 1928*a*, 1928*b*) remains to be settled.

F. DIRECT STIMULATION OF MELANOPHORES

Brücke (1852) pointed out that when a piece of tin foil was put round the body of a dark chameleon in bright light and then after a few minutes removed, the area that had been thus covered was paler than the rest of the animal. This observation has been repeatedly confirmed. Thus Zoond and Eyre (1934) remark that if a small piece of copper sheeting cut in the shape of a letter Y is held close to a chameleon exposed to sunlight so as to throw a sharp shadow on the animal's skin, after two minutes a "print" is obtained, the shaded region being paler than the unshaded. The question then arises, they remark further, whether this response is a reflex, or whether the chromatophores react directly to light; in other words, can these colour-cells play the part of independent effectors?

This question has been put to experimental test by Zoond and Eyre by using an eviscerated chameleon spread out dorsal side up on a wad of cotton-wool soaked in Ringer's solution. On the right side of this lizard parallel to its long axis and close to its vertebral column a longitudinal cut was made severing the spinal nerves of that side of the trunk. The left side remained intact. The operation was carried out in a dark room and the lizard in the beginning was maximally pale. Subsequently it was alternately exposed to daylight and to darkness. The right side, whose nerves had been cut, had quickly darkened and remained continuously black irrespective of its exposure to light or to darkness. The left side changed in tint with the changes in the illumination, being pale in darkness and deep coloured in the light. These conditions led Zoond and Eyre (1934) to state that "the conclusion is unavoidable that the response of the pigment cells of the chameleon to light is dependent upon the integrity of spinal reflex arcs". On removing the autonomic chain of the intact side of their preparation that side also rapidly became black.

Zoond and Eyre doubtless regarded the right side of their original chameleon preparation as darkened in consequence of a paralysis due to the cutting of nerves. The opposite side of their preparation with its changes in tint as a result of changes in illumination they described as the seat of reflex activity. On this side they assumed that light stimulated the nerve terminals in the skin and thus excited impulses which, after passing through the spinal cord, returned to the skin to activate concentration of melanophore pigment.

This view of chromatic action in the chameleon was criticized by Parker (1938 a), who pointed out that in such teleosts as *Fundulus* and *Ameiurus* the darkening of denervated areas of skin was due to the excessive activation of the severed dispersing nerve-fibres and not in any way to paralysis. If this explanation applies to chameleons, as Parker surmised it might, the black half of the chameleon would remain so, not because its melanophores were paralysed and could not respond, but because they were so vigorously activated by the cut nerve that as mild an agent as direct light or darkness could have no effect upon them. Thus from this standpoint Zoond and Eyre's experiment, according to Parker, failed to answer the question at hand and left the problem where it was in the beginning.

The contrast between colour responses due to dermal stimulation and to retinal stimulation in the chameleon was ingeniously investigated by Zoond and Bokenham (1935), who compared the melanophore reactions of these animals under the two following conditions: first, with the eyes eliminated and the skin exposed, and, second, with the eyes exposed and the skin covered. The eyes were eliminated by cutting the optic nerves and the skin (of the trunk at least) was excluded by placing the chameleon in a light-proof sack, out of the mouth of which the head of the animal was allowed to project. The lower threshold of retinal stimulation as contrasted with dermal stimulation was clearly shown and the interplay of these two means of activating the melanophores was described in detail. In this account, however, no observations were made that could not be interpreted as due to direct dermal stimulation of melanophores instead of the reflex stimulation of these effectors as assumed by Zoond and Bokenham. So far then as the matter of direct or reflex stimulation of chromatophores in lizards is concerned this contribution leaves the question quite undecided.

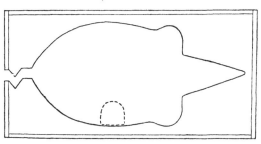

Fig. 108. Plan of a dark box into the cavity of which (inner outline) a *Phrynosoma* could be snugly fitted with head to the left in the drawing and tail to the right. When the top of the box was put on, the lizard was in darkness except for a beam of light which could be admitted through a hole whose position is indicated by the dotted outline. Parker, 1938 a, Pl. 2.

A solution of this problem was attempted by Parker (1938 a) on *Phrynosoma blainvillii*. The final tests were made on hypophysectomized lizards whose pineal eyes had been covered with a light-proof mixture, whose lateral eyes had been removed, and in which a marginal cut (Fig. 106) had been made so as to produce a peripheral lateral band of denervated skin and dentate scales. Of six lizards that were subjected to these preparatory operations, two survived. A week or ten days after the last operation these two appeared to have recovered fully. When put on their backs they righted themselves quickly and in other respects were normally active. Each lizard in turn was placed in a dark box (Fig. 108), so arranged that the animal was restrained from moving about, and a beam of light could be thrown through a hole in the box to impinge on the denervated left edge of the animal. After the lizard had been subjected to this spot of light for a period of time, it was exposed to view and the denervated side was compared with the normal one. The denervated side was always found to be noticeably darker than the normal one. When the denervated left side of the lizard was shielded by a zinc cover (Fig. 109) and the animal as a whole was exposed to a general bright illumination, the covered side became pale as compared with the uncovered one.

In a second test three blinded and laterally denervated lizards, all in the beginning in very close agreement in tint, were placed one in complete darkness, another under a bright electric light, and a third in the diffuse daylight of the laboratory. After two hours they were compared. In each of the three, the denervated side and the normal side were in agreement in tint. The lizard that had been in darkness was slightly paler than the one that had been kept in the diffuse light of the laboratory, and the lizard in bright light was slightly darker than the one from diffuse light. From these several tests, Parker concluded, in agreement with Redfield (1918), that the melanophores of *Phrynosoma* are directly influenced by light and by darkness. The fact that these tests were carried out on horned toads, the nerves of whose denervated areas had been given time to degenerate before the tests were made, precludes the possibility that axon reflexes could have influenced the results.

Fig. 109. Outline of a *Phrynosoma*, showing the attachment of the zinc shield by which the left side of the lizard's body was kept in relative darkness while the rest of the animal was in bright light. Parker, 1938a, Pl. 2.

Tests somewhat similar to those carried out on *Phrynosoma* by Parker were tried on *Anolis* by Kleinholz (1938b), but with results different from those recorded for the horned toad. Kleinholz blanched three anolids by placing them on an illuminated white background, after which a light-proof screen of modelling clay was placed on the back of each animal so as to cover most of its trunk following the plan used by Redfield (1918). The lizards were then put on an illuminated black background. After the animals had turned brown, the clay screen was removed and the covered area was found to be as brown as the uncovered one. As Kleinholz remarked, such results are in agreement with a hormonal control of melanophore dispersion. In a further test, Kleinholz exposed three hypophysectomized animals to direct sunlight and none of them darkened. The unavoidable conclusion, according to Kleinholz, is that the intact skin of *Anolis carolinensis* is not directly responsive to light. By way of criticism of these results, it may be said that in the first of Kleinholz's experiments the amount of intermedine which was undoubtedly produced as a result of placing the lizards on a black background may have been so great as to have obscured completely any direct effect that the darkness of the screen may have had. In the second test, the absence of a standard control for the original tint of the lizards would have made a close comparison of the colour of the animal before and after exposure to sunlight very difficult. In Parker's experiments on *Phrynosoma*, one side of the animal was in darkness and the other in bright light and a comparison was, therefore, always easily made. In these tests, the difference between the two sides was faint but unmistakable. It is doubtful if it could have been established without the use of one side as a standard for comparison. In these respects, Kleinholz's tests, like those of Zoond and Eyre, are not beyond criticism.

In testing the question of the direct stimulation of chromatophores by light and darkness, two recent investigators have worked with isolated pieces of skin. The relative insecurity of this method has been pointed out by Kleinholz (1938 b). Nevertheless, the following records are worthy of consideration. Hadley (1928, 1931) found that isolated pieces of skin from four species of *Anolis*, *A. equestris*, *A. iodurus*, *A. carolinensis*, and *A. watsoni*, became brown in direct sunlight and remained green in ordinary diffuse illumination. These results were confirmed by D. C. Smith (1929) for isolated pieces of skin from *Anolis equestris*. Kleinholz also obtained results from *Anolis carolinensis* that tended to confirm Hadley's observations. Isolated pieces of skin from this lizard remained green when floated on Ringer's solution. When they were placed in direct sunlight, they darkened slightly though they never became brown as was reported by Hadley (1928) and by Smith (1929) for *Anolis equestris*. When the darkened skin was brought by Kleinholz from bright into diffuse light, it became bright green again usually within 30 seconds. The local nature of this response was demonstrated when one half of the isolated strip of skin was shaded and the other half was exposed to bright sunlight. The shaded portion remained bright green while the brightly illuminated part became darker. Kleinholz concludes that though isolated pieces of skin may respond directly to light by local darkening there is no evidence favouring a similar conclusion for the intact skin of *Anolis*.

As the foregoing presentation shows, the question of the direct stimulation of melanophores by light and darkness in lizards is by

no means settled. It is easier to bring objections against the inter-
pretations and experimental procedure used by Zoond and Eyre and
by Kleinholz than it is against the work of Parker, but the differences
in coloration assumed to be the result of direct stimulation are so
slight that conclusive decisions are by no means easy. At best, colour
changes induced in this way appear to be slight in most lizards with
the possible exception of the chameleons, and probably play a very
minor role in the normal chromatic activities of a majority of these
animals. That they do play some part in these changes is the opinion
of not a few investigators (Brücke, 1852; Redfield, 1918; Hogben and
Mirvish, 1928a, 1928b; D. C. Smith, 1928, 1929; Hadley, 1928, 1931;
Parker, 1938a).

G. Lacertilian Neurohumours

The opinion held decades ago by the older students of colour changes
in lizards that these changes are subject to a simple type of nerve
control has been found to be quite inadequate. Nor does it seem
probable from what has been discovered within recent years that the
hope, very naturally expressed by Zoond and Eyre (1934), for a simple
and general theory of reptilian colour change is likely to be realized.
Diversity in the organization of colour systems of lizards rather than
uniformity in this respect seems to be the rule (Parker, 1936c).

Of the lacertilian colour systems thus far worked out, that of *Anolis*,
and particularly *A. carolinensis*, seems to be the simplest. According
to Kleinholz (1938a, 1938b) this lizard, as already stated, shows no
evidence whatever of possessing chromatic nerves. May's statement
(1924) that grafts of skin on *Anolis* remained green (pale), and failed
to turn brown when the rest of the animal darkened, till after 20 to
30 days was shown by Kleinholz to be inaccurate. Colour changes
were seen by Kleinholz in such grafts in about 24 hours after the
skin had been reattached to the animal. Such an early re-establishment
of the colour change in a graft could be accounted for only in consequence of a returned circulation and not as a result of regenerated
nerves. As has been stated already, direct attempts on the part of
Kleinholz to demonstrate physiologically the presence of chromatophoral nerves in *Anolis* always resulted in failure. Neither von Geldern
(1921) nor May (1924) were able to discover by histological means
any traces of chromatophoral nerves in this lizard. So far as the old
view of nerve control of reptilian melanophores is concerned, *Anolis
carolinensis* yields not a vestige of favourable evidence.

The only effective means for inducing colour changes in *Anolis* that Kleinholz could find was humoral, and consisted of a secretion, probably intermedine, from the pituitary gland. The blanching of the lizard has been attributed by Kleinholz to the disappearance of this substance from the blood. The rapid onset of pallor which followed electric stimulation was interpreted as due to vaso-constriction. Thus *Anolis carolinensis* gives evidence of being a unihumoral lizard in which intermedine or some like substance is the only effective chromatic agent present.

Phrynosoma possesses a chromatic system somewhat more complicated than *Anolis*. It takes on its pale phase by the contraction of its melanophores under the influence of concentrating nerve-fibres (Redfield, 1918; Parker, 1938a). Whether the nerve endings of these fibres act on the colour-cells through a secreted neurohumour, as suspected by Parker (1936c), remains to be ascertained. This lizard also blanches as the result of a hydrohumour, probably adrenaline, carried in its blood (Redfield, 1918; Parker, 1938a). It assumes its dark phase in consequence of a pituitary secretion, probably intermedine, and apparently this is the only factor concerned with the dark phase. All attempts by Parker (1938a) to find evidence of dispersing nerve-fibres or other means of darkening yielded only negative results. Although some of the conclusions concerning *Phrynosoma*, particularly those first stated by Redfield, have been sharply criticized by Hogben (1924), Hogben and Mirvish (1928a, 1928b), Zoond and Eyre (1924) and Sand (1935), they are confirmed in the main by the work of Parker (1938a). *Phrynosoma* is certainly bihumoral or, if it is found to possess a nervous concentrating neurohumour, specifically trihumoral.

Probably the most complicated colour system in lizards thus far investigated is that in the chameleon. Unfortunately, the chromatic changes in this animal, though they have been long and much studied, have not been critically tested in a number of important directions. One feature in them, however, seems well established; that the melanophores of this animal are provided with concentrating nerve fibres. This feature has been accepted by almost all investigators from the time of Brücke (1852) to the present (Sand, 1935). Beyond this meagre conclusion there is little agreement. It is by no means settled whether the excitement pallor of the chameleon is due to the action of concentrating nerves, as believed by Hogben and Mirvish (1928a, 1928b), or to adrenaline or other like hydrohumours as might be inferred from the work of Redfield (1918). That the dark phase of the chameleon should result from the simple inactivity of its concentrating nerves, from the action of dispersing nerves as intimated in Parker's interpretation (1938*a*) of the work of Zoond and Eyre (1934), or even from a pituitary secretion such as intermedine, as is suggested by what is known of a number of other lizards, is entirely uncertain. Even the carefully thought out though somewhat cumbersome scheme of Zoond and Eyre (1934) for the activation of the chameleon's melanophores is not free from difficulties. This scheme has already been discussed in the preceding section. According to it, the normal darkening of a chameleon takes place in consequence of the inhibition of the ordinary tonic influences by which the melanophores of this animal are kept contracted. This inhibition is called forth through the illumination of the animal's dermal photoreceptors or through the stimulation of its retina by light from a light-absorbing background. Stimulation of the retinal elements by light from a light-scattering background is, according to Zoond and Eyre, the means of inhibiting the original inhibition, thus allowing the tonic state to reassert itself. As a result, the melanophore pigment becomes concentrated and the chameleon blanches. This mode of chromatic action is reaffirmed by Zoond and Bokenham (1935) as well as by Sand (1935). It has been discussed critically by Parker (1938*a*), who has attempted to show that it is not the only admissible interpretation of the observed facts of colour changes on which it is based. Parker moreover has presented a markedly different view, much of which has been given in the section of this survey which deals with the innervation of teleost chromatophores and which need not be repeated here.

Elusiveness has long been the characteristic of the chameleon and this feature seems still to enshroud our understanding of the colour changes in this animal. From all that is known this reptile may possess on the one hand an extremely simple chromatic system based on a single concentrating neurohumour, but it may be, on the other hand, an example of a highly complex multihumoral type with as many as four activators, two for blanching and two for darkening. Moreover, all discussion thus far has had to do only with the melanophores of the chameleon. Its complement of other kinds of chromatophores is high and about the control of these really nothing seems to be known. Probably when the full chromatophoral system of this animal is worked out, extreme complexity will be discovered. This, however, remains for future investigation.

Although the response in colour change to an illuminated environment, white or black, ceases in a lizard with the loss of its eyes, the animal does not lose completely its power to change colour. After having been blinded, it is usually dark in bright light and pale in darkness. This has long been known and has been recorded recently again for the chameleon by Hogben and Mirvish (1928 b), and by Zoond and Eyre (1934). Kleinholz (1938 a) has seen similar responses in *Anolis*. Such reactions have also been observed in *Phrynosoma cornutum* by Redfield (1918) and in *P. blainvillii* by Parker (1938 a). After enucleation, *Phrynosoma* assumes a tint rather dark than pale. When such blinded animals are put in darkness they blanch slightly, to darken somewhat after they have been returned to bright light. Although they have a well developed pineal eye, the covering of this eye with opaque paint makes no change in their responses to light and to darkness (Parker, 1938 a). Thus *Phrynosoma* yields no evidence in support of the view expressed by Clausen and Mofshin (1936), who worked on *Anolis*, that the pineal eye may be a photoreceptor. Zoond and Eyre (1934) have attributed the colour responses of blinded chameleons to nervous reflexes which, starting from dermal receptors, are believed by these investigators to be mediated by the spinal cord. But, as was pointed out by Parker (1938 a), this interpretation is not a necessary one, for Zoond and Eyre have not in reality excluded the possibility of the direct stimulation of melanophores by light. In fact their observations that shadow patterns are closely reproduced in the coloration of the chameleon's skin is much more likely to be explained by direct stimulation than by reflex activity. It must, therefore, be admitted that the question of the direct stimulation of the melanophores in the chameleon by light or darkness has not yet been definitely settled.

Kleinholz (1938 b) observed that blinded anolids were pale in darkness and dark in the light. When these lizards were hypophysectomized, they became pale (green) and remained so indefinitely. Kleinholz concluded from this that the darkening of a blinded anolid must be due to the stimulation of dermal photoreceptors which then reflexly excite the pituitary gland.

Kleinholz observed further that hypophysectomized anolids, which were green in consequence of the loss of their pituitary glands, failed to darken in direct sunlight. He was, therefore, led to state as an unavoidable conclusion that the melanophores of the intact skin in *Anolis carolinensis* are not directly responsive to light. This statement implies a comparison of the tints of a blinded, hypophysectomized *Anolis*, first in dim light or darkness and then in bright light. But if the degree of darkening under these circumstances should prove to be very slight, as has been found to be the case in *Phrynosoma* (Parker, 1938*a*), it might be difficult without some basis of comparison, apparently not used by Kleinholz, to make this determination. On the question of the direct stimulation of melanophores by light, the positive evidence from *Phrynosoma* appears to be more convincing than the negative evidence from *Anolis*, but it must be kept in mind that these two lizards may in the end be found to differ fundamentally from each other in respect to direct stimulation. This question can be settled finally only by further investigation.

All these facts have a bearing on the complexity of the neurohumoral systems in lizards. Such systems appear to include concentrating nerve-fibres with their possible lipohumours, adrenaline or some adrenaline-like material, pituitary secretion very probably inter-medine, possibly dispersing nerve fibres also with lipohumours and other like agents including direct stimulation, combined in such ways as to give individual species of lizards that remarkable and diversified display of colours which is peculiar to them as a group.

Chapter VIII

DEVELOPMENT AND GROWTH OF VERTEBRATE CHROMATOPHORES

A. Introduction

THE development of vertebrate chromatophores has been traced in only a few animals, notably in some teleosts, and in a few amphibians. Of the several types of colour-cells in these animals, the melanophores, because of their conspicuousness, have claimed first attention. These dark pigment-cells occur regularly in the derma and less regularly in the epidermis of the vertebrate skin.

B. Dermal Melanophores in Fishes

The origin and initial distribution of the dermal melanophores in fishes have been dealt with by a number of investigators, especially Wagner (1911), Stockard (1915b), Gilson (1926a), Becher (1929b), Duspiva (1931a), Tasker (1933), and Graupner and Fischer (1935). Of these workers, Wagner, who studied the development of chromatophores in the trout, has given an account of this subject which for fullness and accuracy is still of highest rank.

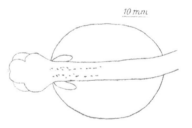

Fig. 110. Dorsal view of an embryonic trout showing the two initial bands of melanophores, one right, the other left, extending posteriorly from the young fish's head. Wagner, 1911, 6.

According to this investigator, the first dermal melanophores in *Salmo fario* appear some 14 days before the fish hatches. They occur in two longitudinal bands one on each side of the mid-dorsal line. These bands stretch from the posterior region of the young fish's head caudally to the hind end of its anterior quarter (Fig. 110). According to Wagner, each melanophore in these bands arises in place and the bands grow by the addition of new melanophores at their ends and not by a migration of such cells from centres of production. The individual melanophores in the trout even at the initial stage of growth show no great regularity in arrangement. In this respect they are unlike the early melanophores in the embryos of *Atherina, Alburnus* and some other fishes, which, according to Bolk (1910), exhibit a distinctly segmental order. A similar metameric grouping had also been noted by Becher (1929b), and by Graupner and Fischer (1935). In *Coregonus*, as in the trout, the first colour-cells, though in bands, are without segmental arrange-

ment (Becher, 1929b). In the embryonic *Fundulus*, the melano-phores of the anterior portion in the forming system exhibit no trace of metameric order, a con-dition in contrast with those of the posterior part which are segmentally grouped (Gilson, 1926a). That these various types of arrangement are dependent upon the attraction of the young melanophores toward the more or less regularly disposed blood-vessels, is a view not favoured by most investigators (Wagner, 1911). It is possible that the characteristic groupings already alluded to may be due to a purely mechanical adjustment of the colour-cells to the available open tissue spaces. However, the question here involved admits at present of no conclusive answer.

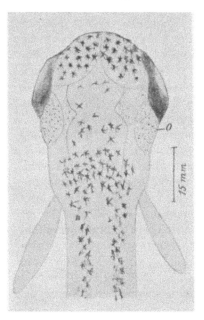

Fig 111 Dorsal view of the head and anterior part of the trunk of a young trout 5 days after hatching or 19 days after the appearance of the first melano-phores. Wagner, 1911, 11.

In the trout, to return to Wagner's account, 19 days after the first appearance of the melanophores or 5 days after the young fish has hatched, the larval colour-cells are abundantly present on the young creature's head and extend as two well-defined bands posteriorly to its tail (Fig. 111). Other bands have also appeared laterally and ventrally. The continued extension and enlargement of these various bands and areas appear to depend, as the original growth did, upon the local formation of new melanophores. In the older parts of the system, the melanophores have come to be very abundant. Nowhere among them is there any sign of a segmental grouping.

Five weeks after the hatching of the young trout, its dermal melanophores are disposed in patches symmetrically placed over the head and the trunk and arranged in such a way as to foreshadow the pigmentation of the young fish (Fig. 112). Meanwhile, step by step, the fins and other parts of the body have acquired their complement of colour-cells and the larval trout thus begins to assume its characteristic bodily markings.

In attempting to discover the embryonic source of the dermal melanophores in fishes the early investigators were obliged to content themselves with an inspection of whole embryos either living or preserved. By this means, they were led to conclude that these and other melanophores were derived from the mesoderm. This view was abundantly confirmed by a number of subsequent workers including Stockard (1915), who saw these colour-cells differentiating from what were believed to be mesenchyme cells between the external ectoderm and the periblast of the yolk sac in *Fundulus*.

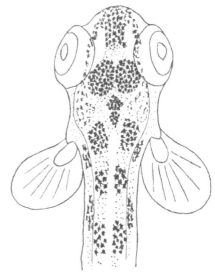

Fig. 112. Dorsal view of the head and anterior part of the trunk of a young trout 5 weeks after hatching from the egg. Wagner, 1911, 16.

The origin of the body melanophores in the young trout was shown by Wagner (1911) to be from the mesenchyme just dorsal to the muscle and the cutis plates of the mesoderm (Fig. 113). Here as early as 14 days before the young fish was hatched single mesenchyme cells were seen to develop granules of melanin and to metamorphose quickly into the melanophores of the first embryonic bands. At this stage the trout embryo is about 10 mm. in length, its circulatory system is functional, and, when freed from the egg envelopes, it is capable of simple swimming movements. This method of origin for dermal melanophores has been confirmed by so many investigators that Becher (1929b) was led to the general conclusion that all chromatophores, irrespective of the kind of pigment that they may contain or of their position in the body, arise from undifferentiated cells of the mesenchyme derived from either the unsegmented mesoderm of the head or the segmented or unsegmented mesoderm of the trunk.

This view has been substantiated by Thumann (1931), Tasker (1933), Graupner and Fischer (1935), and others. The only possible exception to this generalization, according to Becher, is the melanophore system of the epidermis which will be considered later in this account.

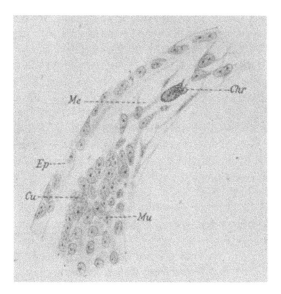

Fig. 113. Left half of a transverse section through an early embryonic trout: *Cu*, cutis layer of a primitive segment; *Chr*, young chromatophore; *Ep*, epidermis; *Me*, mesenchyme; *Mu*, muscle layer of a primitive segment. Wagner, 1911, 7.

Cells destined to become melanophores commonly show as preliminary to their final state an accumulation of colourless granules which eventually take on the character of melanin particles. These colourless granules or plastids have been claimed to be present in fishes and amphibians in general. They have been noted specifically in recent years by Becher (1929 *b*) in *Coregonus*, and by Duspiva (1931 *a*) in *Perca*. Their presence marks the first visible step in the differentiation of a mesenchyme cell into a melanophore. It is often stated that the first actual melanin appears in the fish integument about the time when the embryonic heart begins to beat. This seems to be true of *Fundulus* (Stockard, 1915), *Coregonus* (Becher, 1929 *b*), and *Perca* (Duspiva, 1931 *a*).

C. Epidermal Mélanophores in Fishes

Melanophores are found not only in the derma of fishes but also, though much less constantly, in their epidermis. Epidermal melanophores are as a rule smaller in size, more delicate in form, and more scattered in occurrence than those in the derma. In some fishes the epidermis is entirely devoid of them (Becher, 1929b). They are regularly absent from the epidermis of many if not all young fishes. Thus during the whole period of growth in the trout, *Salmo fario*, from the early embryo even up to a fish 15 cm. in length, no melanophores have been observed in the epidermis (Wagner, 1911), though they are known to be abundantly present in that layer in the adults of this species. The same general condition appears to be true of *Fundulus* according to Gilson (1926a), and of *Coregonus* according to Becher (1929b). The question that naturally arises in this connection is that of the source of these cells. Are they modified epithelial cells differentiated *in situ* and therefore ectodermal in origin, or are they melanophores that have migrated from the derma into the epidermis and hence mesodermic in nature? To these questions, as Becher remarks, no conclusive answers can be given at present. It appears to be beyond doubt that leucocytes, which are very close histologically to melanophores, may migrate from the deeper tissues into the epidermis. But it is equally certain that cells in the external ectoderm of certain animals may of themselves produce melanin. In consequence of these relations and in the absence of any real evidence on the exact source of the epidermal melanophores, it must be admitted that no final conclusion can be drawn as to the origin of these cells.

D. Other Chromatophores in Fishes

The skin of fishes contains, besides melanophores, various other types of chromatophores such as xanthophores, erythrophores, guanophores, and the like. Meagre as is our knowledge of the origin and growth of melanophores in fishes, our acquaintance with these aspects of the other types of piscine chromatophores is much more fragmentary. Of the additional kinds of chromatophores just mentioned, the yellow and red, xanthophores and erythrophores, are certainly very closely related. In fact all transitions from yellow through orange to red may be found in the same fish, a state of affairs which led Schmidt (1918e), Becher (1929b), and others to class both xanthophores and erythrophores under the common heading of lipophores. This step was based upon the assumption that the colouring matter in all these cells, the so-called lipochrome, was one substance or at least a group of closely related substances. The colouring matter of all true lipophores is soluble in alcohol, ether, chloroform and other solvents of fat, and, though it is not well known chemically, it is probably closely related to the vegetable colouring matter, carotin. Notwithstanding this chemical basis for grouping xanthophores and erythrophores together, it must be kept in mind that these two types of chromatophores are not necessarily mingled in all fishes that possess them. Thus in *Fundulus*, xanthophores are present to the exclusion of red cells, and in *Holocentrus*, erythrophores are the only lipophores in the skin. In the trout, however, not only do both xanthophores and erythrophores occur, but also a whole range of other colour-cells (Wagner, 1911).

Though both types of lipophores are of common occurrence in fishes, the xanthophores have received much more attention than the erythrophores. Xanthophores arise early in developing fishes, but not so early as do the melanophores. In *Salmo* the appearance of the melanophores takes place nine days in advance of the xanthophores (Wagner, 1911). In *Coregonus*, according to Becher (1929b), the melanophores are first seen when the embryos are 7 to 8 mm. long, the xanthophores not until they are at least 8 mm. in length. Ginsburg (1929) has shown that in *Lebistes* xanthophores likewise follow melanophores in order of appearance. Thus in all recorded instances the first xanthophores develop later than the first melanophores.

The only extended account of the development of xanthophores is that given by Becher (1929b). According to this author, the xanthophores in *Coregonus* are differentiated mesenchyme cells. After these cells have assumed a spindle shape or have become somewhat branched they show the first signs of differentiation into xanthophores. This step consists in the formation in the cytoplasm of the cell of numerous colourless granules such as have already been mentioned for melanophores. These granules soon take on a yellowish tint, a change first seen in those in the neighbourhood of the cell nucleus where they form a relatively concentrated mass, the xanthome. At this early stage, which is when the embryo of *Coregonus* is about 8 mm. long, the xanthophores are capable of limited amoeboid movement, a capacity which, however, they soon lose. When the young *Coregonus* hatches, which occurs after it has attained a length of from 10 to 11 mm., the deep layers of the skin are richly provided with xanthophores, none of which, however, occurs in the epidermis. Brief as is Becher's account of the development of the xanthophores in *Coregonus*, it nevertheless shows very clearly that, except for the kind of pigment contained in these cells and their lateness in development, their method of origin and growth is much the same as that of the dermal melanophores.

The only other chromatophore, the history of which has been even casually followed and which may be mentioned in a general survey such as this, is the guanophore. This element is concerned with the more brilliant tints and scintillating effects seen in some teleosts. In many species, the skin shows in its deeper parts large areas of a silvery glistening material that often takes on an almost metallic brightness. The layer responsible for this effect is usually designated as the stratum argenteum or simply the argenteum. It is apparently cellular in nature and owes its brightness to deposits of guanin within the bodies of its cells. In addition to the argenteum, many fishes possess small aggregates of crystals whose play of colours is such that under the microscope, particularly with indirect illumination, these aggregates look like the most brightly scintillating jewels. In many instances they are indescribably beautiful. Such small bodies are the iridophores. As in *Fundulus*, they are often associated with melanophores, in fact more or less imbedded in these dark cells, and the complex thus formed has been called a melaniridosome. In such complexes, it is not unusual to have the brightly glistening iridophore temporarily covered or exposed by the movement of the pigment in the melanophore.

From the standpoint of their origin and differentiation, guanophores are very poorly understood. Accounts of certain phases of their history have been given by Becher (1929 b) for *Coregonus*. The guanophores in this fish are modified connective tissue cells which show the first evidence of their final state by a light deposit of guanin near their centres, but external to their nuclei. The guanin is laid down as very fine, doubly refractive granules which sooner or later grow into long crystalline needles or tablets. The guanophore eventually becomes quite gorged with these bodies. In fact as a result of their accumulation the guanophore may be ruptured and the guanin itself may thus come to lie in extra-cellular space. In *Coregonus* the guanophores, according to Becher, are the last chromatophores to be formed, though in *Lebistes* they are said by Ginsburg (1929) to be the first to appear. In a number of fishes many of them eventually come to be attached to melanophores and in fact be enveloped more or less by these cells, thus giving rise to the complex already mentioned, the melaniridosome. Hence, the guanophore is unique in several particulars: it may arise ontogenetically at almost any time in comparison with other chromatophores and it contains a peculiar material, guanin, quite unlike the pigments of the other colour-cells.

E. EGG AND EMBRYONIC PIGMENT IN AMPHIBIANS

The study of the origin and growth of the chromatophores in amphibians is complicated by the fact that in many of these animals the eggs, even at the time of fertilization, are richly provided with melanin granules. As the segmentation of such eggs proceeds, the melanin is handed on from one generation of cells to another and thus establishes in the blastomeres a state of pigmentation which antedates and may overlap the pigmentation of the chromatophores proper. This initial pigment may be designated egg-pigment and is contrasted by certain workers with the pigment which appears subsequently and which has been called embryonic pigment. It is commonly supposed that as development proceeds the egg-pigment gradually diminishes in amount and the embryonic pigment takes its place. This substitution of one kind of pigment for another has been advocated by Weidenreich (1912). It is, however, doubtful whether the rather sharp distinction here implied can really be drawn. Certainly the very circumstantial account of the development of the melanin and the melanophores in *Bufo* as detailed by Goda (1929) gives no support to such a conception. It seems much more probable, as Goda's account shows, that the formation of pigment in those amphibians which may be said to possess egg-pigment is in reality a much more continuous process than is implied in the hypothesis of an egg and an embryonic colouring matter, and, though the formation of pigment in such amphibians is subject to certain fluctuations, it is as a whole relatively uniform. Therefore, the distinction of two classes of melanin in amphibians must be looked upon as essentially artificial. Certainly the majority of recent investigators, though they have recognized the presence of what may be called egg-pigment, have seen fit to pass over as unessential the assumed distinction between this and the so-called embryonic pigment. It must be evident, however, from what has been stated that egg-pigment when, present in amphibians makes it extremely difficult to note the beginnings of melanophores and to follow their early history.

Amphibians in which a well-developed egg-pigment occurs include

Rana, Bufo, Pelobates and the like. A small amount of this pigment is present in *Hyla arborea* and in *Triton taeniatus*. It is absent from the eggs of *Salamandra maculosa* and of *S. atra* as well as from those of *Triton cristatus*. In this respect the eggs of these forms resemble the eggs of fishes. In *Salamandra maculosa* and in *Triton cristatus*, the first melanin to appear is that of the true melanophores and the time of the appearance is when the lens of the eye separates from the external ectoderm in early embryonic life. In this respect the embryos of these amphibians resemble in a striking way those of fishes.

F. INTEGUMENTARY MELANOPHORES IN AMPHIBIANS

Amphibian melanophores may be classed according to their positions under three heads: epidermal, dermal, and visceral (DuShane, 1943), a grouping more direct and convenient than the more detailed ones used by Weidenreich (1912) and by Elias (1939 a). The integumentary melanophores in amphibians are either epidermal or dermal. The embryonic origin of these two classes of colour-cells is a question of much uncertainty. According to Frankenberger (1924), melanophores first appear in the tadpole of *Hyla* when it has attained a length of about 4·5 mm. They are to be seen in two longitudinal rows down the back of the animal, one right, the other left. New melanophores are added to these rows, a number being situated segmentally. A network of melanophores is eventually formed which disintegrates before metamorphosis to give rise to the chromatic system of the adult. That these melanophores are primitive connective tissue or mesenchyme cells has been accepted by a large number of investigators including Kornfeld (1919), Fischel (1920), Schmidt (1920), Berweger (1926), Elias (1932, 1934), and others. This view is in agreement with the conclusion that has already been stated for the origin of similar cells in fishes.

The melanophores whose source is thus accounted for are the dermal melanophores. As already stated, the epidermis of amphibians also contains melanophores and the origin of these colour-cells has led to much discussion (DuShane, 1943). The more important hypotheses that have been advanced for the origin of these cells may be brought under the three following heads. According to the first of these, epidermal melanophores, like dermal colour-cells, are believed (Kornfeld, 1919) to be derived from the mesenchyme and to have migrated from the regions in which they arose into the epidermis. The chief evidence in favour of this view is based upon the occurrence of cells whose bodies were found to be in the derma and some of whose processes extended into the epidermis as though they were in the act of migrating from one layer to the other. It has also been pointed out that many of the epidermal melanophores have a striking histological resemblance to those of the derma and are unlike other epidermal cells.

The second view on the origin of the epidermal melanophores is that they arise locally in the epidermis, where they remain. This opinion, which has been advocated by Weidenreich (1912), Fischel (1920), and others, is based like the preceding one upon histological resemblances and lacks as the other also does any really conclusive basis.

The third general hypothesis on the origin of these colour-cells is one that has been espoused by Winkler (1910b). According to this view, all melanophores originate in the epidermis and those found in other parts of the body have migrated to these parts from this superficial source. This view is exactly the opposite of the first one, and within recent years has received very considerable support though it has undergone more or less modification at the hands of those who have favoured it.

When these several views are compared, the most fundamental differences that they show pertain without doubt to the germ-layers. From this standpoint, there may be said to be three possibilities: first, the integumentary melanophores may be derived exclusively from the mesoderm (mesenchyme or primitive connective tissue); secondly, they may come exclusively from the ectoderm (epidermis or still more primitive ectodermic sources); and, finally, some may arise from a mesodermic source and others from an ectodermic one. If all melanophores come from the mesoderm as advocated by many of the older workers, some must reach the epidermis, where they are found in most adult amphibians, by migration. If all come from the ectoderm, those in other parts must reach the derma and deeper regions also by migration. If these colour-cells come from the two embryonic sources, some from the mesoderm and others from the ectoderm, they may remain where they are formed and migration is not a necessary step in their evolution. It is, therefore, important to determine the exact embryonic source or sources of these colour-cells, and such a determination should be tested experimentally and not left to a decision based on unassisted observation. A new attack upon this problem seems therefore reasonable and as a matter of fact has been carried out from the standpoint just indicated by several very recent investigators, including Goda (1929), Holtfreter (1933), Blanchard and McCurdy (1934), and DuShane (1934–43).

The source of the integumentary melanophores in *Bufo* has been exhaustively studied by Goda (1929). According to this author, all the cells of the early larval toad, excepting those that form the lens of the eye and its covering cornea, contain melanin granules, the so-called egg-melanin. As development proceeds, this melanin gradually disappears except in such places as the pigment layer of the eye, the nerve cells, and the melanophores, where it is further increased. At very early stages in the development of *Bufo* the ectoderm consists of two layers of cells, one superficial, the other deep. From the deep layer, cells can be seen to escape into the space between this layer and the subjacent entoderm. Subsequently, when the neural tube is being formed, numerous such ectoderm cells make their way inward from the region of the neural crests to the subjacent mesenchyme, where they mingle with the cells of this tissue and eventually become melanophores.

Other cells from the deep layer of the ectoderm develop large amounts of melanin and differentiate into epidermal melanophores. Still others migrate inward and give rise to the melanophores of the derma and of other deeper parts. Goda is persuaded that all larval melanophores in *Bufo* are derived from the ectoderm. Although this investigation was carried through with much greater fullness and care than were those of the earlier advocates of the ectodermic origin of amphibian melanophores, and though the evidence presented in it for an ectodermic origin of these cells is very strong, it must be admitted that Goda's methods, most of which were of the older observational type, could not be expected to yield conclusive evidence on this point. An experimental test of this problem naturally suggested itself as much more promising and this was attempted by Holtfreter (1933), Blanchard and McCurdy (1934), and especially by DuShane (1935, 1938). Holtfreter (1933) observed that when pieces of medullary tissue were transplanted from one amphibian to another, melanophores were formed in the transplant and under conditions that led him to suspect that these colour-cells were derived from the ganglionic ridges of the lip of the medullary plate; that is, from ectoderm. This observation induced Blanchard and McCurdy (1934) and DuShane (1934a) to make extended experimental studies of the subject in part by means of chemical responses of the tissues and in part by transplantation of tissues.

Blanchard and McCurdy (1934) traced the growth of the melanophores in the young of *Triturus torosus* by means of a method specific for these cells—Laidlaw's modification of the Bloch 'dopa' technique. The stages in the development of the newt which they studied extended from the time of the partial closure of the neural tube to that of the first appearance of dendritic melanophores. At the very early stages, cells positive to the 'dopa' reaction were found in the entire ectoderm of this animal. At later stages such cells occurred only in the regions of the neural crests. From these regions they could be traced to positions where they metamorphosed into definitive melanophores. Hence it was concluded that these colour-cells had an ectodermic origin.

DuShane (1934a, 1935) carried out a research on this general problem by extending and amplifying the method used by Holtfreter. *Amblystoma* embryos, from which the neural crests in the trunk region had been removed, were found to develop into larvae that were normal in all respects except that they lacked completely the fin fold and the melanophores in the region of extirpation. When prospective ganglion crest cells were transplanted to the same or to another species of salamander, they always gave rise in their new location to pigment cells. As a check on this type of observation, ectoderm from the flank of a young salamander and fragments of its mesoderm were transplanted into another individual and it was found that no melano-phores were produced in these transplants. Furthermore, trans-plantations of appropriate pieces of ectodermic tissue from *Amblystoma punctatum* to the flank of a white axolotl induced pigment formation in the axolotl, a form not normally pigmented. These several tests show that the integumentary melanophores in *Amblystoma* arise not from the connective tissue or epidermis of this animal, but from its neural crests. They relegate these colour-cells not only to a purely ectodermal source, but exclusively to the neural crests.

These most recent pieces of work on the origin of the integumentary melanophores in amphibians set the entire question in a wholly novel light. The evidence from them points very conclusively to an ecto-dermal origin of melanophores and we can no longer look upon the problem of the embryonic source of these cells in amphibians as a problem the solution of which is still very uncertain (Elias, 1934). That the several accounts already noted are not in themselves wholly in harmony does not militate against the general conclusion here drawn. DuShane declares that in *Amblystoma* the melanophores are derived from the neural crests and from these parts alone. Blanchard and McCurdy on the other hand intimate that in *Triturus* the general ectoderm is involved, and Goda points very specifically in *Bufo* to the deep epidermis of the skin as a centre of melanophore production. All these claims may be correct, for it is entirely possible that such minor differences may well exist among the various species studied. All the accounts agree in one particular; namely, that amphibian melanophores both epidermal and dermal are ectodermic in origin. This whole field has been ably reviewed recently by DuShane (1943).

Granting the correctness of this view and keeping in mind the novel methods by which it has been attained, it is not unnatural to inquire whether the conclusion concerning the origin of the melano-phores in fishes, as stated in the earlier part of this summary, may not be open to revision. It must be obvious from what has just been stated that the evidence on which the mesenchyme origin of piscine melanophores is based can no longer be said to be final. These cells were first identified in the mesenchyme of fishes by their own natural pigment. What their history was before they became pigmented and thus identifiable no one can claim to know. So far as our present knowledge of them is concerned they may as well have wandered into the mesenchyme from the neural crests as to have arisen in the mesenchyme itself. For this reason the real origin of the melanophores in fishes may really be said to be in doubt and the subject therefore to be thrown open to reinvestigation. Steps in such a new research would naturally be guided by modern methods such as experimental embryology is now bringing forward.

G. Other Integumentary Chromatophores in Amphibians

In addition to melanophores, the skin of amphibians contains xantho-phores, erythrophores, guanophores and other like colour-cells. The origin of these cells has not been as exhaustively investigated as that of the melanophores. As a rule their early phases have been treated by the various authors who have been mentioned as an adjunct to what has been reported on the melanophores. The origin of the xanthophores or yellow lipophores has been noted by Berweger (1926) in *Salamandra*, and by Elias in *Rana* (1932) and in *Bufo* (1934). In all three investiga-tions, these cells are said to arise from connective-tissue sources. A similar origin is ascribed to the guanophores in amphibians by Schmidt (1920), who studied the early stages of these elements in the embryos and larvae of *Salamandra* and in the tadpoles of the common frog. The first evidence that a cell is to differentiate into a guanophore, as observed by this author, is the appearance of minute, crystalline particles of guanine in its cytoplasm. This early stage in the development of guanophores has also been fully described in the fish *Lebistes* by Ginsburg (1929), who calls attention to the fact that these particles of guanine show Brownian movement in the living cells. In the amphibians studied by Schmidt, the guanine particles in course of time increase in size, take on the form of platelets, and eventually gorge the cell to such an extent as to project partly on its periphery and thus give to its outer margins a roughly angular contour. Schmidt's statement that the guanophores in *Salamandra* arise from connective tissue cells has been reaffirmed by Berweger (1926) and has been declared by Elias to hold true also for the frog (1932) and the toad (1934). All these references to the origin of special types of chromatophores were published before the experimental work on the amphibian melanophores already referred to had been made known. In none of these reports is the evidence so complete as to preclude the possibility of an ectodermic origin for these special colour-cells such as has been ascribed recently to the melanophores.

In work published in 1938, DuShane showed that when the neural folds in the future trunk region of *Amblystoma* embryos were removed, not only melanophores but also xanthophores and guanophores were lacking from the corresponding regions in later development. The visceral chromatophores were also absent. Hence, he concluded that in amphibians all chromatophores (epidermal, dermal, and visceral) have a common origin from the ectoderm (DuShane, 1943).

H. Migration of Chromatophores

The earlier students of chromatophores were so impressed by the remarkable capacity of these cells to disperse and concentrate their pigment, a capacity that resembles in a striking way the projection and withdrawal of pseudopodia by an amoeba, that many of them unhesitatingly ascribed to these cells amoeboid locomotion as a means of migration. This position, however, was rather uncritically assumed. Fuchs (1914), after surveying this field, was forced to admit that though ample evidence had been brought forward in favour of much protoplasmic activity in chromatophores, no conclusive proof was given that chromatophores moved from place to place. Of course, those workers who advocated an exclusive origin for colour-cells either in the ectoderm or in the mesoderm were forced to admit a migration of these cells from one layer to the other, but such an admission is in no sense a proof. Bolk (1910), who was the first to call attention in a precise way to the segmental arrangement of the embryonic melanophores in such fishes as *Atherina* and *Alburnus*, was opposed in general to the idea of migration, though he recorded evidence in favour of this activity in what he regarded as an exceptional instance in *Lophius*. In recent times Ginsburg (1929), in his studies on the colour-cells in *Lebistes*, has also expressed himself as opposed to the idea of the locomotion of colour-cells.

When the evidence on this question is carefully sifted, it becomes clear that chromatophores differ greatly, depending upon their relative age in development and the part of the animal in which they are located. There appears to be no doubt whatever that in many early stages in the development of fishes the melanophores on the yolk-sac move about with considerable freedom. Such a migration was fully confirmed by Stockard (1915) and by Gilson (1926 a) on the same fish, *Fundulus*.

Fig. 114. A black and a brown chromatophore on the yolk-sac of *Fundulus* 72 hours old; 11, the two melanophores in contact; 12, 15 minutes later, the two melanophores separated; 13, 20 minutes later than 12, the two melanophores still farther apart. Stockard, 1915 b, 542.

Fig. 114, taken from Stockard's monograph, illustrates a mutual change in location as shown by two migrating melanophores on the yolk-sac of *Fundulus*. The migration of chromatophores, including xanthophores, on the yolk-sac of *Coregonus* has been recorded by Becher (1929 b); and Duspiva (1931 a) has observed a similar activity among the melanophores in the perch and the saibling. From these records it is clear that the chromatophores, and particularly the melanophores of the yolk-sacs of very young fishes, are capable of migration, though the extent of their movement is distinctly limited. Somewhat different is the question of the migration of chromatophores within the body proper of a developing fish or amphibian. A number of the earlier investigators believed that the embryos of such animals possessed a limited number of centres in which chromatophores were produced in quantity and from which these colour-cells migrated to different parts of the animal there to take up final and permanent positions. These centres were usually located on the flanks of the young creature sometimes anteriorly and at other times posteriorly. But the evidence that was gained from the growth of the chromatophoral system as a whole did not favour this view. Bolk (1910) pointed out that as lines and bands of melanophores were formed on the growing embryo the new colour-cells were added to the series, not by the migration of young cells from a preformed centre, but by a local production of melanophores step by step at the end of the line. Such cells, moreover, remained where they had been formed. It was observations of this kind that led to the opinion that though melanophores on the yolk-sac of fishes might show a real though limited capacity for migration, those in the embryonic body proper were without this ability.

An attempt to test this matter from an experimental standpoint was made by Wagner in 1911. This investigator cut off parts of the edge of the tail of young salmon just taking on their pigmentation and made incisions into this organ so as to sever its fin rays. He then looked for the possible heaping up of melanophores at the cut ends of the rays over which they were suspected of migrating. Such a heaping up was to be observed, though the results were not always as clearly shown as might have been wished. However, so far as they went they were distinctly in favour of the migration of melanophores lengthwise over the fin rays of this fish, and Wagner believed that his observations called for the acceptance of this assumption. Dederer's observation (1921) that melanophores made their way out of explants of embryonic tissue from *Fundulus* and crept about for short distances substantiated Wagner's general view.

Gilson (1926a), in testing this question in *Fundulus*, turned to the older observational methods and recorded from hour to hour the positions of melanophores in the body of the embryonic fish. He found on comparing these records that the melanophores at this stage actually did creep about. Fig. 115 represents successive sketches from a group of melanophores in a *Fundulus* embryo some six days old and

made over a period of about three hours. It is perfectly obvious from
these records that at this stage of development in *Fundulus* melanophores in the body of the embryo move about with much freedom.

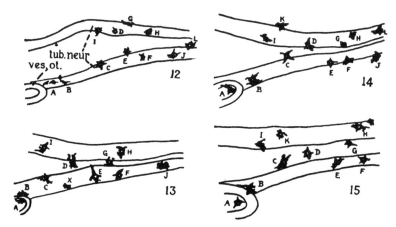

Fig. 115. A group of melanophores in an embryo of *Fundulus* 6 days and 9 hours
old. The group is above the medulla oblongata. 12, at the beginning of observation;
13, 2 hours, 40 minutes later; 14, 43 minutes later than 13; 15, 47 minutes later
than 14. Each individual melanophore is designated by the same letter. Gilson,
1926a, 453, Figs. 12–15.

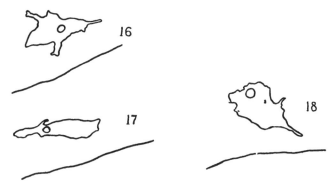

Fig. 116. Three outlines of the same melanophore from an embryo of *Fundulus
heteroclitus* 6 days old: 16, initial outline; 17, outline 50 minutes later; 18, outline
4 hours and 45 minutes after 17. Gilson, 1926a, 453, Figs. 16–18.

In a second illustration from Gilson's work (Fig. 116), the change of
form undergone by a single melanophore is given in enlarged outline,
and when the records here presented are combined with those given
in Fig. 115, it must be admitted that the comparison of a very young

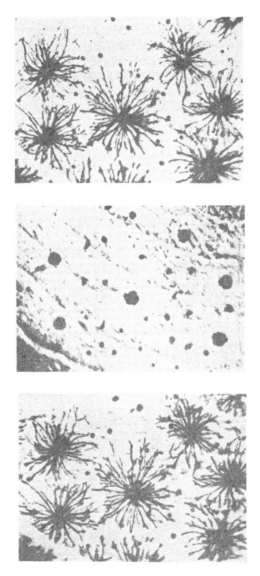

Fig 117. A group of melanophores in a scale from *Fundulus heteroclitus*, with pigment dispersed (upper figure), with pigment concentrated (middle figure), and with pigment redispersed (lower figure) Spaeth, 1913a, 521–3, Figs 1–3

melanophore with an amoeba, so far as its powers of locomotion are concerned, is not so far-fetched as has been maintained by some of the critics of this view. It seems, therefore, definitely established that at least in very early stages of chromatophore development the individual cells may exhibit powers of locomotion that compare very favourably with those of a leucocyte or even with an amoeba and that this activity, though not as extensive as was implied by many older workers, is nevertheless a real factor in placing the melanophore in its definitive location.

The locomotor activity just described is apparently characteristic of chromatophores only in their very early stages of growth. After they have become members of a general pattern, even in larval individuals, they retain their places in that pattern and their original locomotive powers seem to subside almost completely (Franz, 1935 a). Their activities in the later stages are limited to the dispersion and concentration of their pigment, an operation which they retain throughout the remainder of their life and which is well known to be quite stereotyped in character. If photographs are taken of them during these changes, they can be shown to return to their identical forms again and again. This was well illustrated by Spaeth (1913) in *Fundulus* (Fig. 117). In summary it may be stated that the evidence at hand shows that the older workers were correct in their assumption that chromatophores can move from place to place, but they were wrong in their belief that this activity was extensive. Apparently it is limited in amount and restricted to the very early stages in the life of these cells.

I. Time of Onset of Colour Changes

The most important chromatic neurohumour in the vertebrates is without doubt the pituitary secretion intermedine. According to Kleinholz (1940a, 1940b) this neurohumour first appears in the developing pituitary glands of the larvae of *Rana pipiens* when these young animals are from 4 to 7 mm. long. In making this determination Kleinholz used the *Anolis* method (Kleinholz and Rahn, 1940), by which he was able to show that the hypophysial primordium of the gland contained at this early stage some 0·3 of an *Anolis* unit of intermedine. Etkin (1941 a), who worked upon the developing toad *Xenopus*, found evidence of intermedine in the developing glands of this animal 48 hours after fertilization. In *Amblystoma*, Drager and Blount (1941) noted the first traces of the melanophore-expanding hormone in this salamander at stage 38 (Harrison's series) and clearly definite evidence of its presence at the next stage, 39. These determinations were made by the Kleinholz and Rahn *Anolis* method checked by that of Keaty and Stanley (1941). The first appearance of the functional activity in the pars intermedia of the pituitary gland in *Amblystoma* is estimated by Drager and Blount to be relatively later than that in the frog as reported by Kleinholz, and in *Xenopus* by Etkin. Nevertheless, all these evidences of the presence of functional amounts of intermedine in the three amphibians thus far tested come from very early stages, stages that in general mark the beginning of the first colour-changes.

The times at which the chromatophoral systems in developing vertebrates first take on functional activity has been determined in comparatively few forms and of these almost all are fishes. *Fundulus* has been reported on in this respect by three investigators, Bancroft (1912), Spaeth (1913) and Gilson (1926a), all of whom agree that the functional activity of the melanophore system in this fish sets in immediately after the creature has hatched and that the colour changes that ensue are essentially the same as those in the adult. The evidence for this determination is most fully presented by Spaeth, who states that many observations on twelve-day embryos failed to show any change in their melano-

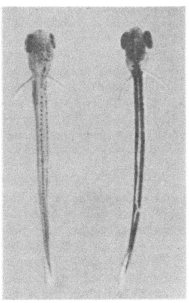

Fig. 118. Two larvae of *Coregonus*, the left one adapted to a white background, the right to black surroundings. Koller, 1934, 422.

phores so long as their egg-membrane remained intact, even though the embryos were watched over white and black backgrounds. But after the young fishes had left the egg-cases, which occurred in a little over 12 days after fertilization, and had been placed some in white-bottomed and others in black-bottomed vessels, those in the white vessels became pale and those in the black ones remained dark as they were when first hatched. Thus at about six hours after hatching, *Fundulus* can first adapt itself to the tint of its environment.

Becher (1929 b) records that the chromatophores of the trout, of the saibling, and of *Coregonus* (Fig. 118) become active also immediately after the young fishes have hatched from the egg. He remarks, however, that it is quite possible to excite the system to action by removing well-grown embryos prematurely from the egg-shell. Duspiva (1931 a) notes that in *Salmo salvelinus* the streaming of melanin granules can be observed four days after the first of this pigment has appeared and that the same is true of *Perca fluviatilis* three days after that event. In *Perca*, there is so much variation in this particular that Duspiva is inclined to believe that the beginnings of pigment movement may be somewhat earlier than three days after pigment formation. He points out that before such changes can appear two events must transpire; namely, the development of cell processes from the differentiating chromatophore into which the pigment may spread and from which it may retreat, and the production of the pigment itself in the colour-cells. After discussing these matters at some length, Duspiva concludes that a melanophore is in all probability able to function as soon as it has formed its supply of pigment, a conclusion that would set the fate of first activity some time ahead of that at which he actually noticed it.

The observations which have thus been somewhat categorically recorded all refer to oviparous fishes and amphibians. Apparently only one record has been made of the state of the chromatophoral system in a viviparous vertebrate. *Mustelus canis* is a dogfish that may bring forth in one litter as many as six or eight young fishes or pups. At birth each pup may measure as much as 30 cm. in length. These young pups are born with a fully developed melanophore system, the individual colour-cells of which are at the moment of birth in an intermediate condition so far as their pigment is concerned. This condition is the same as that of an adult dogfish which has been a day or so in the dark. If a litter of such newly born pups is separated and some are put in an illuminated aquarium with white walls and others in an illuminated aquarium with black walls, their colour begins to change quickly. The melanophores of the fishes in the black-walled vessel disperse their pigment and those in the white-walled vessel concentrate theirs. These operations agree in almost all particulars with those of the adults. Moreover, the newly born pups blanch when injected with adrenaline and darken when injected with pituitary extract as do the adults. Viviparous dogfishes then are born with an immediately effective and functional melanophore system (Parker, 1936 *b*).

From this instance, as from those noted in the earlier part of this account, it is obvious that in those vertebrates which have been examined in this particular the chromatophore system takes a very early start in life and is at once fully effective either at birth or, on hatching as the case may be. In fact, it may be functional even before these events. In this very early assumption of its activities, the vertebrate chromatophoral system resembles the blood system of these animals, a condition that suggests for chromatophores some initial use quite unlike that which is finally assumed by these cells. What these relations are it is difficult to say; they may be more metabolic in nature than otherwise. This problem will be discussed in a subsequent part of this work.

J. EARLY SYSTEMATIC RESPONSES OF
CHROMATOPHORES (PATTERNS)

As will be recalled from chapter IV of this survey, the chromatophoral systems in a number of lower vertebrates show during their early stages of development two strikingly different reaction patterns. The first of these characterizes the very earliest period of chromatophoral response in embryos and early larvae, and the second the period of later larval development and adult life. The two patterns are almost the reverse of each other. In the first or primary type of response, the young animals, as already pointed out, become pale in darkness and dark-tinted in bright light. In the secondary type so-called, the animals are usually dark in darkness but in bright light pale in white surroundings and dark in a black environment (Babák, 1910 *b*). Many vertebrates in the course of their early growth pass through these stages, primary and secondary, in the order given (*Perca, Salmo, Macropodus, Rana, Axolotl,* etc.). Others seem to omit the primary stage and begin their chromatic activities in the secondary stage (*Mustelus, Fundulus, Coregonus, Oncorhynchus, Amblystoma, Triton*). Such an abbreviation in development as the omission of the primary stage is in no sense surprising, for occurrences of this kind are well known in the early growth of many animals. On being blinded, all these forms revert to the primary type of response in which the individual is pale in darkness and dark in full light. Apparently there is no form known in whose history the state here described as secondary precedes the primary, not would such a condition be anticipated from the standpoint presented in this survey.

After a chromatic vertebrate has reached either indirectly or directly its secondary stage of chromatic organization, it begins to take on its final colour pattern; namely, that characteristic of its adult life. The assumption of the mature chromatophoral pattern is a complicated process and efforts at its understanding have led to decided differences of opinion. Is the colour pattern of a given animal determined by the chromatophores themselves or does it depend upon the tissue environment in which the chromatophores find themselves? This question was tested by DuShane (1935) on two species of *Amblystoma* whose coloration patterns were sufficiently different for satisfactory experimentation. In his review of this subject, DuShane (1943) states that when the prospective pigment cells of *Amblystoma tigrinum* are transplanted to the embryos of *A. punctatum*, the resulting pigmentation of the region of the trunk affected by the graft is entirely that of the donor. Hence, the type of colour pattern is to be attributed to the intrinsic activity of the colour-cells themselves and not to the embryonic environment in which these cells are to be found. A similar conclusion is to be drawn from the investigations of Twitty (1935, 1936) and of Twitty and Bodenstein (1939), whose studies on transplantation were carried out on three species of *Triturus*. These three newts have significant differences in pigmentation. When transplantations of neural crest cells were made between their embryos, the patterns assumed by the transferred colour-cells were always those of the donor individuals. Costello (1942) has also reported like findings in transplantation experiments between the anuran *Hyla* and the urodele *Triturus*. *T. torosus*, the species of newt worked upon by Costello, has a marked dorsal pigment band. When this newt was supplied with chromatophores from *Hyla*, the resulting pigmentation showed no suggestion of a band but the chromatophores remained scattered, thus recalling the state of the donor species. All these investigations give support, therefore, to the view that in some way not yet clearly understood the colour pattern is an intrinsic property of the chromatophores concerned.

In contrast with this conclusion, other investigators have obtained evidence that the determination of pattern is dependent upon the individual into which the prospective colour-cells have been transplanted. Thus Rosin (1940) transferred a neural fold from the axolotl to *Triton palmatus* and found that the grafted melanophores tended to distribute themselves largely according to the pattern of the recipient rather than to that of the donor. Somewhat similar results were obtained by Balzer (1941), who transplanted neural folds from *Hyla* to axolotls and to *Triton*. Here grafted melanophores and guanophores migrated extensively and accumulated at the border of the yolk, thus simulating in part the pattern of the recipient *Triton*. De Lanney (1941), who worked with transplants of mesoderm in *Triturus*, found that this material exerted a limiting control over the migration of melanophores, thus demonstrating an environmental influence in the final arrangement of these colour-cells. Barden (1942, 1943) sought evidence on this general question by transplanting and retransplanting the eyes in several species of salamanders and noting the changes in the iris pigment in transplanted and in normal eyes. As a result, he declared that "the changes in the iris chromatophores which lead to darkening of the eye, whether by fading of the guanophores and xanthophores as in (*Triturus*) *rivularis*, or by the blanketing of this bright pigment through the emergence of an obscuring layer of melanophores as in (*Amblystoma*) *punctatum*, seem to be stimulated by some influence originating in the other tissues of the eye", a conclusion which points to the tissue environment rather than to the chromatophores themselves as the occasion of such changes. Thus the contributions referred to in this paragraph emphasize the surrounding tissues as determiners of pattern and not the chromatophores themselves. It is hardly to be supposed, however, that the contrast between these two views as developed in this and the preceding paragraph is as rigidly maintained in nature as is implied in the two sets of contributions quoted. It seems more likely that in all chromatic vertebrates both the colour-cells themselves and the environment in which they are found have to do with pattern and that the contrast here noted arises in that in certain species the chromatophore influences predominate in large part and in others the influences of the organic surroundings. Such at least might be a reasonable attitude to assume as a step toward understanding the discordant views here presented. This whole problem, however, is far from settled. It has been ably discussed by DuShane (1943), whose survey has been freely drawn upon in the preparation of this section.

K. Production and Loss of Chromatophores and Pigment in Adult Animals

The chromatophoral system of an adult animal is by no means a fixed one. Chromatophores from time to time disappear and new ones take their places. These gradual changes have a certain resemblance to the loss and renewal of many integumentary organs such, for instance, as the hairs on the human body. In the case of the hairs, however, the change is probably much more considerable than is that of the colour-cells. In fact, the loss and renewal of chromatophores is such an insensible operation that it can be demonstrated only by what are obviously experimental means.

Apparently the first to advance the idea of a production and loss of chromatophores in adult animals were Keeble and Gamble (1904), who offered this as an explanation for the more or less permanent adaptation of *Hippolyte varians* to its environment. Their view of the method by which such an adjustment might be brought about may be stated in the following way. If a part of the body of this shrimp remains in shadow, its chromatophores, according to these authors, would assume the active state of expansion, one favourable to growth. In consequence, the shaded region would be a location in which an increase in the number of chromatophores would occur. If another part of the shrimp's body is in bright light, the chromatophores of this part would shrink, a state of relative inaction and in consequence one unfavourable for growth. Hence, in this region the colour-cells would gradually diminish in number. In this way Keeble and Gamble attempted to explain the differences in chromatophoral numbers of animals that had been in different environments, an explanation that was based upon the commonly accepted belief that activity excites growth and inactivity retards it. This view remained with Keeble and Gamble a purely hypothetical one. It received, however, the general approval of van Rynberk (1906).

In 1910, Franz pointed out an instance which seemed to be a good example of the kind of response just described. He called attention to the fact that flatfishes (*Pleuronectes platessa*) from the Baltic Sea, the bottom of which is darkened by much seaweed, were deeper in tint than those from the North Sea where the bottom is lighter in colour. This darker shade in the Baltic fishes persisted even when they, as marked individuals, were transferred to the North Sea. Franz attempted to explain this situation more or less on a trophic basis. In applying such an explanation, he accepted the common assumption, the reverse of that made by Keeble and Gamble, that the light phase of the fish (contracted melanophores) was the active one. He pointed out, however, that in such a case it would be impossible to explain the increase that took place in melanophores on a dark background. He was, therefore, led to suggest two phases of activity, one associated with a dark environment and the other with a light one. In his opinion dark surroundings excited in excess dark melanophores (an excitation which as a trophic influence induced them to increase in numbers) and light surroundings in a similar way activated the light chromatophores. Certain insufficiencies in this attempt were pointed out by von Frisch (1911c), who from a study of the trout advanced evidence in favour of the original idea of Keeble and Gamble and pointed out the importance of melanophore counts in considering any such problem.

Babák (1912, 1913) demonstrated quite conclusively that the larvae of *Amblystoma* when reared under different conditions of illumination showed consistent increases or decreases in counted melanophores. He found that in normal larvae on a dark background or in darkness, and in blinded larvae in the light, in all of which conditions the larval melanophores were expanded, an increase in the number of these cells took place. He also found that in normal larvae on a white background and in blinded larvae in darkness, in both of which conditions the larval melanophores were contracted, there was lack of increase and even a decrease in the numbers of these cells. Babák concluded that light and darkness exerted a direct and an indirect influence (through the eyes and nerves) on the integumentary chromatophores. The latter he believed to be trophic. He further maintained that chromatophores exhibited two states of activity, one of complete contraction and the other of complete expansion, and that it was probable that they possessed with these states an appropriate double innervation.

These observations and other advances in chromatophoral physiology led Kammerer (1913), as already described, to carry out an extended series of experiments on the young of the fire-salamander with the view of testing certain aspects of the general theory of evolution. Some of Kammerer's experiments were repeated by Herbst (1919) with contradictory results and the subject became one of general discussion. In consequence, von Frisch (1920) worked over again the chromatophoral responses of the salamander and declared that in his opinion an explanatory basis of all the facts newly accumulated and brought forward was still to be found in the original hypothesis of Keeble and Gamble.

Meanwhile Kuntz (1917) had published on the colour changes in *Paralichthys albiguttus*. This flatfish when on a white background suffered a decrease in the number of its melanophores to the extent of about 30 per cent in 11 days. When such a light-adapted fish was transferred to a black background, it increased the number of its melanophores though not always to full recovery. The melanin from the disintegrated melanophores was, according to Kuntz, discharged to the exterior through the skin. Erythrophores followed the same general course as the melanophores did.

Some three years later, Murisier (1920-21) published an extended account of the chromatophores in the trout. The light and the dark phases in this fish were found to be pronounced and the degeneration of its melanophores could be seen when it was kept on a white background. Murisier believed that this reduction depended upon inanition rather than upon the background on which the fish rested and he interpreted the control of these colour changes as due to metabolic factors. In this, he was probably in error, for even in his own studies he recorded the absence of melanophore degeneration in trout that had been kept on a black background and insufficiently fed. He made clear that in colour changes of the kind under consideration, it was important to separate two elements, changes in the amount of melanin and changes in the number of chromatophores.

At about this time, it became customary to distinguish between what were called morphological and physiological colour changes. The latter were the ordinary quick changes brought about by a dispersion or concentration of chromatophoral pigment in immediate response to the environment. The morphological changes were those of a more permanent character in which by an increase or decrease in the quantity of pigment or in the number of colour-cells a given animal would gradually assume a permanently deeper tint or the reverse. These two activities were examined by Himmer (1923), who declared them to be in reality the same. After an extended study of the melanophores of *Salamandra*, he asserted, in accordance with the general view already expressed, that continual expansion of melanophores was favourable for their multiplication and continual contraction for their reduction.

The common European dab, according to Hewer (1927), increases its epidermal melanophores in numbers after it has been two to three weeks on a black background, but its dermal melanophores under these circumstances remain unchanged. On a white background, the number of its erythrophores are said by this author to be reduced and that of its iridophores to be increased.

In 1933, two important contributions to this subject appeared, one by Sumner and Wells and the other by Odiorne. Sumner and Wells worked upon one of the guppies, *Lebistes*. They showed that two days after these fishes had been transferred from a white background to a black one, minute, dark-brown spots, smaller than ordinary melano-phores, began to appear throughout the central areas of their scales. Three days after the transfer, these spots would respond to adrenaline and to urethane as melanophores do. Such young melanophores continued to enlarge so that after the fifth day an obvious increase in the number of melanophores could be clearly made out. Melanin as such also increased in amount and the new melanophores reached their definitive condition in about three to four weeks. The source of these new melanophores was not ascertained by Sumner and Wells. There was no evidence that the new colour-cells migrated from elsewhere for to all appearances they arose *in situ*. Nor did they seem to result from the division of other melanophores, though such cells have been shown to divide mitotically. After 30 days, on a white background, dark guppies became quite pale again, but not so pale as those reared from birth on such a background. A most active extrusion of melanin took place during the first two weeks. Beyond this the loss of pigment was relatively slow. A similar degeneration of melanophores with loss of pigment to the exterior has been recorded by Graupner and Fischer (1935) for *Atherina*.

The work done by Sumner and Wells on the growth in numbers of melanophores in *Lebistes* was subsequently repeated by Sumner (1940*b*) on the same fish with much improved technique. In the illumination of the aquaria in these new tests, special attention was given to the albedo. By albedo is meant the proportion of light reflected from a given environment as compared with that coming directly from the source of the illumination. Guppies were exposed for six weeks to light in aquaria, the walls of which were either white, pale grey, medium grey, dark grey, or black. After this sojourn, the fishes were measured, killed, and prepared so that the numbers of melanophores on definite measured areas of skin could be counted and recorded. These areas were either on the upper lip or the tail. The average numbers of melanophores for the two areas under the five types of illumination are given in Table 14. The records from the tail areas were far less satisfactory than those from the upper lips. When the records for the lips were plotted against the several albedos, that for white being taken as 100 and that for black as 0, the average numbers of melanophores, 33·88 to 60·98, were found to conform fairly well to the logarithms of the albedo. Sumner therefore con-cluded from this more recent study that direct light had little if any effect upon numbers of melanophores in this fish, but that these colour-cells tend to vary in number inversely as the logarithm of the albedo. Thus this more recent work of Sumner supports with some greater detail the main conclusion of the earlier investigations by Sumner and Wells. The question of light effects from the standpoint of the ratio of the reflected light to that directly from the source was first discussed by Keeble and Gamble (1904) in their work on crustacean colour responses. The subject was revived and taken up subsequently by Sumner (1911), by Mast (1916), and by Sumner and Keys (1929) in their account of flatfish responses.

TABLE 14. Average numbers of melanophores in uniform areas of skin from the upper lip and the tail of *Lebistes* exposed for six weeks to continuous illumination in aquaria whose walls were either white, pale grey, medium grey, dark grey, or black (Sumner, 1940 b, p. 335)

Tint of wall	Average number of melanophores	
	Upper lip	Tail
White	23·88	12·78
Pale grey	33·48	21·45
Medium grey	39·48	24·39
Dark grey	49·78	40·65
Black	60·98	76·54

Incidentally, it may be mentioned that the relation of the total light intensity and the shade of a given background to the degree of dispersion of the melanin in the melanophores of the silver-mouthed minnow, *Ericymba buccata*, has been investigated by Brown (1936) with results that point to the intensity of light as a significant factor in melanophore activity. Brown found that on a black background illuminated with from 0·000053 to 1·75 foot-candles of light the average diameter of the pigment masses was directly proportional to the logarithm of the light intensity. At an intensity below 0·000053 foot-candle, the fish became pale even on a black background. At 1·75 foot-candles, the fish was completely dark. Here the diameter of the melanin masses appeared to vary directly with the ratio of the incident light to the reflected light. In this fish, according to Brown, colour changes within certain limits are determined by the total light intensity as well as by the shade of the background.

Odiorne's first work (1933 a) on changes in the numbers of melanophores was carried out on *Fundulus heteroclitus*. When this fish was placed on a white background, its melanophores soon contracted and remained so. In a few days, evidence of melanophore degeneration appeared; clumps of pigment were visible and counts revealed a decrease in the number of melanophores. These counts were made on given areas on the backs of fishes, the areas being each of 0·75 sq. mm. and capable of re-identification with certainty so that the same area could be recounted from time to time. In all, 23 fishes were subjected to these counts over a period of about four weeks. The numbers in Table 15 are the totals from the areas on all 23 fishes under consideration as recorded at the intervals of time given in the table.

TABLE 15. Decrease in the numbers of melanophores seen in identical areas on 23 fishes, *Fundulus*, kept on white backgrounds over a period of 28 days. In each number, the individual counts for the 23 fishes are added together (Odiorne 1933a, p. 330)

At the beginning			1835	melanophores
After 7 days on white			1714	,,
,,	11	,,	,,	1621	,,
,,	18	,,	,,	1478	,,
,,	25	,,	,,	1283	,,
,.	28	,,	,,	1224	,,

Under the conditions of this test, it is clear that the melanophores in *Fundulus* were reduced in number to almost exactly two-thirds of their original total.

When *Fundulus* was placed on a black background, no such reduction in its melanophores appeared; instead the number of these cells remained nearly constant. If fishes that had been for some time on a white background, and which had thereby lost a number of their melanophores, were placed on a black background, the decrease was not only checked, but a gradual increase in melanophores took place. Low temperatures (10° C.) retarded these operations; higher ones facilitated them.

In 1936, Odiorne published an extended account of his researches, accompanied by photographic views showing the degeneration of melanophores. A pair of illustrations from this source is reproduced in Fig. 119. A close comparison of these will show the granular disintegration and complete disappearance of several pigmented cells. In 1937, Odiorne reported on the changes in the numbers of chromatophores in several other fishes as a result of lengthy sojourns on white or black backgrounds. *Fundulus majalis*, like *F. heteroclitus*, when on a white background, suffered a loss in the number of its melanophores. On a black background, the numbers tended to increase. This was especially true of the catfish *Ameiurus*. Young specimens of *Macropodus* and of *Gambusia* showed a retardation of pigmentation on white and a marked increase on black. Thus, the independent researches of Sumner and Wells and of Odiorne have yielded essentially similar results. In the fishes that these investigators have worked upon, a white background favours a decrease in the number of melanophores and a black one an increase.

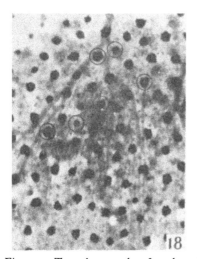 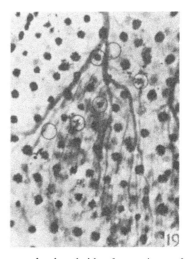

Fig. 119. Two photographs of a selected area on the dorsal side of a specimen of *Fundulus heteroclitus* kept on a white background and showing the disappearance of melanophores: 18, at the beginning of the experiment; 19, 10 days later. Two of the melanophores enclosed in circles in 18 have disappeared in 19, two others are represented by scattered particles, and still two others are beginning to degenerate. Odiorne, 1936, 37.

How these changes are induced is not wholly clear. The reduction of melanophores appears in the form of degeneration with disintegration, and the discharge of the refuse, particularly the melanin, on the surface of the fish. The increase of melanophores is apparently not due to the migration of formed melanophores from other parts of the animal's body into the region concerned nor to the division of melanophores already present. It seems to result from the differentiation of reserve cells in the tract under examination.

What controls the differentiation of melanophores and their final disintegration? Murisier favoured inanition as the occasion of melanophore disintegration, but in *Fundulus* on a white background melanophores will disintegrate in well-fed as well as in poorly fed individuals (Odiorne, 1937). Moreover, melanophores are abundantly produced in underfed fishes on a black background as Murisier's own observations show. Functional activity and its converse also probably play no important part in chromatophore changes. Keeble and Gamble (1904) regarded a chromatophore with dispersed pigment as in an active state and one with concentrated pigment as in a condition of rest. The majority of workers, however, have taken the opposite view; namely, that the condition of concentrated pigment is the active state and that of dispersed pigment the resting one. Babák (1913) opposed both of these opinions in that he held "expansion" and "contraction" each to be a state of activity. These declarations, however, are based more or less tacitly on a comparison of chromatophores with muscle cells and this was true even in Babák's discussion of the case. Such a comparison, however, is essentially misleading. It has recently been pointed out (Parker, 1935f) that the active state of a chromatophore is not one of immobility, but one in which this cell is in process of change. The immobile conditions, full concentration or full dispersion of pigment, however, are just those in which the decrease and the increase of chromatophores are induced. If these are quiescent states of chromatophores, their functional activity cannot be a factor concerned with their numerical changes.

Babák (1913) pointed out very clearly that light and darkness had two very distinct effects on chromatophores, one direct and the other indirect (through the eye and nervous system). That there is a direct effect of light and of darkness on chromatophores is beyond doubt. In vertebrates, the direct effect of light is a dispersion of melanophore pigment and of darkness a concentration of it. How light and its absence accomplishes these changes is not known, but that light or darkness can directly affect protoplasm is obvious, for ordinary vision is dependent upon just such a change. The indirect effect of light through the eye and the nervous system on chromatophores is also well established and here the activities are naturally much more complicated for they involve ordinarily an intricately illuminated environment. The melanophore responses to this indirect activation are nevertheless necessarily simple. They are either dispersion or concentration of pigment. These two conditions are the states under which melanophores increase or decrease in numbers. Thus a nervous control of melanophores must in some way be responsible for their growth and destruction; in other words, they are subject to trophic influences favourable and unfavourable to growth. Such a general view has already been suggested by Franz (1910) and by Babák (1913) and, though these authors have not especially enlarged upon it, they have pointed out its obvious application to the system of vertebrate melanophores.

What is meant by trophic influences under these circumstances is perhaps somewhat vague. If, however, a neurohumoral interpretation is applied to the situation, a meaning is not difficult to find. From this standpoint a melanophore in such an animal as *Fundulus* is brought to disperse its melanin by the action of a substance, a specific neurohumour, which, produced by the appropriate nerve terminals, passes from them to the melanophore and excites it to appropriate reaction. Such a neurohumour, if produced in sufficient quantity and over a long period, may be the means of inducing differentiation in dormant cells and thus of increasing the supply of melanophores. In this way, a trophic influence may be interpreted in terms of neurohumours. In a like manner the nerve terminals in *Fundulus* concerned with the production of a concentrating neurohumour may dispense this substance in such a way as to check the growth of new melanophores or even induce their disintegration, and thus an opposing trophic influence may be brought to bear on melanophore activities. These suggestions are purely hypothetical, yet they show clearly how the production and destruction of melanophores may have a trophic background based on neurohumours. Such an interpretation has already been suggested by Odiorne (1937).

If this interpretation fully represents what actually takes place, it follows that the chromatophoral neurohumours have a close resemblance to growth hormones. They not only excite chromatophores to normal action, the dispersion and concentration of their pigment, but they also induce these cells to grow and to disintegrate. This resemblance of neurohumours to growth hormones can be seen well in the chromatophoral neurohumours from the eye-stalks of crustaceans. When such eye-stalk extracts are prepared and applied with proper precautions to the decapitated coleoptiles of the young oat plant, *Avena*, a perfectly definite acceleration in growth can be demonstrated (Navez and Kropp, 1934). In other words, the crustacean neurohumour acts as a growth hormone to growing plant tissue and thus exhibits a property which has already been inferred for it and other like substances in their effects upon chromatophores. The observations upon which these general conclusions are based are avowedly extremely scanty and yet so far as they go they point to a trophic control of the indirect effects of light and of darkness on the growth and destruction of chromatophores—processes which may well rest in the end upon neurohumoral activities.

Chromatic neurohumours not only give evidence of being activators of normal growth, but they appear also to play a part in pathological processes. They may be concerned in the excessive melanosis in certain fishes, as for instance in the Mexican *Platypoecilus* (Reed and Gordon, 1931), in wound healing in the goldfish (G. M. Smith, 1931, 1932 b, 1934), and in the abnormal darkening of the tail in trout and in salmon (Nigrelli, 1934) where the unusual melanic development is said to resemble the neurohumoral darkening due to nerve cutting. Astwood and Geschickter (1936) report positive responses in the chromatophores of *Phoxinus* to blood from a patient suffering from melanosarcoma. Sumner and Doudoroff (1938 b) describe an infectious diseased condition involving emaciation with lesions of the skin and fins, particularly in the tail, of *Gambusia*. This state occurred in 36 per cent of the fishes in black bowls and in only 2·5 per cent of those in white bowls. After a survey, they ascribed this state to a possible diffusion hormone such as a neurohumour. None of these instances shows in a conclusive way that such hormones are really excitants of pathological states, but they point to the suggestions of Nigrelli and of Sumner and Doudoroff as worthy of further consideration.

The general question of the experimental increase and decrease in numbers of chromatophores in fishes has been fully reviewed by Sumner in several recent publications (1939 a, 1939 b, 1940 a).

L. Production of Pigments

Of the several types of vertebrate chromatophoral pigments (Verne, 1936), the only ones whose formation has been followed to any considerable extent are melanin and the carotenoids. The formation of melanin from a histological standpoint has been described in amphibians by Goda (1929). In the degenerating epidermal cells in the tail of *Bufo* at metamorphosis, the nuclei contain several large nucleoli in which eventually a few melanogranules appear together with special granules. As degeneration proceeds, the nucleus ruptures and the special granules, called by Goda nucleolar bodies, pass into the cytoplasm, there to form many melanogranules. Thus melanin particles are believed by Goda to be traceable to nuclear sources. Graupner (1933), who worked upon the goldfish, associated the formation of melanin with cells having large nuclei, much nucleolar substance, and finely divided chromatin. Melanin in the opinion of Graupner is an end product of cellular metabolism.

Kudo (1922) investigated juices extracted from the skins of a number of fishes, *Esox, Carassius, Phoxinus, Gobius,* and *Nemachilus.* These juices darkened in the air as a result of the formation of completely darkened melanin. Juices from the skin of fishes in the dark phase yielded much stronger deposits of melanin than those from the skins of fishes in the pale phase. In consequence, Kudo concluded that fishes darkened not only because of the dispersion of dark pigment in their melanophores, but also as a result of the formation of new supplies of this pigment.

TABLE 16. Milligrams of melanin in 100 g. of original weight of fishes (*Gambusia*) from black or grey backgrounds and under different intensities of light (Sumner and Doudoroff, 1938*a*, p. 462)

Intensity of light	Black		Grey	
	Bowl 1	Bowl 2	Bowl 1	Bowl 2
0·25 foot-candle	—	—	84	88
1·5 foot-candles	94	84	77	86
10·0 ,,	91	102	85	84
90·0 ,,	98	93	75	88
Means	93·7		83·4	

This question was taken up by Sumner and Doudoroff, whose recent paper (1938*a*) on this subject gives their results from the mosquito fish *Gambusia.* In testing these fishes, the amounts of melanin contained in their bodies were compared under different intensities of direct light and under different degrees of albedo. The intensities of the direct lights used were in foot-candles 0·25, 1·5, 10 and 90. The backgrounds were white, black, and three shades of grey whose albedos were estimated in per cents as follows: light grey, 33; medium grey, 19; and dark grey, 12. Black was assumed to be zero. Fishes were subjected to these conditions for about two months, then killed, prepared, and their melanin extracted and measured. One set of tests in which gambusias from a black background were compared with those from a grey one is given in Table 16.

From Table 16 it will be seen that there is no correlation between the intensity of illumination and the amount of melanin finally present. Thus in the grey section, bowl 2, 88 mg. of melanin were produced under both 0·25 and 90·0 foot-candles of light. But the amount of melanin present in fishes after some two months on a black background (93·7 mg.) was about 12 per cent greater than that from fishes on a grey background (83·4 mg.). Thus the albedo and not the direct illumination appears to be the significant factor and an illuminated black background is clearly not only favourable for an increase in the number of melanophores, as pointed out previously, but also for an increase in the absolute amount of melanin. This increase in melanin is probably a response to the neurohumour intermedine which has already been shown to be a substance with growth-exciting qualities. These results from the work of Sumner and Doudoroff on *Gambusia* substantiate their earlier findings on *Gillichthys* (1937), as does the very recent work by Sumner (1943) on *Girella* and *Fundulus*. *Girella*, after four months in illuminated black-walled containers, showed two and a half times as much melanin as did fishes in pale-grey containers, whereas *Fundulus* under similar circumstances showed an increase of only one and a half times as much of this material. An improved method of assaying melanin has been put forward by Sumner and Doudoroff (1943).

Like melanin, the yellow or carotenoid pigments have been made the objects of study so far as their sources are concerned. These pigments have been identified in both vertebrates and invertebrates by Lönnberg, who in a long series of brief articles (1930–36) has recorded his results. The question of the origin of these colouring matters has been assailed from the quantitative side by Sumner and Fox (1933, 1935*a*, 1935*b*). In their first paper, these investigators reported results obtained from the three fishes *Girella*, *Fundulus*, and *Gillichthys*. No carotene was obtained from these forms, but all their yellow pigment seemed to belong to the xanthophyll series. The retention of *Fundulus* and *Gillichthys* for one or two months on backgrounds of black, white, red, or yellow led to no changes in the amount of xanthophyll present. Since these fishes, when on a yellow background, were strikingly yellow, this tint must be attributed to the dispersion of their xanthophore pigment and not to its increased production. Newly captured specimens of *Girella* contained more than twice as much xanthophyll as those that had been kept four months in the laboratory, especially if they had been retained on a white background. Hence, it was believed that in the laboratory this fish not only failed to add to its store of xanthophyll, but actually suffered a reduction in its supply of this colouring matter.

In a second paper on this subject, Sumner and Fox (1935*a*) substantiated their earlier finding on *Girella* and showed that in 30 days this fish could suffer a reduction of 70 per cent of its xanthophyll, and in one lot of fishes kept in the experimental tanks for five months the reduction reached 85 per cent. The greater amount of xanthophyll in fishes on a yellow background as compared with those on a white one is not due to an increase of this substance in the "yellow" individuals, but to a retardation of the loss of this material in such forms as compared with what occurs in "white" fishes. In Sumner and Fox's tests, a neutral grey background was found to be as effective as a yellow one in retarding the loss of the xanthophyll, and a black one was even more effective than a yellow one in this respect.

In a third paper on this subject, Sumner and Fox (1935 b), working upon *Fundulus*, showed that these fishes when fed on a diet rich in xanthophyll gained in 14 weeks more than 22 per cent in the absolute amount of this substance in their bodies. On a diet free from carotenoid pigments, no change in xanthophyll was observed, showing that the original store of this substance had been retained nearly or quite intact. When carotene was added to a diet containing no significant amount of xanthophyll, the fishes so fed increased their store of xanthophyll, thus justifying the provisional conclusion that in some way these creatures converted carotene into one or more carotenoids of the xanthophyll group. Sumner (1937) concluded that xanthophyll must be rapidly lost from certain fishes if it is not present in sufficient amounts in their diet and that this loss is most considerable on a white background, less so on a yellow one, and least on black. Evidently these slow xanthophore changes in fishes take place in a way very different from that shown by their melanophores.

The chemistry of melanin formation has been discussed by Du-Shane (1935, 1936, 1939, 1943), Twitty and Bodenstein (1939), Twitty (1942), and Figge (1938–40). Figge's work took its start from a discovery made by Lewis (1932), that melanin disappears from tadpoles when they are reared in solutions of indophenol dyes. Figge was thereby led to study the type of pallor thus produced. As stated by Dawes (1941 b), such blanching "seemed to imply the destruction of melanin, but was found (by Figge) to arise in consequence of the inhibition of tyrosinase action by the dye. Tyrosinase is regulated by oxidation-reduction potentials, and in a mixture with redox indicators, sodium hydrosulphite not only decolorizes the dye but also alters black melanin to a light brown colour. The subsequent addition of potassium ferricyanide to the mixture reverses the change and restores to the pigment its former black colour. Figge concluded that melanin is a natural reversible oxidation-reduction system." This conclusion by Figge, together with work done previously by Dawes (1941 a), led

Dawes to declare (1941 b) that the usual conception of the nature of background colour responses in amphibians is inadequate and must be extended. Accordingly, the following revised view was expressed by Dawes. When an amphibian is kept on an illuminated black background for several weeks, two events occur: first, melanin granules are dispersed in the melanophores of the animal, and secondly, the amount of melanin in its skin is increased. Conversely, when the animal is kept on a white background, the melanin granules are aggregated in the creature's pigment cells and the amount of melanin in its skin is decreased. In short, beside pigment migration, additive and subtractive processes are operative. These processes, whereby melanin is intensified or reduced in colour, take place according to Dawes on the basis of the natural reversible oxidation-reduction system worked out by Figge, and thus a new factor is introduced into the chromatic physiology of amphibians and probably of other vertebrates. This idea is by no means novel. As early as 1917, Ruth and Gibson, who worked on the colour changes in certain Philippine house-lizards, were led to conclude that in these animals their integumentary colour-cells neither expand nor contract but the pigment in these cells fades and re-forms and thereby induces the changes in tint shown by these reptiles. No pigment migration was ever seen in these lizards by Ruth and Gibson, who attribute such colour alterations as these animals show to intracellular chemical changes. Thus Ruth and Gibson, in their interpretation of reptilian chromatics, took a step much more radical than that taken by Dawes, for they denied that pigment migration played any part in such colour changes as they had studied, and attributed the whole of these changes to the fading and the re-forming of pigment within the melanophores themselves. Whether the extreme position assumed by these investigators is justifiable or not, and what part this novel process of pigment fading and re-formation may play in vertebrate colour changes, must now be considered.

That the melanin content of a number of chromatic vertebrates is increased one- or two-fold by the retention of these animals in an illuminated black-walled receptacle is unquestionable. The extended researches in this field of Sumner and his associates have placed this matter beyond doubt. All these researches show, however, that the formation and elimination of melanin is a process of many days, usually weeks. The ordinary darkening or blanching of most animals to environmental changes is a matter of relatively brief intervals, of hours, minutes, or even seconds. Moreover, these changes are well known, through the microscopic inspection of living melanophores by innumerable investigators, to be accomplished through the actual migration of melanin particles. In such dark cells, the melanin particles can be seen in joggling motion, suggestive of Brownian movement, making their way in one direction or the other through the processes of a given melanophore. In view of these facts, the statement of Ruth and Gibson that in the lizards studied by them no pigment migration could be discerned becomes extremely questionable (Parker, 1938a). Thus far no reputable observer has given evidence to show that melanin migration is completely absent from any chromatic vertebrate. That such migration may be accompanied by the re-formation of melanin as claimed by Dawes is another question and may very well be an actual occurrence in actively dispersing dark cells. But judging from the very slow rate at which melanin is accumulated in chromatic vertebrates as indicated by the measurements of these matters given by Sumner and his associates, the amount of this substance formed in the time usually taken for most chromatic vertebrates to change colour would be very small. It is not likely that it would amount in most instances to more than 1 per cent or so of the total quantity of melanin present in the colour-cells of a given animal at the beginning of pigment dispersion. It is, therefore, improbable that the view advanced by Dawes as to the formation and obliteration of melanin in connection with pigment migration can be of much significance in the physiological darkening or blanching of an animal. The rate of the increase and the decrease in the melanin content of the dark cells of such creatures is much too slow to make operations of this kind important in physiological colour changes. In the morphological colour changes, on the other hand, where the blanching or darkening of the individual animal takes on increasing permanency in the course of one or more weeks, such a process as that conceived of by Figge and seconded by Dawes may very well represent real steps by which these changes are brought about. It is to the morphological rather than to the physiological colour changes then that the ideas of Dawes and of Figge seem to apply. The view that colour changes are not due to pigment migration but strictly to the fading and re-formation of pigment, as advocated by Ruth and Gibson, is probably quite erroneous.

A summarized statement of the present discussion may be given in the following form. The dark phase in vertebrate colour changes is dependent chiefly upon intermedine and acetylcholine. This phase favours the two operations of the differentiation of new melanophores and the formation of additional melanin. Whether the trophic processes involved in these operations are determined by intermedine, by acetylcholine, one, both, or neither, remains to be ascertained, though the probabilities point to intermedine as an agent of first importance in this respect. The pale phase in vertebrate chromatics is determined chiefly by adrenaline. This phase favours a decrease in the differentiation of melanophores and a reduction in the formation of melanin. The inhibition of growth activities in these two directions may well be due to adrenaline, though this too is uncertain. Nor is it known what agent, if any, is concerned with the degeneration and discharge of melanophores and their debris as it occurs on the outer surface of the skin of chromatic vertebrates. This and many other like questions of importance in animal colour changes must remain for future investigation.

Chapter IX

GENETICS OF VERTEBRATE CHROMATOPHORAL SYSTEMS

A. Introduction

THE chromatophoral colours of the lower vertebrates have been convenient indicators to those students who have worked upon the genetics of these animals. It is not within the scope of this survey to consider the general genetic problems raised by such studies. Only that part of this field which throws light on the chromatophoral systems as such will be discussed. Students of this subject appear to have limited themselves entirely to experimentation on teleost fishes and among these chiefly to the domesticated or semi-domesticated stocks. Here about ten species of the so-called aquarium fishes have received almost all the attention devoted to this matter. Wild species, however, have not been entirely neglected, for Bancroft (1912) pointed out over a quarter of a century ago that hybrids from two killifishes, *Fundulus heteroclitus* and *F.·majalis*, gave evidence in their coloration of Mendelian inheritance. Thus, in the hybrids from these fishes, the large red type of yolk chromatophore found in *F. heteroclitus* dominated over the small one characteristic of *F. majalis*, and the black type of yolk colour-cell common to *F. heteroclitus* was found to replace that which occurred in *F. majalis*. These and other chromatophoral traits in such hybrids were found by Bancroft to exhibit not a mixed condition but true Mendelian segregation. Newman (1918) also worked with success on hybrids between wild teleosts. He was able to fertilize the eggs of *Fundulus* with sperm from *Scomber* and discovered that the chromatophoral system in *Fundulus*, which is very different from that in *Scomber*, dominated in the resulting hybrids. Newman's results have been confirmed by Russell (1939), who remarks, however, that in the *Fundulus-Scomber* cross no two embryos are alike in detail. The colour patterns of certain wild types of Tortugas fishes were studied experimentally by Goodrich (1935a, 1935c), who showed that when a dark scale of *Hallichoeres* was re-implanted in a pale region on this fish and vice versa the transferred scale took on a coloration appropriate for its new location. This was not found to be true for *Thalassoma*. Such conditions reported for wild teleosts are at best only suggestive of hereditary connections. The actual work on the genetics of fish coloration has been carried out with forms acclimatized to aquaria. These include a number of types among which are the goldfish, *Carassius*, and the cyprinodonts, *Oryzias* or *Aplocheilus*, as it was formerly designated, *Lebistes*, *Platypoecilus*, and *Xiphophorus*. The majority of these fishes are shown in outline in

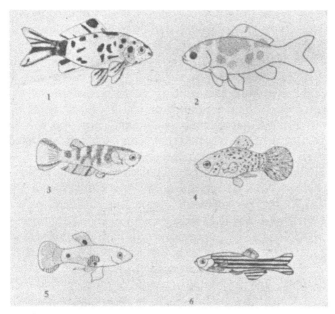

Fig. 120. Outlines of six species of teleost fishes commonly used in genetic studies: 1, the common shubunkin, a variety of the goldfish, *Carassius auratus*, showing the pattern of the melanophores (solid black) and the gold areas (stippled); 2, a gold and white goldfish, *C. auratus*, gold areas stippled; 3, *Oryzias latipes*, showing the pattern of the melanophores (coarse stippling) on a yellow body (fine stippling); 4, *Platypoecilus maculatus*, showing areas of macromelanophores (black) and of micromelanophores (stippled); 5, *Lebistes reticulatus*, dark areas (black), yellow (stippled), and red (cross-barred); 6, *Brachydanio rerio*, striped. Goodrich, 1935 b, 268.

Fig. 120. The chemical identification of the pigments in *Platypoecilus* and *Xiphophorus* has been accomplished by Goodrich, Hill and Arrick (1941), who have found in these and other tropical fishes melanin, carotenoids, lubein, zeaxanthin, violaxanthin, and erythropterin. This whole subject has been well summarized in three papers by Goodrich (1929, 1935 b, 1939) and these summaries will form the basis of the following account.

B. *CARASSIUS AURATUS*

The goldfish, *Carassius auratus*, a native of China, is of brown colour in its wild state. The red-scaled variety was known some fifteen hundred years ago, but was not domesticated till much later. The coloration of this fish and its genetics have been discussed by Koketsu (1915), Hance (1924), Chen (1928), Berndt (1925, 1928), Fukui (1930), Goodrich and Hansen (1931), Goodrich and Nichols (1933), and Goodrich and Anderson (1939). Hance noted the probable dominance of red and black over white in *Carassius*. Chen (1928) and Berndt (1928) found that when the ordinary goldfish was crossed with an unpigmented type a heterozygous form was produced which showed irregular mottling, the ordinary shubunkin (Fig. 120, 1). Berndt (1925) and Fukui (1927) reported that common goldfishes, which are usually dark brown for the first three months of their lives, assume their final red colour by the destruction of colour-cells. Their initial brown colour is due to melanophores and xanthophores and the change in tint is the result of the disintegration of the dark colour-cells. Parti-coloured red and black fishes are those in which this process has not been completed. Later the red cells may be destroyed, though as a rule not fully. Thus there is left a red and white pattern. In colourless goldfishes, the depigmentation starts as early as one week after hatching. In the colour changes associated with growth, melanophores are produced as well as destroyed (Goodrich and Hanson, 1931). The destructive steps can be taken only at certain periods in the development of the fish. If scales bearing melanophores are transplanted to red areas during depigmentation, the melanophores will disintegrate (Goodrich and Nichols, 1933), a process which will not occur in fishes whose colour pattern is completed. Although the three common types of goldfishes differentiate after hatching by the destruction and emergence of chromatophores, individual melanophores have been known to persist for as long as 19 months (Goodrich and Anderson, 1939).

C. *ORYZIAS (APLOCHEILUS) LATIPES*

Oryzias latipes (Fig. 120, 3) is a small fish some 3 cm. long native in the fresh-water ditches and pools of Southern Japan. Its genetics have been studied by Toyama (1916), Ishiwara (1917), Aida (1922, 1930), Goodrich (1927, 1933) and Winge (1930). Certain colour varieties, brown, red (yellow), white, variegated, and blue, have been subjected to genetic analysis by Aida (1922). The fishes possess melanophores and xanthophores. The brown variety, which is the wild form, depends for its tint on both these kinds of colour-cells combined. In the yellow type of fish, the xanthophores are normally developed but the melanophores are apparently few in number. Goodrich (1927), however, has shown that melanophore cells are present in this type in normal numbers but are incapable of developing the usual supply of melanin. This recessive character then differs from its allelomorph in the absence of melanin but not of the colour-cells. Goodrich (1933) found that these cells were deficient in the chromogen, probably tyrosine, but not in oxidase and were, therefore, colourless. A sufficient amount of chromogen is apparently dependent on the dominant gene in contrast to the double recessive condition where an inadequate amount is laid down. As these are local effects in the fish, they cannot be ascribed to hormones, but must be associated with particular colour-cells or groups of such cells.

D. LEBISTES RETICULATUS

Lebistes reticulatus, the guppy (Fig. 120, 5), is a native of Venezuela and certain neighbouring West Indian islands. It has been spread even to Africa as an anti-mosquito fish. Sexual dimorphism is pronounced. The females are some 5 cm. in length and coloured a uniform olive-brown. The males are about half this size and possess a highly varied combination of brilliant hues (Winge, 1922 b). The fish is viviparous and reaches sexual maturity in from three to four months. All in all, it is a most favourable form for genetic investigation. The colours of *Lebistes* and their inheritance have been studied by Winge (1922 a, 1922 b, 1923, 1927, 1930), Blacher (1926, 1927) and Demerec (1928).

The female of *Lebistes*, as already stated, is ordinarily brown. Occasionally on ageing it shows chromatic sex reversal, as has been observed by Blacher (1926) and by Winge (1927). This assumption of male coloration by old females is believed to be due to changes in their gonads and to indicate an invisible chromatophoral pattern in the females open to development under certain conditions. It is known that females, though without observable pattern themselves, will transmit to their male offspring the pattern of the particular male with which they have been paired.

E. *Platypoecilus maculatus*

The Mexican killifish, *Platypoecilus maculatus* (Fig. 120, 4), is a native of the freshwater streams of southern Mexico. The female attains sexual maturity in some eight months. The species is viviparous, the period of gestation being about thirty days. It exhibits numerous colour patterns of which Gordon (1927) has listed fifteen. The genetics of its coloration has been studied by Gerschler (1914), Bellamy (1923, 1924, 1928, 1936), Gordon (1926a, 1926b, 1926c, 1927, 1928a, 1928b, 1931a, 1931b, 1931c, 1933, 1934, 1936a, 1936b), Fraser and Gordon (1928a, 1928b, 1928c, 1929), Kosswig (1929a, 1929b, 1929c, 1931, 1937), Gordon and Fraser (1931), and Reed and Gordon (1931).

According to Gordon (1927), *Platypoecilus* possesses erythrophores, xanthophores, and two kinds of dark colour-cells, macromelanophores and micromelanophores. The patterns known as one-spot, twin-spot, crescent, and moon are due to aggregations of micromelanophores in well defined areas at the base of the caudal fin. The black pattern, nigra, is formed by macromelanophores. These fishes appear to have no colourless melanophores, such as have been found in *Oryzias*, but the production of their dark colour-cells seems to suffer some inhibition during their embryonic stages.

F. *Xiphophorus helleri*

The swordtail killifish, *Xiphophorus helleri*, is a native of Mexico and is noteworthy because of its ability to hybridize with *Platypoecilus maculatus*. It has been studied especially by Kosswig (1929a, 1929b, 1929c, 1931, 1937), by Gordon (1931b, 1931c, 1933, 1935a, 1935b, 1936a, 1936b, 1937, 1938), by Breider and Seeliger (1938), and by Reed and Gordon (1931). *Xiphophorus helleri*, according to Gordon (1938), varies in colour from light orange to deep orange-red and is spotted with large black dots. These are formed by macromelano-phores and are arranged in a variable number of lateral bands. This species is distinct from what has been called *Xiphophorus montezumae* which is brownish olive in tint and dull in appearance. This species is not to be confused with the montezuma variety of *X. helleri*. When this variety is crossed with the golden variety of the same species, four types of offspring result in which the genetic factor for the montezuma type controls the inheritance for red body colour and large black spots (macromelanophores) and a second dominant factor controls that for extremely small black dots (micromelanophores).

In the matings between *Xiphophorus* and *Platypoecilus*, it is convenient to pair a female *Xiphophorus* with a male *Platypoecilus* (Fig. 121). The coloration of the resulting hybrids shows evidence of

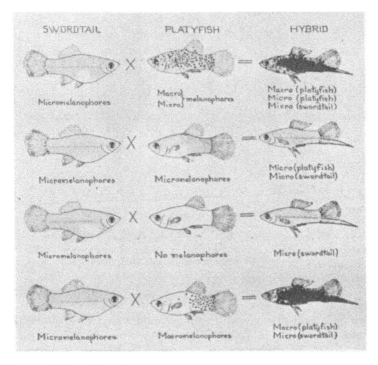

Fig. 121. Relationship of macromelanophores to melanosis in hybrids of the swordtail, *Xiphophorus helleri*, and the platyfish, *Platypoecilus maculatus*. Melanosis occurs in those crosses only where macromelanophores are involved and never where only micromelanophores are present. Gordon, 1937*b*, 363.

segregation. Excessive melanosis is of common occurrence. When the *Platypoecilus* used in the cross carries micromelanophores, melanosis is not produced, but when it is provided with macromelanophores the darkening in the skin of the hybrid may be considerable (Fig. 121). In fact, in such hybrids, the melanosis may be accompanied with an extreme neoplastic overgrowth in the form of a tumour (Fig. 122). Such growths are destructive to the fishes. The genetics of these melanomas have been studied by Gordon and Flathman (1943).

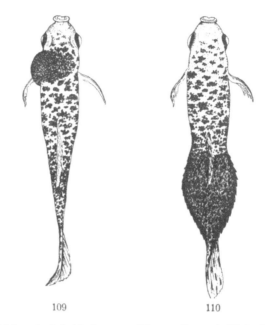

109   110

Fig. 122. Melanotic hybrids between *Platypoecilus* and *Xiphophorus*, showing neoplastic overgrowths or tumours in the head region (109) and in the pedunculus region (110). The dark cells are macromelanophores. Reed and Gordon, 1931, 1529.

G. Other Fishes

Other fishes, the genetics of whose colorations have been studied, are *Rivulus*, *Brachydanio*, *Betta*, *Macropodus*, and the trout. *Rivulus urophtalmus*, according to Constantinescu (1928), exhibits two colour phases, one black and the other red. These are believed to be allelomorphs with the black dominant over red. The zebra fish, *Brachydanio rerio* (Fig. 120, 6) is striped. Goodrich and Nichols (1931) have shown that in the beginning the pale stripes are covered by a sparse supply of melanophores which later disintegrate. Thus the definitive pattern is only gradually acquired. The Siamese fighting fish, *Betta splendens*, has been shown by Goodrich and Mercer (1934) to be represented by two forms, one green and the other blue, probably genetically different. The paradise fish, *Macropodus opercularis*, studied by Goodrich and Smith (1937), is represented by either normal dark specimens or by albinos. The dark individuals show vertical, alternating red and green stripes, the albinos have faint orange and blue-green stripes. The colour patterns are produced by melanophores, xanthophores, erythrophores, and iridophores. Of these, the albinos lack melanophores. The dark coloured type has been shown by Goodrich and Smith to be dominant over the albino. In a brief communication, Pomini (1938) has discussed the relation of the melanophores in crosses of two races of the trout.

After a survey of the elements that enter into the coloration of teleost fishes, Goodrich (1939), on reviewing briefly the genetics of the situation, has discussed with much insight the mutual significance of pigments, chromatophores, and patterns. Of these three, the pigments are the least diversified, consisting of a few melanins, a possible hundred or so of carotenoids and the like. Chromatophores are assuredly much more diversified than the pigments. In the group of the melanophores, there are not only macro- and micromelanophores, but on the basis of branching and other morphological aspects a diversity may be discovered among these cells which may allow a close student to distinguish a given species even by the form of its melanin-bearing cells alone. Finally colour patterns are the most diversified of the three elements, for it is on the basis of colour patterns that the multitude of races, the phenotypes of the geneticists, are distinguished. These three, slightly variable pigments, moderately variable colour-cells, and very variable patterns, are the elements involved in the genetics of coloration, a field the complexity of which is to be glimpsed in what is beginning to be discovered in this aspect of the colour genetics in teleost fishes.

Chapter X

SPECIAL ACTIVITIES OF VERTEBRATE CHROMATOPHORES

A. PULSATION OF CHROMATOPHORES

CHROMATOPHORE pulsations were first observed by Ballowitz (1913i) in the erythrophores of the excised skin from the trunk of *Mullus* after the skin had been immersed in 0·75 per cent NaCl. They were independently described by Spaeth (1916a) in the melanophores of the scales from the killifish, *Fundulus heteroclitus*. If a living scale containing melanophores is taken from *Fundulus*, rinsed in 0·1 N NaCl,

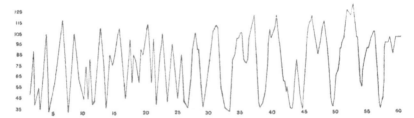

Fig. 123. Rhythmic pulsations of a melanophore in a scale of *Fundulus heteroclitus* after treatment with BaCl₂. The successive diameters of the mass of pigment in the melanophore were measured by an ocular micrometer at various times and these measurements (vertical column) were plotted against time (horizontal). Maximum diameter of pigment mass 130 units, minimum 33. Spaeth, 1916a, 206.

transferred for ten minutes to 0·1 N BaCl₂, and finally returned to 0·1 N NaCl, the melanophore pigment after half an hour or so will show active and characteristic pulsations. Each pulsation lasts about a minute and during that time the pigment spreads from the body of the melanophore into its processes and then returns (Fig. 123). Such pulsations, after having been initiated, will continue with great regularity over a period of as much as four to five hours, whereupon they will gradually cease.

Spaeth's original observations were confirmed by Gilson in 1926, who found a similar condition in a second species of fish *Gambusia affinis*. D. C. Smith in 1930 also substantiated Spaeth's accounts of these responses in *Fundulus* and described four new methods by which melanophore pulsations could be induced. In the first of these, scales were treated with 0·1 N KCl at hydrogen ion concentrations between 7·5 and 6·6 till the pigment in the melanophores was dispersed, after which the scales were transferred to o·1 N KCl at pH 6·0. In from 5 to 10 minutes following this transfer, pulsations began and lasted some 2 hours. In the second method, scales were immersed in o·1 N NaCl buffered with potassium phosphate and with concentrations of hydrogen ions ranging between 8·6 and 6·1, after which pulsations began in from 35 minutes (pH 7·6) to 2 hours (pH 7·0). In the third method, scales were submerged in o·1 N SrCl$_2$ for 5 minutes and then transferred to o·1 N NaCl. Pulsations could be seen in 30 minutes. The fourth method was distinctly the most novel. It consisted of immersing the scales in o·2 M dextrose and then after 49 to 90 minutes transferring them to NaCl, presumably o·1 N, after which pulsations began in about half an hour. These new methods added very materially to the means by which this interesting phenomenon could be induced.

In 1931, D. C. Smith reported on the effects of differences of temperature on melanophore pulsations. In *Fundulus*, these changes could be observed at as low a temperature as 10° C. Below this, pulsations failed to appear. The upper limit for these movements was found to be between 30° and 34° C. Two features were studied by Smith between these extremes of temperature, the extent of pigment migration and its rate. The extent of migration, that is, the length of the melanophore process that filled with pigment and emptied, was different for different temperatures. In the higher ranges, the processes were filled over only some 4 units of length, but at the lower temperatures the pigment invaded the processes over as much as 25 to 30 units. Above the highest temperatures, the pigment in the melanophores was completely concentrated about the centre of the cell and motionless, the melanin mass having assumed the punctate form. Below the lowest temperatures, the pigment was fully dispersed and also motionless.

Smith likewise observed that temperature differences were accompanied by a change in the frequency of pulsation. At the lower temperatures, there was roughly half a pulsation or less per minute. These increased to two or more per minute at about 27° C., above which there was a rapid falling off of the rate. These are the obvious results to be drawn from a rather scattered series of readings on the relations of melanophore pulsations to temperature (D. C. Smith, 1939).

Yamamoto in 1933 showed that the melanophores of the freshwater fish *Oryzias latipes* could be made to pulsate by a method essentially the same as that first used by Spaeth. Yamamoto pointed out the importance of ascertaining the effects of the ions of Na, K, and Ca on the pulsations. Following the early work of Ringer and particularly of Loeb, in which it had been shown that the quiescent or active state of nerve and of muscle depended upon a particular balance of these ions, Yamamoto applied this idea to melanophores and sought especially to ascertain the effects of altering the proportions of Na and of Ca in the fluids surrounding these colour-cells. In general, two sets of salts of Na were tested, those whose anions formed relatively insoluble compounds with Ca and those whose anions formed soluble compounds with this element. On treatment with these various salts, the percentages of melanophores that showed pulsations were fairly distinctive and formed two rather well defined categories. These are clearly indicated in Table 17 from Yamamoto's paper.

It can be seen from Table 17 that all Na-salts tested by Yamamoto brought about rhythmic pulsations in the melanophores of *Oryzias*.

TABLE 17.   Table of percentages of melanophores of *Oryzias* that showed pulsations to solutions of different Na-salts (Yamamoto, 1933, p. 126)

Na-salts whose anions are Ca-precipitants		Na-salts whose anions are not Ca-precipitants	
Na_2SO_4	80	NaCl	10
NaF	70	NaBr	10
Na_2HPO_4	80	NaI	10
Na_2CO_3	80	$NaNO_3$	50
Na-oxalate	100	Na-formate	10
Na-citrate	10	Na-acetate	20
Na-tartrate	90	Na-propionate	30
		Na-valerate	10
		Na-succinate	20

Those whose anions formed insoluble compounds with Ca were as a rule much more vigorous in this respect than those whose anions formed soluble compounds with this element. The former class brought about an increase in the ion quotient $\dfrac{\text{Conc. Na}}{\text{Conc. Ca}}$ of the fluid that bathed the melanophore, the latter a decrease. From this it may be concluded that Na-ions excite melanophore pulsations and that Ca-ions retard them. Hence any readjustment that reduces the Ca-ions in the presence of Na-ions tends to excite pulsations, a general condition that brings these melanophores closely into line with nerve and especially with muscle.

Smith and Smith (1934), who worked on the erythrophores of

Scorpaena, showed that these cells, as well as the melanophores of other fishes, could be excited to pulsate rhythmically for some two to three hours by the means first devised by Spaeth. These authors (1935) subsequently demonstrated chromatophore pulsations in the squirrel fish *Holocentrus*, where not only the erythrophores but also the xanthophores could be induced to concentrate and disperse their colouring matter rhythmically. Matsushita (1938) has been led to suspect that the melanophores of *Parasilurus* may also be capable of pigment pulsation. All these records, from the time of Ballowitz and of Spaeth to the present, tend to show that the migration of chromatophoral pigment back and forth in colour-cells is a process that is not necessarily associated with the environment of the cell, but when once initiated may continue apparently as an operation entirely within the colour-cell itself. This view implies that a chromatophore on dispersing its pigment assumes a physiological state quite unlike that peculiar to it when its pigment is concentrated. It would appear, if this interpretation is correct, that the dispersed state when completed initiates concentration and that, in turn, concentration brings about dispersion. This opinion is in a measure emphasized by the work of Parker and Pumphrey (1936), who have shown that the melanophores in the scales of *Fundulus* can be made to pulsate long after their activating nerve terminals have degenerated, again indicating that the complete mechanism for pigment migration is contained within the colour-cell itself.

B. The Resting and Active States in Chromatophores

Which is the active and which the resting state of chromatophores is a question that has been a matter of speculation for over a century. The majority of workers on this subject have declared in favour of the view that chromatophores with concentrated pigment are in an active state and that those with dispersed pigment are at rest (Brücke, 1852; Keller, 1895; von Frisch, 1912; Giersberg, 1930, and others). This opinion was reversed by Carlton (1903), whose work on *Anolis* led him to conclude that the condition of pigment concentration is that of rest and of dispersion that of activity. Babák (1913), who worked on amphibians, assumed a third position; namely, that both the dispersed and the concentrated states are active ones and that the resting condition is between these two. That the resting state is intermediate to those of complete concentration or dispersion appears to be implied, if not always so stated, in the writings of Abramowitz (1936), of Osborn (1938) and of Wykes (1938).

The opinion of the older workers that the state of pigment concentration is that of melanophore activity was based upon a comparison of colour-cells and their nerves with skeletal muscle and its motor innervation as first suggested by Brücke. After it had been shown by Pouchet (1872) that the innervation of vertebrate chromatophores was from autonomic sources and not from the cerebro-spinal system, this comparison was no longer appropriate, for, in accordance with this type of innervation, chromatophores should be compared with smooth muscle-fibres and not with skeletal muscle-elements. The propriety of this change was seen in the fact that in the action of chromatophores as in that of smooth muscle, tonus was found to play a very significant role. This more consistent interpretation was adopted by Zoond and Eyre (1934), Zoond and Bokenham (1935), and by Sand (1935), who modified in this respect Brücke's original conception. Another position (Parker, 1935 b) was at variance with all these opinions and came from a suggestion by Wyman (1924 a) that the active state of a chromatophore was one in which the pigment particles were in migration and the passive state one in which they were at rest. According to this view, chromatophores may be in action or at rest at any point in the course of their changes, be such a point terminal or intermediate.

This problem has been worked upon experimentally in the catfish *Ameiurus* (Parker, 1940 g). The colour changes in *Ameiurus* are to be seen chiefly on the dorsal side of the fish and range from a coal-black to a pale yellowish green. The principal receptors concerned in these changes, as already pointed out, are the eyes and the skin. The controlling effector agents are intermedine from the pituitary gland and two other neurohumours, a dispersing one, acetylcholine, from the dispersing nerve-fibres and a concentrating one, adrenaline, from the concentrating fibres (Parker, 1940 c). The action of these three chromatic neurohumours may be eliminated by two steps, blinding the fish and putting it in complete darkness. In this situation, the skin of the fish is not stimulated by light, and consequently intermedine is no longer liberated in significant amounts from the pituitary gland, and with the loss of the eyes the two sets of chromatophore nerves, concentrating and dispersing, are rendered inactive. Under these conditions, the whole melanophore system lapses into a state of quiescence. What is the tint of a catfish in this state, and what is the condition of its melanophores? For these are the only kind of chromatophores that this fish possesses.

All investigators agree that blinded *Ameiurus* in the light are fully dark (Van Heusen, 1917; Bray, 1918; Pearson, 1930; Parker, 1934e; Abramowitz, 1936c; Odiorne, 1937; Osborn, 1938b; Wykes, 1938). When, however, eyeless catfishes were put into darkness, the reports on the resulting colour changes were extremely diverse. They were said to be deep in tint but perceptibly paler than full dark (Parker, 1934e), a statement substantiated by Odiorne (1937). Abramowitz (1936c) declared that *Ameiurus*, even after 12 days in darkness, was pale. Blanching in darkness was also reported by Osborn (1938b) and by Wykes (1938). Abramowitz (1936c) pointed out further that a pale catfish blinded and put immediately into the dark would be found pale after a day. He also showed that pale, blinded catfishes when removed from darkness to light became dark in tint in a few minutes. Osborn (1938b) declared that blinded, dark catfishes when kept in darkness became pale. Thus a considerable difference of opinion has been expressed as to the outcome of placing blinded catfishes in situations without light.

It is noteworthy that the workers whose conflicting reports on the effect of darkness on catfishes have been summarized paid little or no attention to the tint of the fishes previous to their transfer to the experimental dark-room. A new set of tests was, therefore, devised (Parker, 1940g) in which the colour of the fish was recorded before it was put into the dark. In these tests, four classes of fishes were used: dark and blinded; pale, blinded and put in the dark room before they had begun to darken; blinded and intermediate in tint; and, finally, fishes normal in all respects and used merely as controls. All experiments extended over periods of at least one week. During the periods of the tests, the dark-room in which the fishes were kept was never opened and the fishes remained in running water at about 19° C. As a means of checking the conditions of the melanophores of a given fish before and after the test, one pectoral fin was removed from the fish before the test, spread, fixed in water at 60° C., and preserved in formol-alcohol, and after the test the other pectoral fin from the same fish was similarly preserved for comparison. This very satisfactory technique was first used by Wykes (1938), and by means of it a close examination of the states of the melanophores before and after a given test could be made.

In all, eleven sets of such fishes were subjected to this kind of test. After a set of fishes had been a week in darkness, they were brought into the light, the opposite pectoral fins prepared, and the fishes immediately compared with others, fully dark, intermediate, and fully pale, held as controls. To the eye the two pale fishes, the experimental one and its control, in each set were indistinguishable in tint. The same appeared to be true of the intermediate pairs though here the comparison could be less close because the two fishes in a set were not uniformly tinted but somewhat mottled. The darkest pairs of the test specimens were slightly but perceptibly paler than the outside controls. Nevertheless, in the main, all three sets of fishes emerged from darkness with essentially the same tint as that with which they entered. In no instance did any fish originally pale or dark approximate even an intermediate tint not to mention an opposite extreme. In these tests, the individual fishes were marked for identification by clipping one of their barbels in an appropriate way. From such tests it was concluded that, so far as inspection with the unaided eye is concerned, blinded catfishes in darkness for a week or more retain over that period precisely or almost precisely the tints which they exhibited when they were put into darkness. This conclusion was also supported by a close comparison of the melanophores in the pairs of pectoral fins from each set of fishes tested. Hence it may be stated that blinded catfishes whose melanophore pigment is in the position for pale, intermediate, or dark before the fishes are put into darkness emerge after a week in darkness with this pigment in essentially its original position. This constancy of the several initial pigmentary phases in darkness may well be the explanation of the diverse conditions reported by the earlier workers. Except one or two of the most recent of these, none took into account the colour of the fishes when they were put into darkness. Abramowitz (1936c) was one of those who observed this feature, and he was much puzzled at the diversity of the outcome of his tests. Wykes (1938), who also worked on *Ameiurus*, gives no definite statement as to the tints of the fishes she subjected to darkness.

The conlusion to be drawn from these studies on the melanophores in blinded *Ameiurus* in darkness is that the colour-cells of these fishes may remain in any given position for many days without important change in their pigment. This immobility may last as long as the blinded fish can be kept in an environment without light. Such a fish is no longer open to the really effective stimuli for change of colour. As stated before, its chromatophoral system under these conditions lapses into a state of quiescence. At whatever stage of change its colour-cells may have reached as the fish goes into darkness, there they rest. As these observations indicate, such cessation of activity may occur anywhere between full dispersion and full concentration of pigment. In other words, the melanophores of *Ameiurus* and probably of all other related animals may assume a resting condition at any stage of pigmentary changes. This is the view that was originally suggested by Wyman (1924*a*). In such a condition the melanophore pigment granules cease to show movement, even Brownian movement. If an active state is to be ascribed to these colour-cells, it may be said to be that in which their pigment particles are in motion, usually migratory motion. This is a conception of rest and activity as applied to colour-cells that is radically different from that entertained by almost all preceding investigators from the time of Brücke to the present.

The active condition of melanophores as thus understood appears to be associated with a very fluid state of their cytoplasm, a sol state, and conversely their resting condition one of greater gelation, as first suggested by Redfield (1918). These studies imply that chromatophores organized like the melanophores in *Ameiurus* are very closely related to smooth-muscle fibres though probably not so intimately as Spaeth (1916*a*) believed. This subject, however, as well as that of the relation of cytoplasmic motility in the resting and active states of melanophores, is scarcely ripe for discussion, but is open rather for further investigation (Parker, 1940*g*).

If rest and activity of chromatophores are to be understood in the sense of the present discussion, what can be said about exhaustion in such colour-cells? Can these cells be brought to a functional standstill by incessant activity as ordinary skeletal muscle can? A test of this kind on melanophores was attempted by Parker and Brower (1937). It is well known, of course, that fishes may remain indefinitely in the pale phase or in the dark phase without any evidence of exhaustion in their chromatic systems. But if what has been stated in this section is correct, these phases are not conditions of activity but of rest, and consequently should not be expected to result in exhaustion. To produce exhaustion the colour-cells should be kept shifting their pigment incessantly from the position for the dark phase to that of the pale one and the reverse. To accomplish this a simple, automatic device was arranged to change at regular intervals the colour environment of a fish contained in a small tube through which a current of water was running. The tests were carried out on the killifish, *Fundulus heteroclitus*. In the most extreme test, a killifish, which previous to exercise changed from pale to dark in an average period of 61 seconds and from dark to pale in 118 seconds, was subjected to environmental colour changes one every two minutes for 48 hours (in all, approximately 1440 successive changes), and was then found to change from pale to dark in 62·3 seconds and from dark to pale in 120 seconds. These two sets of time records, one before the excessive colour exercise and the other after it, are so nearly in agreement that they may be said to give no evidence in favour of exhaustion. The colour changes in *Fundulus* appear to take place without signs of fatigue. In this respect, they must be very inexpensive operations for the fish. They resemble the activity of smooth muscle rather than that of skeletal muscle. Perhaps their activity may best be compared to that of the vertebrate heart whose recovery is involved in its functional processes to such a degree that its normal operation may be said to be practically continuous. This state of affairs is what would be expected from a system of parts such as the chromatic system in vertebrates whose rest and activity are really such as have been described in this section.

C. Responses of Chromatophores to Temperature Differences

The generalization is often made concerning chromatophore responses to heat and to cold that high temperatures induce concentration of melanophore pigment, low temperatures its dispersion, and intermediate temperatures allow these changes to be subordinate to illumination (Hogben, 1924; von Buddenbrock, 1928; Parker, 1930). This rule, though of considerable application, is not without important exceptions. One of these is seen in the colour responses of the minnow *Phoxinus* to different temperatures, as recorded by von Frisch (1911 a, 1912 b).

Von Frisch placed newly killed minnows snugly between two vertical sheets of glass. On the outside of one sheet, cold water was run, on the other warm water. On the warm side, the fish's skin after 7 to 25 minutes blanched. On the cold side it also blanched, but somewhat later. These changes were eventually interpreted by von Frisch as due to lack of oxygen. He therefore repeated his experiments with living minnows held between glass and provided with an abundant supply of fresh water. The conclusion that he reached was that the local application of heat led to melanophore expansion and of cold to contraction. The contrast between these results and those of the earlier workers, as stated in the opening paragraph of this section, was left by von Frisch for the time being as unsettled.

This question was taken up by D. C. Smith (1928), who worked on *Fundulus heteroclitus*. Smith found that when the living scales from this fish were put into o·2 N NaCl at 20° C., their melanophores soon showed maximum expansion and these melanophores remained expanded at temperatures of 1° to 30° C. When living scales were put into tap water between 5° and 41° C., they remained contracted, but between 1° and 5° C., they expanded. Smith concluded from these observations that with isolated scales high temperatures induced melanophore contraction and low temperatures expansion. The same was found to be true when whole fishes were immersed in warm or in cold water. The melanophores of the tail, either denervated or innervated, also contracted in warm water and expanded in cold. Local tests on the trunk led to different results. Denervated melanophores on the trunk contracted to heat, 32° to 42° C., but expanded to cold, 1° to 5° C. Innervated melanophores of the trunk, however, did exactly the reverse, expanding to heat, 32° to 42° C., and contracting to cold, 1° to 5° C. Thus in *Fundulus* all denervated melanophores and all caudal melanophores, denervated or innervated, follow the rule of contracting at high temperatures and expanding at low ones. Only the innervated trunk melanophores do the opposite; in this respect they are like the trunk colour-cells of *Phoxinus* as observed by von Frisch. Smith made the interesting observation that when two whole fishes, one of which had been in cold water and the other in warm, were simultaneously put in water of intermediate temperature, 20° C., the fish from the cold environment blanched and the one from the warm water darkened; in other words, an intermediate temperature of 20° C. acted as warmth to the cold fish and as cold to the warm one. Smith's observations not only show the relations of the temperature responses of melanophores in *Fundulus* to the rule already stated, but they offer a possible explanation of von Frisch's apparently anomalous instance. Innervated melanophores on the trunks of the two teleosts concerned responded to high and low temperatures differently as compared with melanophores which were denervated or in other situations.

The teleost *Jenynsia*, according to Muzlera (1934), follows the general rule in that its melanophores are expanded at low temperatures, 2° to 3° C. Melanophore contraction in this fish increases from 17° to 30° C., again in conformity with what might be expected. At 35° C., however, the melanophores are said to expand, though under circumstances that appear to be beyond the range of normal response.

Cole and Schaeffer (1936, 1937) and Cole (1939) have worked on the influence of temperature differences on the colour changes of *Fundulus* in response to black and to white backgrounds and to the different kinds of water in which the fishes may be tested. The time of response in *Fundulus* diminishes regularly as the temperature increases from 5° to 30° C., and may be expressed by the Arrhenius equation. In sea water, blanching on a white background occurs 2·7 times more rapidly than darkening under similar circumstances on a black background. In fresh water, blanching is very slightly slower than darkening. The temperature characteristics were found to be as follows: for blanching in sea water, 9700 units; in fresh water, 11,400 units; for darkening in sea water, 10,900 units; in fresh water, 11,400 units. It is obvious that the aqueous environment plays an important part in these reactions. This may be an explanation of the symmetrical and the unsymmetrical curves of chromatic responses recorded for *Fundulus* by Spaeth (1916a), by Hill, Parkinson and Solandt (1935), and by D. C. Smith (1936a).

All amphibians whose melanophore reactions at low temperatures have been tested have shown a dispersion of pigment with a consequent darkening of the animals. This has been demonstrated for the frog (Hooker, 1912; Hogben and Winton, 1923; Hewer, 1923), for the young of *Amblystoma* (Laurens, 1915), and for *Diemyctylus* (Collins and Adolph, 1926). In *Necturus*, which was reported by Dawson (1920) as showing pigment concentration at 9° or 10° C., full dispersion takes place, according to Field (1931), at 2° C., a temperature lower than any tried by Dawson.

In amphibians, concentration of melanophore pigment commonly though not invariably results from high temperatures. This has been observed in the frog (Hooker, 1912; Hogben and Winton, 1923), in the tree frog (Hargitt, 1912), and in the young of *Amblystoma* (Laurens, 1915). Hewer (1923) reported frogs at high temperatures as of intermediate tint and Collins and Adolph (1926) stated that *Diemyctylus* was not pale between 34° and 37° C. Dawson (1920) recorded that the melanophores of *Necturus* are expanded at high temperatures, an observation confirmed by Field (1931). These instances appear to be valid exceptions to the rule that high temperatures are associated with a blanched state.

Among lizards, a low temperature is very commonly accompanied by a pale skin as has been shown in *Phrynosoma* (Redfield, 1918; Parker, 1938a, Atsatt, 1939), in the chameleon (Hogben and Mirvish, 1928b), in *Anolis* (D. C. Smith, 1929; Hadley, 1929, 1931), in the gecko *Coleonyx* and in more than a dozen iguanids (Atsatt, 1939). The temperature range for the assumption of the pale phase in the iguanids tested by Atsatt was from about 35° to 43° C. *Uta*, however, blanched at 25° C. The lizards reported by Atsatt as dark at low temperatures were found by this investigator to be pale at high ones. This is in agreement with what had previously been found for *Phrynosoma* (Weese, 1917; Redfield, 1918; Parker, 1938a), for *Anolis* (D. C. Smith, 1929; Hadley, 1929) and for the chameleon (Hogben and Mirvish, 1928a, 1928b; Zoond and Eyre, 1934). Noteworthy exceptions to this very uniform set of temperature responses have been pointed out by Atsatt (1939) in the two lizards, *Xantusia vigilis* and *X. henshawi*. At high temperatures, illuminated, these lizards are dark, but neither low temperatures nor darkness will induce blanching in them. Thus among lizards, as among most other chromatic vertebrates, the rule of darkening at low temperature and blanching at high holds, but not invariably. Fuchs (1914) has pointed out that low temperatures may be purely inhibitory in their effects on colour change, and Zoond and Eyre (1934) have declared that, since a chameleon will become pale in a dark, cold refrigerator, low temperatures may not be the real agents in darkening. This may be true for the chameleon, but the evidence from other lizards favours the view that low temperature is a true means of exciting melanophore expansion in these animals. The relations of high and low temperatures to adaptive coloration, as seen especially in lizards, will be considered in the section on the ecology of colour changes later in this chapter.

D. Pressure, Atmospheric and Hydrostatic

Pressure, atmospheric and hydrostatic, has relatively little influence on organisms except in limiting their vertical distribution. Man has no difficulty in moving about on the surface of the earth under an atmospheric pressure of 14·7 pounds per square inch. Animal life reaches down to the greatest depths of the sea, yet deep-sea creatures five miles below the surface are as little inconvenienced by pressure as is man, though they are under a pressure exceeding 11,400 pounds per square inch. All animals are seriously inconvenienced, however, when they attempt to rise out of their normal level or sink below it. The consequences of changes of this kind make themselves felt in the condition of the substances, particularly the gases, dissolved in the water of the given animal's body. With decrease of pressure, especially hydrostatic pressure, gases appear as such in the tissues of a creature with the result that it may be physically torn to pieces. As long, however, as an animal remains near its normal pressure level, no such disturbances occur. It would, therefore, seem probable that physical pressure would have very little to do with chromatophoral activity. Marsland (1940, 1942 a, 1942 b) has shown nevertheless that pressure has a significant effect on the melanophores of *Fundulus*.

This investigator subjected the melanophores in the isolated living scales of *Fundulus* in a strong microscope-pressure chamber to various degrees of compression up to 540 atmospheres (nearly 8000 pounds per square inch). The expanded condition of the melanophores in the scale, as induced by the sodium choride solution in which the scale was immersed, was maintained at the highest pressure. Pulsations of the melanophore pigment, initiated by Spaeth's barium chloride method, continued to occur at pressures lower than 400 atmospheres (some 5880 pounds per square inch), but the amplitude of the ebb and flow of the pigment was progressively reduced as the pressure increased. When the pressure reached a point where the flow ceased, the melanophore pigment was always fully dispersed. The pulsations, abolished by high pressure, began again about one minute after decompression. These several phenomena were observable on denervated melanophores.

From such observations, Marsland was led to conclude that the processes of the melanophores in *Fundulus* are not like the pseudopodia of the amoeba, which collapse at pressures higher than 350 atmospheres. The changes that occur in the melanophore processes are to be understood from the gel-sol condition of their protoplasmic contents. Gelation is fostered by increased temperature and by decreased pressure. Concentration of pigment is accomplished by a contraction of the plasmagel whereby the plasmasol is squeezed into the outlying branches of the colour-cell, thus replacing the pigmented plasmagel as it retreats from these branches. Thus the pigment of the colour-cell is transported toward the centre of this cell and its processes left filled with transparent plasmasol.

E. ECOLOGY OF COLOUR CHANGES

Chromatophores and their attendant colour changes are so intimately involved in the lives of many animals that they have become associated with a number of general activities of chromatic creatures. Supposed in the beginning to be merely aggressively favourable or protective to their possessors, they have since been suspected of playing a role in sexual affairs, in temperature regulation, and the like. These several kinds of ecological relations of the chromatic system will be considered in this section.

Aristotle, in his *History of Animals*, as already noted, remarks that when the octopus comes into the neighbourhood of small fishes it may change its colour so as to resemble the surrounding stones and that it reacts in a similar way when it is alarmed. According to him, the sepia is reported to behave in much the same manner. These few comments by the Father of Natural History contain what may be said to be the germs of the modern theories of aggressive and of protective coloration. Comments such as these have been repeated from time to time by the older naturalists, who without detailed discussions seem to have taken it for granted that animal coloration was attuned to the environment. Over a century ago, Stark (1830) noticed that the English minnow was pale in a white receptacle and dark in a black one. He was thereby led to experiment with this and other freshwater fishes and discovered that not only the minnow (*Leuciscus phoxinus*), but the stickleback (*Gasterosteus aculeatus*), the loach (*Cobitus barbatula*), and the perch (*Perca fluviatilis*) exhibited similar chromatic powers. This capacity of matching colour with background was in Stark's opinion a means by which the fish could escape with greater success from its enemies. In short, Stark described the conditions met with in these fishes in the terms of simple protective coloration. As a result of this and other like statements, the subject of colour changes became involved in the long series of discussions which have on the whole obscured rather than clarified this subject and which have reached from the time of Darwin almost to the present. These discussions have led to no satisfactory conclusion, partly because of the extreme positions assumed by certain advocates of the theory involved and partly because of the absence of satisfactory and sufficient data. Within the last few years, the deficiencies here noted, at least so far as colour changes are concerned, have been met in some degree by the experimental investigations of Cuénot (1927) and particularly of Sumner (1934b, 1934c, 1935a, 1935b, 1937) and Brown and Thompson (1937).

Cuénot (1927) exposed in aquaria certain shrimps, including the variegated *Hippolyte*, to attacks from fishes. The shrimps were adapted to the colours of the sea-weeds with which they were associated or to the bottoms on which they rested. Cuénot put them, living or dead, on sea-weeds of their own or contrasting colours and on sand. From these tests, he concluded that fishes were as likely to catch and devour shrimps which were adjusted in colour to their backgrounds as they were to take those that were not so adjusted. He believed that these crustaceans, regardless of tint, were protected from fishes as long as they remained motionless and that colour as such had little or nothing to do with their safety.

Fig. 124. Various colour phases of *Gambusia patruelis* for use in experiments on the efficiency of protective coloration 1, four specimens, one pair of which had been in a black tank for about 7 weeks, and the other pair in a white tank for an equal period; 2, the same fishes after 24 hours on a pale grey background; 3, four specimens of the same history as those shown in 1, 20 seconds after their transfer to a black dish (the pale fishes in this photograph are too pale), 4, the same fishes as in 3, after 27 hours in a black dish. Sumner, 1935a, 254

Sumner (1935a) experimented with mosquito fishes, *Gambusia*, in their pale and in their dark phases (Fig. 124). These fishes were exposed in tanks with white or black bottoms and walls to the Galapagos penguin, *Spheniscus mendiculus*. Both pale and dark fishes were put into white or black tanks and the number of fishes of each tint eaten by the birds was recorded. In all, 1726 fishes were used, 1056 in the white tank and 670 in the black one. Of these, 746 or about 43 per cent were eaten by the birds. In the white tank, 278 dark fishes (61 per cent) and 176 pale fishes (39 per cent) were captured, and in the black tank 78 dark fishes (26 per cent) and 217 pale ones (74 per cent) were taken. Thus pale fishes (39 per cent) had the advantage in the white tank and dark ones (26 per cent) in the black tank. These figures are the combined results of eleven sets of tests in each of which the trend of the outcome for the same tank was always the same. Similar experiments were conducted with a night heron in place of the penguins and with results quite as indubitable as in the first set of tests, for with the heron as with the penguins the more conspicuous fishes were captured in excess.

In a second series of experiments conducted by Sumner (1935 b), pale and dark specimens of *Gambusia* were exposed, on appropriate backgrounds and under particular conditions, to attacks from the sunfish, *Apomotis cyanellus*. In these tests, the numbers of pale and of dark fishes were varied and the activity of the fishes used as prey was limited by drugs. As a result of these trials, it was shown that conspicuousness in a given environment, far more than availability in a spatial sense, determined survival under attacks from a piscine predator and that total quiescence was a handicap rather than an advantage to fishes in escaping capture. Sumner concluded from these experiments that the chromatic adjustment of fishes to their backgrounds may be of vital importance to them as a protection from predators despite the circumstantial evidence and theoretic objections that have been offered in disproof of this conclusion.

The general problem of the protective significance of colour changes to animals was approached by Brown and Thompson (1937) from another standpoint. These investigators sought to ascertain whether fishes in the dark state would tend to gather in a black rather than in a white environment, and conversely, whether those in the pale condition would move into white rather than into black surroundings. For this purpose they tested, in all, eight species of freshwater fishes. Individuals in one or other state were liberated in an aquarium on a line between its black and its white halves. The subsequent distributions of these fishes, black-adapted and white-adapted, in the two halves of the aquarium are given in percentages in Table 18.

It is evident from Table 18 that each of the eight species of fishes when in its dark phase tends to move into the black half of the aquarium in considerably greater numbers than into the white half. In this respect, the outcome would appear to be on the whole advantageous to the fish. Among the white-adapted individuals on the other hand, no such consistent tendency is to be seen and the different species show in their pale phase much diversity. It was found further that these various differences in response disappeared rapidly as the fishes readjusted their bodily tints. Although these results are admittedly preliminary, they show, at least in the responses of the dark fishes to black, an obvious protective trend in harmony with what had been found in this general field by Sumner. That chromatophoral changes in the fishes tested by these several workers play a significant part in the protection of these animals against predators is hardly open to doubt. Hence, chromatic adaptation as part of the general scheme of protective coloration appears to have a real place in nature.

TABLE 18. Percentages of individuals of eight species of teleosts which after adaptation to white or to black backgrounds sought on liberation white or black environments (compiled from Brown and Thompson, 1937, pp. 174-5)

Name of fish	Black-adapted choice		White-adapted choice	
	Black	White	Black	White
Ericymba	87	13	43	57
Hyborhynchus	82	18	70	30
Notropis whipplii	85	15	54	46
N. deliciosus	76	24	30	70
N. umbratilis	77	23	33	67
Semotilis	88	12	63	37
Campostoma	90	10	61	39
Allotis	86	14	79	21

That chromatic changes are concerned in aggressive coloration whereby a predatory species may stalk its prey with increased success, as was intimated even by Aristotle in his comments on the octopus, appears never to have been subjected to experimental test. But in view of what Sumner has demonstrated for chromatic changes and protective coloration, there is good reason to suspect that aggressive coloration may be aided by chromatic change. This suspicion, however, is purely hypothetical and the question of the importance of colour changes in aggressive activities is one open for future investigation.

During the breeding season of the lower vertebrates, many of these animals exhibit remarkable and striking alterations in patterns and tints, the so-called nuptial markings and colorations. These features usually appear with the onset of reproductive activities, reach their height in the midst of these performances, and gradually disappear with the close of mating.

The relation of these changes to the sexes is by no means uniform. In the common eel, as sexual maturity approaches, both males and females change from a dirty greenish yellow to a more silvery hue due to an increase of melanophores and guanophores (Fosi, 1934). In *Tribolodon*, according to Sato (1935), the nuptial colours of the males, reddish orange and black, also occur in the female but in less marked degree. In the female *Amia*, there is a potential caudal ocellus; this is always clearly seen in the male, but it becomes intensified in the female during the breeding season (Zahl and Davis, 1932).

In the majority of fishes with nuptial colours, however, these features are found exclusively in the males. The nuptial dress of the European stickleback, *Gasterosteus aculeatus*, has been fully described by Titschack (1922). The females of this species are marked throughout the year by greenish black spots on a lighter background. This is the colour of the males till about March when the throat in this sex begins to take on a reddish hue which in April and May as breeding comes on spreads upward over the head and posteriorly over the trunk. In July or somewhat later, after breeding has ceased, this bright reddish coloration fades and the males take on the tones shown by the females. These tints they retain till the next breeding season. Such nuptial colorations, according to Titschack, involve melanophores, guano-phores, xanthophores and erythrophores, especially the last. The details of these changes and their relations to the fighting of the males, nest building, egg laying and the care of the young, have been discussed by Wunder (1930) and by Leiner (1930). Osterhage (1932) has given descriptions of the nuptial colorations not only of *Gasterosteus* but also of *Phoxinus* and *Rhodeus*, on the latter of which Wunder (1931, 1934) has likewise worked. The Japanese stickleback, *Acheilognathus*, has been studied by Tozawa (1929). In this form, the nuptial colours of the male are red for the head, operculum and antero-lateral surface of the trunk, with much black on the trunk proper. With the oncoming of breeding, these colours develop in some 8 to 10 days, maintain a maximum over 5 to 10 days, and then fade very slowly. The nuptial colours in this fish may be visible for a period of from 30 to 70 days.

In addition to the sticklebacks, many other fishes show nuptial coloration and marks. These features have been described by Blacher (1926) in *Lebistes*, in which the colours are black and especially red and yellow. In *Tribolodon*, black, orange, and red predominate and the erythrophores are said to increase in size and in number (Sato, 1935). Such nuptial colours, which are slightly noticeable in females, appear once a year and last some 50 days. In *Helichoeres* the young fishes, both male and female, are described by Kinoshita (1934) as uniformly coloured and the females while mating still retain this juvenile tint. The males on the other hand assume as breeding approaches a strikingly novel nuptial coloration. Parker and Brower (1935) have described an ocellus-like mark on the dorsal fin of the male *Fundulus* (Fig. 125) which is present from April to November, a period during which breeding occurs. This spot cannot be seen during the rest of the year and is always absent from females.

Fig. 125. Nuptial spot on the dorsal fin of a male *Fundulus*: *A*, dark phase; *B*, pale phase. Absence of spot in the dorsal fin of the female: *C*, dark phase; *D*, pale phase. Parker and Brower, 1935, 5.

It is due to an aggregation of melanophores which are so abundant that even in the pale phases of the males the spot is easily visible.

In all such fishes as *Fundulus*, the nuptial colorations occur in the males and are absent from the females. In *Chloea sarchynnis*, according to Kinoshita (1938), nuptial markings are limited to the mature females. During the breeding season, the females assume a nuptial black coloration of the lower jaw, throat, head, and ventral, anal, and dorsal fins. These markings last about 60 days in each individual and may appear between the end of February and the middle of May. They arise suddenly and reach full development in a few days and are caused by an increase and expansion of melanophores.

It is noteworthy that in *Chloea*, whose nuptial colours are thus reversed so far as sexes are concerned, there should also be a reversal of the fighting instinct, for in this fish the females, not the males, fight one another.

Numerous experiments have been made on fishes with nuptial colorations to ascertain the means by which these features are excited. The seasonal arrival of the change of tint is probably dependent upon increase of temperature or light. The males of *Gasterosteus pungitius* assume nuptial colours in the spring. Van Oordt (1924a) found that if these fishes were subjected to an artificially warmed environment (12° to 20° C.), they would develop their characteristic colorations two to three months earlier than usual. Attention was naturally directed toward the testis as a gland whose internal secretions might possibly be concerned with nuptial colour changes. Castration was found to prevent or seriously interfere with the development of nuptial colours in *Phoxinus* (Kopéc, 1918, 1927), *Gasterosteus* (Courrier, 1921, 1922a, 1922b; Van Oordt, 1923; Bock, 1927), *Lebistes* (Blacher, 1926), *Acheilognathus* (Tozawa, 1929), *Amia* (Zahl and Davis, 1932), and *Helichoeres* (Kinoshita, 1935). It was noted by Kinoshita that, though the loss of the testis by the male of *Helichoeres* was followed by an inability on the part of this fish to develop nuptial coloration, the loss of the ovaries by the female resulted in nothing so far as colour changes were concerned. As an outcome of a study of the seasonal states of the testis in *Gasterosteus*, Courrier (1921, 1922a, 1922b) was led to conclude that the internal chromatic secretions of this gland came from its interstitial cells. This conclusion was opposed by Van Oordt (1923, 1924a, 1924b), who declared in favour of the reproductive tissue of the testis as a source of the nuptial hormone. The majority of subsequent writers have sided with Courrier (Kinoshita, 1934, 1938).

The injection of a number of extracts, hormones, and the like has been found to be successful in inciting nuptial coloration. Thus Wunder (1931, 1934) obtained this type of chromatic response on injecting into *Rhodeus thelygan*, testiglandol, testogan and the vegetable alkaloid yohimbine. This substance, yohimbine, has also yielded positive results when injected into various fishes by Osterhage (1932), by Glaser and Haempel (1932) and by Kinoshita (1938). Osterhage (1932) noted that nuptial colours could be made to appear on injecting both yohimbine and atropine, but that these substances acted in this way only during the breeding season. At other times of year, they induced an expansion of chromatophores, but without the production of a specific nuptial coloration. Finally, it may be mentioned that

Saito (1936) succeeded in calling out a nuptial coloration in *Archeilognathus* by the injection of vitamin E.

In their investigation of the mating behaviour of lizards, Noble and Bradley (1933) have dealt with the nuptial colorations of these reptiles. Judging from the accounts of such features given by students and particularly field students of lizards, these animals as a class are as striking in their sexual colour differences as are the teleosts. If the work on the excitation of nuptial colours in teleosts is slight, that in lizards is almost nil. Regamey (1935) observed that the males of the common European lizard, *Lacerta agilis*, are greenish during the warm part of the year, especially in May and June, and that they lose this tint in the winter. They also lose it after they have been castrated. Evans (1935), who studied the mating and fighting behaviour of *Anolis carolinensis*, discovered that an injection of antuitrin S into a quiet winter male excited its mating proclivities. Such males were green instead of dark in colour, assumed dominant attitudes over other males, chased females, and paired homosexually as normal males are known to do during the breeding season.

The notes on nuptial colorations and markings thus gathered point indubitably to the actual presence of such types of coloration, but give almost no evidence of their significance. That a particular animal with highly developed nuptial coloration should be more successful in mating than another with a less striking chromatic display has apparently never been shown. In this respect, the ecology of nuptial chromatophoral coloration is less fully understood than that of protective chromatophoral coloration. So far as the present state of knowledge is concerned, it is quite conceivable that nuptial colorations may serve more as a means of driving off other members of the same sex and may thus act as warning colorations than that they may be of use in attracting the opposite sex (Leiner, 1930; Noble and Bradley, 1933; Evans, 1935).

Colour changes in the lower vertebrates have been suspected by a number of workers to be associated with heat regulation. It is believed on good grounds that animals with a dark skin readily absorb radiant energy and thereby become warm, and those with a pale one reflect such energy and remain cool. An animal that can change its tint from light to dark and the reverse may thus be able to assume conditions favourable for warming or not warming itself. This hypothesis so impressed Fuchs (1912, 1913) that he was led to declare that though chromatic changes doubtless played some part in protective coloration they were primarily concerned with temperature regulation.

This question was ably discussed by Bauer (1914) who, after an adequate account of the historical sources of the idea, proceeded to examine its applicability. He pointed out that in consequence of the great heat capacity of water as compared with air, about 3000 to 1, the heat thus obtained by an aquatic animal would be taken up almost at once by the surrounding water and thus the creature would fail to profit by such a source. Hence, as a means of obtaining heat a dark skin would be of significance only to air-inhabitant creatures, lizards and possibly amphibians. He also indicated that so far as temperature was concerned cold-blooded animals are not particularly inconvenienced by cold. They are, however, seriously disturbed by overheating as in fact all animals are. Thus, Bauer put very considerable limitations on the question as set by Fuchs, who sought to make this hypothesis apply to all chromatic animals irrespective of the medium, water or air, in which they lived. These limitations have been recognized and accepted by almost all subsequent investigators.

Kruger and Kern (1924) regarded chromatophores as possible heat regulators in amphibians and reptiles, and Kruger (1929), who worked on the absorption of radiant energy by the skins of poikilothermous vertebrates, showed that this absorption was complete for violet and indigo and less so for blue and extreme red. He showed further that animals on the earth were greatly warmed by heat rays from the hot ground. At 1·5 cm. above the earth, they were heated 0·54° C. per minute; at 0·5 cm., 0·66° C. per minute; and in contact with the earth, 1·8° to 2·0° C. per minute. Plainly, heat conduction must play a considerable part here, convection being largely insignificant.

H. W. Parker (1935), in his description of a melanic lizard from Transjordania, collected some interesting facts concerning the relations of such creatures to their environments. He pointed out the striking differences between the life of lizards on sandy deserts and on rocks. The heating of desert sand during the daytime is extremely superficial. At 50 mm. below the surface of the desert sand at full heat, the temperature may be 14° to 15° C. lower than at the surface, and at 100 mm. below, some 25° to 26° C. lower. *Lacerta*, which cannot stand a temperature higher than 40° C., may burrow during extreme heat 10 cm. into the sand and thus remain sufficiently cool. But rock-dwelling lizards, such as *Agama stellio*, cannot thus escape excessive heat. Parker suggests that since the unusually dark skin of the species described by him is a good absorber of radiant energy it may also be an efficient means of dispersing it, but from a physical standpoint this seems scarcely possible.

Mosauer and Lazier (1933) called attention to the remarkable reputation that lizards have for living in extreme heat, and pointed out that *Uma notata* plays and thrives on sand at 55° to 60° C. This toleration of excessive heat by lacertilians was worked on experimentally by Mosauer (1936). He subjected *Uma notata* and *Dipsosaurus dorsalis* to overheating by radiation from electric heat-filaments. These lizards were killed at environmental temperatures of 55° to 60° C. with body temperatures ranging at death from 44° to 53° C. *Uma notata* was exposed to sunshine in its native habitat with an air temperature of 25° C. and a sand temperature of 45° to 48° C. It showed discomfort within 12 minutes and was dead after about an hour and a half. Its body temperature was approximately 47° C. In a temperature cage ranging from 25° to 60° C., *Uma* usually buried itself and came to rest in sand at 37° C. The optimum temperature for the activity of this lizard ranges from 35° to over 40° C.; the lethal body temperature is from 45° to 50° C. Thus, contrary to the common belief, these desert lizards are not very unlike other cold-blooded vertebrates in their actual heat tolerance. The temperature of the sand surface in the Colorado Desert exceeds 50° C. between morning and evening and may reach 60° C. for two to three hours at noon. This kind of environment is what has led to the belief that a dark coloration would be favourable for these animals early in the morning and late in the evening when the desert is relatively cool and that a pale one would be advantageous during much of the daytime. Klauber (1939) has discussed this question in a speculative way and favours to some degree Parker's idea of cooling by radiation from a dark skin.

L. C. Cole (1943) made a study of the influence of temperature on ten species of lizards, all of which with one exception came from the desert regions of Utah and Arizona. The animals were subjected to heat in a closed chamber provided with thermocouples for the measurement of temperatures. In these tests, atmospheric humidity was measured as well as the temperatures of the substratum and of the air. By means of an anal thermocouple, the temperature of the lizard could be known. Irrespective of the habits of the lizards, the lethal temperatures for all species were much the same and averaged 45·6° C. In this respect, Cole's results confirm those of Mosauer. The lizards received heat by conduction from the air, from their own metabolic processes, by conduction from the substratum, and as radiant energy. Of these four sources, the first two were shown to be negligible. The significant supplies of heat were from the substratum and radiation. Cole made the interesting discovery that lizards could cool themselves by evaporation from their lungs. He also pointed out the advantage of small size for quick and effective cooling. He agreed with Mosauer that the death of desert lizards in heated chambers was a true thermic death. Colour changes in these animals, though often very pronounced, appear to have comparatively little to do with their bodily temperatures, but Cole showed, as others had done before him, that dark-coloured lizards are more readily overheated by the absorption of radiation than are pale-coloured ones. Klauber expresses what is probably a correct opinion that chromatic changes in lizards are more concerned with protective coloration and other like states than with temperature control.

Closely related to the idea of heat regulation by chromatophores is the suggestion that these colour-cells, particularly melanophores, may afford a protection to certain internal organs by acting as a screen to cut off injurious radiations. Such chromatophores could well be deep-seated and in this situation form sheet-like coverings for organs. This possible function for colour-cells does not necessitate their motility, though should certain rays be at one time advantageous and at another time not so, motility might well be of significance. All anatomists are familiar with the fact that vertebrates show in many of their internal parts considerable aggregations of melanophores (Gordon, 1933). These dark colour-cells have been studied in certain teleosts by Ballowitz (1920a, 1920b), who pointed out that in deep situations there are not only melanophores of the ordinary kind but pigment spots also. Such spots are irregular areas of pigment larger as a rule than the melanophores, but without regular outlines or branches, and containing numerous nuclei (Fig. 126). They were suspected by Ballowitz (1920b) of being centres for the formation of melanophores. This author appears to have taken no steps to ascertain whether deep-seated melanophores were really motile.

K. Yamamoto (1931) studied internal melanophores in the cyprinoid fish *Acheilognathus*. This worker demonstrated the motility of such melanophores and declared that they reacted to light in a way the opposite of that of the melanophores in the skin. When they were illuminated, the peritoneal melanophores generally contracted while the dermal ones as a rule expanded. In the matter of reaction time, the peritoneal colour-cells were found to be slower than the dermal ones.

Independently of Yamamoto and without knowledge of his results, Sears (1931, 1935) carried out investigations on the internal melanophores of the frog, the tadpole, and the killifish In the frog, these dark cells were found in the fascia of the muscles of the leg, the back, and the body-wall, and on the mesenteries and the pericardium. They responded appropriately when the animal was placed on a white or

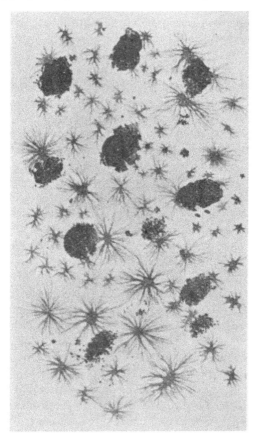

Fig 126   Melanophores and pigment spots from the peritoneum of
the teleost *Leuciscus rutulus* Ballowitz, 1920a, 408

a black background or when it was injected with adrenaline or pituitrin, but about six times more slowly than did the melanophores in the skin. Much the same was true of the deep-seated melanophores in the tadpole, except that they were still slower in response than those in the adult frog In the killifish, however, the external and internal melanophores responded synchronously Thus, through the independent work of Yamamoto and of Sears, the presence of true motile melanophores in the interior of vertebrates may be said to have been established. The peritoneal melanophores in the tadpole of the frog were subsequently studied by Yamamoto (1937). They are said by this investigator to contract and expand with decrease and increase of light. In this respect, they agree with the peritoneal melanophores of *Acheilognathus*, but not with its dermal melanophores. Under narcotization, all melanophores in the tadpole, internal as well as external, expand. They are not particularly influenced by adrenaline, but pituitrin is an effective means of dispersing their pigment. These contributions taken together show beyond a doubt that deep-seated, true melanophores are present in the vertebrates and that they respond usually though not always as do the integumentary melanophores.

Bauer (1912) was led to believe that the formation of dermal fat in shrimps was influenced by the relation of the chromatophores to rays in these animals. Titschack (1922) noted that when the skin of *Gasterosteus* was removed, the fish was quite transparent, an observation that led him to conclude that the chromatophores of this creature must be a very effective screen against rays. Kruger and Kern (1924) in their consideration of chromatophores as heat regulators called attention to the exclusion of rays from the interior of animals by colour-cells, and Kruger (1929) in particular emphasized the nature of these cells as filters for radiant energy eliminating the blue and the extreme red ends of the spectrum as already mentioned. Finding it difficult to ascribe a function to the deep-seated melanophores of the vertebrates, K. Yamamoto (1931) intimated that these cells might protect certain internal parts against injurious rays, though P. C. Foster's work (1935) on the effects of infra-red on human skin gave little support to this notion. Finally Rass (1937) pointed out that larval pigmentation in herrings appears to be related to habitat. When the larvae are pelagic, melanophores appear first over the dorsal portion of the young fishes. When the young are from the deep sea, the colour-cells occur laterally and ventrally. Deep-sea young are not especially exposed to rays, but pelagic ones are so exposed and their early melanophores were believed to be arranged to protect the developing central nervous organs. Suggestions such as these, and they are at best only suggestions, serve to indicate the possibilities of the view that chromatophores may be the means of admitting to an animal's interior or excluding from it certain kinds of radiant energy, but that such a function is a real one in any creature remains still to be demonstrated.

Finally chromatophores have from time to time been assumed to serve as excretory elements. May not then colour-cells be the means of accumulating waste materials in organisms and, as occasion offers, of discharging this waste to the exterior? Such a view has been advocated for the colour-cells in leeches by Iuga (1931). It has long been known that the melanophores of certain teleost fishes are more or less continuously produced and continuously shed. This view has recently been placed beyond dispute by the work of Sumner and Wells (1933) and particularly of Odiorne (1936). In *Fundulus*, according to Odiorne, melanophores regularly differentiate, and after a period of activity they disintegrate and pass as accumulations of melanin to the exterior. Fuchs (1913) long ago described this discharge of melanin as an excretion of a specific product of metabolism, an activity which Kuntz (1917) regarded as a normal and continuous physiological operation. Gordon (1931a) described it as a means of eliminating protein waste from the bodies of fishes, and Sumner and Wells (1933) classed it as an obvious excretion. Much the same view appears to have been held about guanin, but in this instance the material appears to form and accumulate in guanophores and never to be discharged to the exterior. If melanin is an end product of the metabolism of chromatic vertebrates, its excretion might well be associated with the disintegration of melanophores. But that this function is really to be attributed to these colour-cells is purely gratuitous.

Thus this survey of the ecology of chromatophores may be said to lead to the view that the colour-cells in chromatic animals may have a number of functions. Of these, protective coloration seems to be the most assured, but no one can deny that aggressive and nuptial coloration, heat regulation, protection against injurious rays, excretion of certain metabolic end-products and the like may not also be listed under the functions of these cells. Moreover, it is quite conceivable that any colour-cells may assume at one time or another several of these functions. This multiple significance of chromatophores has already been stressed by Bauer (1914). Any one familiar with the life history of *Fundulus*, for instance, would be led at once to suspect that the chief function of its chromatic system is protective coloration, but in addition to this its chromatophores may well play an important part in its nuptial activities and its melanophores may equally well serve as a means of excretion and of protection against noxious rays.

Thus, the functions of the colour-cells may differ from one animal to another with the result that diversity rather than uniformity may dominate the picture of the system as a whole.

F. CONCLUSION

During the period covered by this survey, 1910 to 1943, more radical proposals as to the nature and activities of animal chromatophores have been advanced than in all previous times. Before 1910, and in fact for almost a decade after that date, all workers in this field were persuaded that the natural responses of active colour-cells were excited exclusively by nerves. These chromatic nerves, as already stated, were assumed to act upon the chromatophores as motor nerves do upon muscles. But after the work of Redfield (1916, 1918) on lizards and of Huxley and Hogben (1922) and Hogben and Winton (1922) on amphibians it became increasingly evident that at least certain chromatophores were normally excited either exclusively or partly by hormones or, as they have been called in this volume, neurohumours. In this way such agents as adrenaline, intermedine, and acetylcholine in the vertebrates and the eye-stalk extract in the crustaceans took their places as essential elements in the chromatic mechanism. For a time it seemed as though the activation of colour-cells as a whole was a mixed phenomenon, in some instances purely nervous, in others purely humoral, and in still others in part nervous and in part humoral. It soon appeared, however, that nerves acted upon colour-cells not so much by virtue of changes in their electric potentials, as had been held to be true by the older workers for the nervous system in general, but in consequence of substances discharged at the nerve terminals and in very close proximity to the colour-cells themselves. These substances induced in the chromatic cells a dispersion or a concentration of their pigment precisely as the neurohumours did. In fact, except for the closeness of their regions of origin to the colour-cells, such substances were in no wise to be distinguished from true neurohumours and they were therefore included under this designation. Thus in the activation of colour-cells the distinction between nervous stimulation and that by neurohumours really disappeared. From this standpoint all excitation of chromatophores, be it nervous or humoral, was believed to be of a single type and dependent upon substances, neurohumours, which from near or far eventually impinged upon the colour-cells and thus called them into action. This modern interpretation of chromatophore excitation brought the system of colour-effectors into line with that of nervous units in general where receptor cells, neurones, and the various types of effectors are believed to be activated in sequence by material excitants which emanate from one unit and act upon the next, thus handing on the nerve impulse from cell to cell. This view of the progressive humoral activation of one unit by another, a view vigorously espoused by such workers as Loewi, Cannon, and especially Dale and his associates, though by no means fully established, has proved to be a remarkably successful working hypothesis. From this general standpoint, and yet with chromatic effectors always in view, what can be said of the excitation of nervous units and their appended effectors by what have often been called chemical activators?

In speculating on the transmission of nervous impulses over the interruptions between units in the nervous system, be they receptor-cell and neurone, or one neurone and the next, or neurone and effector-cell, we are prone to think of the process as time-consuming. To exude a neurohumour from a discharging terminal, to have this neurohumour diffuse across the internuncial space, and to impinge finally upon a receiving neurone or upon an effector and thus stimulate such an element cannot be accomplished instantly. Yet it is well known that in not a few cases these steps must be taken with incredible swiftness. Such instances are clearly illustrated by those repetitious movements in animals that are accomplished in extremely brief intervals of time. In man the repeated tapping of a single finger can be brought about at a rate of somewhat more than ten beats per second. From the standpoint of neurohumours this implies that the motor-nerve terminals must discharge, check, and redischarge exciting neurohumours about once every tenth of a second. Other examples, however, call for even higher rates of activity as, for instance, the scratch-reflex of the hind leg of the mouse, 20 per second (Williams, Barness and Sawyer, 1943), the beat of the fins of the seahorse, 35 per second (Breder and Edgerton, 1942), of the wings of the hummingbird, 50 to 75 per second (Blake, 1939), or again the vibrations of the tail of the rattlesnake which, according to Chadwick and Rahn, may reach as high as 100 per second (Williams, Barness and Sawyer, 1943). Although these rates, all of which come from vertebrates, are relatively rapid, especially those of the rattlesnake's tail, they are exceeded by what occurs in the wings of some insects. The movements of the wings of the fruitfly, *Drosophila pseudo-obscura*, were recorded by Reed, Williams and Chadwick (1942), who used a stroboscopic technique and found that these parts beat as high as

191 double strokes per second. This rate, however, is exceeded by the wing movements in several other insects as mentioned by Wigglesworth (1939): the bumble-bee, 240 per second; the common bee, 250 per second; the mosquito, 307; the fly, 330. All these rates, except that from the common bee, date from workers of thirty or more years ago. How reliable the technique of these workers was cannot be stated, but if all these earlier records are discarded, there still remains the fully reliable count of 191 beats per second for the fruitfly (Reed, Williams and Chadwick, 1942). Can this remarkably high rate be accomplished by successive discharges of humoral substance at a frequency as high as that of the wing beat? It is such high rates as these that put a strain on neurohumoralism as applied to rapid responses. But even the electric interpretation of such activities implies in such transfers the passage of charged particles and there can be no more concern in conceiving the rapid passage of chemical activators than in imagining those of particles carrying electric charges. Hence the difficulty introduced by the brief time in which many of these nervous operations are accomplished is a common one to both the electric and the humoral interpretation of synaptic and other like types of transmission. It must be clear from what has been repeatedly stated in this survey that the neurohumoral interpretation of chroma-tophore activities meets with none of the difficulties that the stimulation necessary for extremely rapid muscle response does. With colour-cells the most rapid rate of change is a matter of seconds, a time interval amply sufficient for interunitary transmission by neurohumours as implied in this modern view of chromatic activation, but how this transmission, electric or humoral, is to be understood for muscle movements, some 200 per second, is a question for future speculation. This question, though not directly concerned with chromatophores, is one of the many raised by the study of these colour-cells, and illus-trates how inviting to investigation is this field.

BIBLIOGRAPHY OF COLOUR CHANGES

THE bibliography of colour changes is assembled under the three following heads: Surveys, Historical Publications, and Publications which have appeared between 1910 and 1943 inclusive.

A. SURVEYS

Publications which present a general survey of the structure and function of chromatophores and which have appeared at any time in the past.

ABRAMOWITZ, A. A. (1939). Colour changes in animals. *Tabul. biol., Berl.*, **17**, 267–337.
BACQ, Z. M. (1937). *L'acétylcholine et l'adrénaline, leur rôle dans la transmission de l'influx nerveux.* 114 pp. Paris.
BALLOWITZ, E. (1931). Die Pigmentzellen, Chromatophoren und ihre Vereinigungen (chromatische Organe) in der Haut der Fische, Amphibien, und Reptilien im Hinblicke auf Färbung und Farbenwechsel der Haut. In L. Bolk, E. Göppert, E. Kallius und W. Lubosche, *Handb. vergl. Anat. Wirbelt.* **1**, 505–20.
BIEDERMANN, W. (1926). Vergleichende Physiologie des Integuments der Wirbeltiere. *Ergebn. Biol.* **1**, 1–342.
BRECHER, L. (1938). Pigment-Bildung (Farbwechsel und Farbanpassung) bei Wirbellosen und Wirbeltieren. *Tabul. biol., Berl.*, **16**, Pars 2, 140–61.
VON BUDDENBROCK, W. (1928). *Grundriss der vergleichenden Physiologie.* 830 pp. Berlin.
CUNNINGHAM, J. T. (1921). *Hormones and Heredity.* 246 pp. New York.
ERHARD, H. (1929). Farbwechsel und Pigmentierungen und ihre Bedeutung. In A. Bethe, G. v. Bergmann, G. Embden, A. Ellinger, *Handb. norm. path. Physiol.* **13**, 193–263.
FUCHS, R. F. (1914). Der Farbenwechsel und die chromatische Hautfunktion der Tiere. In H. Winterstein, *Handb. vergl. Physiol.* **3**, 1189–656.
GIERSBERG, H. (1930). Der Farbwechsel der Tiere. *Forsch. Fortschr. dtsch. Wiss.* **6**, 450–51.
GIERSBERG, H. (1931). Über Farbwechsel der Tiere. *Jber. schles. Ges. vaterl. Kult.* **103**, 25–35.
GIERSBERG, H. (1934). Physiologie des Farbwechsels bei Tieren. *Verh. dtsch. zool. Ges.* 1934, 96–126.
GIERSBERG, H. (1937a). Hormone. *Fortschr. Zool.* N.F. **1**, 311–33.
GIERSBERG, H. (1937b). Hormone. *Fortschr. Zool.* N.F. **2**, 363–414.
HANSTRÖM, B. (1939). *Hormones in Invertebrates.* 198 pp. Oxford.
HOGBEN, L. T. (1924). *The Pigmentary Effector System.* 152 pp. Edinburgh.
HOGBEN, L. T. (1926). *Comparative Physiology.* 219 pp. New York.
HOGBEN, L. T. (1927). *The Comparative Physiology of Internal Secretion.* 148 pp. Cambridge.
JEENER, R. (1933). L'année zoologique. IV. 1931–2. *Rec. Inst. zool. Torley-Rousseau* (Bruxelles), **4**, 211–23.
KLEINHOLZ, L. H. (1942). Hormones in Crustacea. *Biol. Rev.* **17**, 91–119.
KOLLER, G. (1938). *Hormone bei wirbellosen Tieren.* 143 pp. Leipzig.
PARKER, G. H. (1930). Chromatophores. *Biol. Rev.* **5**, 59–90.

PARKER, G. H. (1943). Animal colour changes and their neurohumours. *Quart. Rev. Biol.* **18**, 205–27.

VAN RYNBERK, G. (1906). Über den durch Chromatophoren bedingten Farbenwechsel der Tiere (sog. chromatische Hautfunktion). *Ergebn. Physiol.* **5**, 347–571.

SCHARRER, B. (1941). Endocrines in invertebrates. *Physiol. Rev.* **21**, 383–409.

TRENDELENBURG, P. (1929–34). *Die Hormone, ihre Physiologie und Pharmakologie.* **1**, 351 pp. (1929); **2**, 505 pp. (1934). Berlin.

VERNE, J. (1926). *Les pigments dans l'organisme animal.* 603 pp. Paris.

WARING, H. (1942). The co-ordination of vertebrate melanophore responses. *Biol. Rev.* **17**, 120–50.

WEIDENREICH, F. (1912). Die Lokalisation des Pigmentes und ihre Bedeutung in Ontogenie und Phylogenie der Wirbeltiere. *Z. Morph. Anthr.* (Sonderheft), **2**, 59–140.

WELSH, J. H. (1939). Diurnal rhythms. *Quart. Rev. Biol.* **13**, 123–39.

VON DER WENSE, T. F. (1938). Wirkungen und Vorkommen von Hormonen bei wirbellosen Tieren. *Zwangl. Abh. Geb. Inneren Sekretion*, **4**, 80 pp.

B. PUBLICATIONS, CHIEFLY HISTORICAL, BEFORE 1910

These publications are of historical importance and have been referred to in the text from this standpoint.

ARISTOTLE (1883). *History of Animals.* Trans. R. Crosswell. 326 pp. London.

BALLOWITZ, E. (1893*a*). Die Innervation der Chromatophoren. *Verh. anat. Ges. Jena*, **7**, 71–6.

BALLOWITZ, E. (1893*b*). Die Nervenendigungen der Pigmentzellen, ein Beitrag zur Kenntnis des Zusammenhanges der Endverzweigungen der Nerven mit dem Protoplasma der Zellen. *Z. wiss. Zool.* **56**, 673–706.

BERT, P. (1875). Sur le mécanisme et les causes des changements de couleur chez le caméléon. *C.R. Acad. Sci., Paris*, **81**, 938–41.

BIEDERMANN, W. (1892). Ueber den Farbenwechsel der Frosche. *Pflug. Arch. ges. Physiol.* **51**, 455–508.

BRÜCKE, E. (1852). Untersuchungen über den Farbenwechsel des afrikanischen Chamaleons. *Denkschr. Akad. Wiss. Wien*, Math.-nat. Kl. **4**, 179–210.

BUCHHOLZ, R. (1862). Ueber die Mikropyle von *Osmerus eperlanus*. *Arch. Anat. Physiol. wiss. Med.* 1863, 71–81.

CARLTON, F. C. (1903). The colour changes in the skin of the so-called Florida chameleon, *Anolis carolinensis* Cuv. *Proc. Amer. Acad. Arts Sci.* **39**, 259–76.

CHUN, C. (1902). Über die Natur und die Entwicklung der Chromatophoren bei den Cephalopoden. *Verh. dtsch. zool. Ges.* **12**, 162–82.

CORONA, A. e MORONI, A. (1898). Contributo allo studio dell' estratto di capsuli surrenali. *La Reforma Medica*, **14**. (Cited from van Rynberk, 1906.)

CUVIER, G. (1817). *Mémoire sur les céphalopodes et sur leur anatomie. Mémoires pour servir à l'histoire et à l'anatomie des mollusques.* 466 pp. Paris.

EBERTH, C. J. (1893). Die Nerven der Chromatophoren. *Verh. anat. Ges. Jena*, **7**, 70–1.

EBERTH, C. J. und BUNGE, R. (1895). Die Nerven der Chromatophoren bei Fischen. *Arch. mikr. Anat.* **46**, 370–8.

GAMBLE, F. W. and KEEBLE, F. W. (1900). *Hippolyte varians*; a study in colour-change. *Quart. J. Micr. Sci.* **43**, 589–698.

HOFMANN, F. B. (1907). Histologische Untersuchungen über die Innervation der glatten und der ihr verwandten Muskulatur der Wirbeltiere und Mollusken. *Arch. mikr. Anat.* **70**, 361–413.

KEEBLE, F. W. and GAMBLE, F. W. (1900). The colour-physiology of *Hippolyte varians*. *Proc. Roy. Soc.* **65**, 461–8.

KEEBLE, F. and GAMBLE, F. W. (1902). The colour-physiology of higher Crustacea. *Proc. Roy. Soc.* **71**, 69–71.

KEEBLE, F. and GAMBLE, F. W. (1904). The colour-physiology of higher Crustacea. *Philos. Trans.* B, **196**, 295–388.

KEEBLE, F. and GAMBLE, F. W. (1905). The colour-physiology of higher Crustacea. Part III. *Philos. Trans.* B, **198**, 1–16.

KLETT (1908). Zur Beeinflussung der phototropen Epithelreaktion in der Froschretina durch Adrenalin. *Arch. Anat. Physiol.*, *Lpz.* (Physiol. Abt.) Suppl. pp. 213–18.

KRÖYER, H. (1842). Monographisk Fremstilling af Slaegten Hippolyte's Nordiske Arter. *K. danske vidensk. Selsk. Skr.* **9**, 209–361.

VON LENDENFELD, R. (1883). Über Coelenteraten der Südsee. II. Mittheilung. Neue Aplysinidae. *Z. wiss. Zool.* **38**, 234–329.

LIEBEN, S. (1906). Ueber die Wirkung von Extrakten chromaffinen Gewebes (Adrenalin) auf die Pigmentzellen. *Zbl. Physiol.* **20**, 108–17.

LISTER, J. (1858). On the cutaneous pigmentary system of the frog. *Philos. Trans.* **148**, 627–43.

PLINY, SEC., C. (1887). *Natural History*. Trans. J. Bostock and H. T. Riley. 6 vols. London.

POUCHET, G. (1871). Sur les rapides changements de coloration provoqués expérimentalement chez les poissons. *C.R. Acad. Sci.*, *Paris*, **72**, 866–9.

POUCHET, G. (1872a). Du rôle des nerfs dans les changements de coloration des poissons. *J. Anat. Physiol.* **8**, 71–4.

POUCHET, G. (1872b). Sur les rapides changements de coloration provoqués expérimentalement chez les crustacés et sur les colorations bleues des poissons. *J. Anat. Physiol.* **8**, 401–7.

POUCHET, G. (1872c). Sur les rapides changements de coloration provoqués expérimentalement chez les crustacés. *C.R. Acad. Sci.*, *Paris*, **74**, 757–60.

POUCHET, G. (1872d). Sur les colorations bleues chez les poissons. *C.R. Acad. Sci.*, *Paris*, **74**, 1341–3.

POUCHET, G. (1873). Recherches anatomiques sur la coloration bleue des crustacés. *J. Anat. Physiol.* **9**, 290–307.

POUCHET, G. (1874a). Sur les changements de coloration que présentent certains poissons et certains crustacés. *C.R. Soc. Biol.*, *Paris*, **24**, 63–5.

POUCHET, G. (1874b). Note sur l'influence de l'ablation des yeux sur la coloration de certaines espèces animales. *J. Anat. Physiol.* **10**, 558–60.

POUCHET, G. (1875a). Note sur le mécanisme des changements de coloration chez le caméléon. *C.R. Soc. Biol.*, *Paris*, **26**, 86–8.

POUCHET, G. (1875b). Lésion du grand sympathique chez le turbot. *C.R. Soc. Biol.*, *Paris*, **26**, 350–1.

POUCHET, G. (1876). Des changements de coloration sous l'influence des nerfs. *J. Anat. Physiol.* **12**, 1–90, 113–65.

POUCHET, G. (1877). Note sur un changement unilatéral de couleur produit par l'ablation d'un œil chez la truite. *C.R. Soc. Biol.*, *Paris*, **28**, 364–5.

POUCHET, G. (1880). Nouvelle note sur le changement unilatéral de couleurs produit par l'ablation d'un œil chez la truite. *C.R. Soc. Biol.*, *Paris*, **30**, 344–5.

RABL, H. (1900). Über Bau und Entwickelung der Chromatophoren der Cephalopoden, nebst allgemeinen Bemerkungen über die Haut dieser Thiere. *S.B. Akad. Wiss. Wien*, Math.-nat. Kl. **109**, Abt. 3, 341–404.

ROESEL VON ROSENHOF, A. J. (1758). *Historia naturalis ranarum nostratium.* 115 pp. Nürnberg.

SANGIOVANNI, G. (1819). Descrizione d'un particolare systema di organi cromoforoespansivo-dermoideo e dei fenomeni ch' esso produce, scoperto nei molluschi cefalopodi. *Giornale enciclopedico di Napoli*, anno 13, no. 9.

SARS, G. (1867). *Histoire naturelle des crustacés d'eau douce de Norvège.* 145 pp. Christiania.

VON SIEBOLD, K. T. E. (1863). *Die Susswasserfische von Mitteleuropa.* 431 pp. Leipzig.

SOLGER, B. (1889). Zur Structur der Pigmentzelle. *Zool. Anz.* 12, 671–3.

STARK, J. (1830). On changes observed in the colour of fishes. *Edinb. New Phil. J.* 9, 327–31.

STEINACH, E. (1891). Ueber Farbenwechsel bei niederen Wirbelthieren bedingt durch directe Wirkung des Lichtes auf die Pigmentzellen. *Zbl. Physiol.* 5, 326–30.

VON UEXKÜLL, J. (1896). Vergleichend sinnesphysiologische Untersuchungen. II. Der Schatten als Reiz für *Centrostephanus longispinus*. *Z. Biol.* 34, 319–39.

VALLISNIERI, A. (1715). *Osservazioni intorno alle rane.* Venezia. (Cited from van Rynberk, 1906.)

VOGT, C. (1842). Embryologie des salmons. In L. Agassiz, *Histoire naturelle des poissons d'eau douce de l'Europe centrale.* Neuchatel.

WENCKEBACH, K. F. (1886). Beiträge zur Entwicklungsgeschichte der Knochenfische. *Arch. mikr. Anat.* 28, 225–51.

WILD, G. (1903). Einige Mittheilungen uber Fisch und Fischerei in Heilbronn. *Jh. Ver. vaterl. Naturk. Wurttemb.* 59. (Cited from Bronn, *Tierreich*, 1924, 6, 1, 656, 665.)

C. PUBLICATIONS FROM 1910 TO 1943 INCLUSIVE

This list contains the titles of all publications on chromatophores that have come to the notice of the author of this survey and that have appeared between 1910 and 1943 inclusive. The list is intended as a supplement to that contained in the survey by Fuchs (1914), which was based on the earlier one by van Rynberk (1906). The year 1910 was taken so as to overlap safely the date of Fuchs' publication. Although the present list cannot claim to be complete, an effort has been made to render it as full as circumstances would permit. The subject of chromatophores has received in recent years greatly increased attention, with the result that publications concerning these effectors have greatly multiplied. The lists given by Fuchs, which extend from antiquity to 1914, contain a total of some 500 titles. The following list (1910–43) contains some 1200 titles.

ABEL, J. J. (1930). On the unitary versus the multiple hormone theory of posterior pituitary principles. *J. Pharmacol.* 40, 139–69.

ABOLIN, L. (1925a). Beeinflussung des Fischfarbwechsels durch Chemikalien. I. Infundin- und Adrenalinwirkung auf die Melano- und Xanthophoren der Elritze (*Phoxinus laevis* Ag.). *Anz. Akad. Wiss. Wien*, Math.-nat. Kl. 61, 170–2.

ABOLIN, L. (1925b). Beeinflussung des Fischfarbwechsels durch Chemikalien. I. Infundin- und Adrenalinwirkung auf die Melano- und Xanthophoren der Elritze (*Phoxinus laevis* Ag.). *Arch. mikr. Anat.* 104, 667–97.

ABOLIN, L. (1925 c). Beeinflussung des Fischfarbwechsels durch Chemikalien. II. Annahme männlicher Erythrophorenfärbung durch das infundinisierte Weibchen der Elritze (*Phoxinus laevis* Ag.). *Anz. Akad. Wiss. Wien*, Math.-nat. Kl. **61**, 172–3.

ABOLIN, L. (1926). Beeinflussung des Fischfarbwechsels durch Chemikalien. III. Einfluss zentraler und peripherischer Nervengifte auf das gesamte Chromatophorensystem der Haut der Elritze (*Phoxinus laevis* Ag.). *Anz. Akad. Wiss. Wien*, Math.-nat. Kl. **63**, 32–5.

ABOLINS, L. (1929). The sexual specificness of the skin pigments of the fishes of the genus *Crenilabrus* colorimetrically investigated. *Latv. biol. Biedr. Rakati*, **1**, 135–49.

ABRAMOWITZ, A. A. (1935 a). Degeneration of xanthophores in *Fundulus majalis*. *Proc. Nat. Acad. Sci.*, *Wash.*, **21**, 132–7.

ABRAMOWITZ, A. A. (1935 b). Regeneration of chromatophore nerves. *Proc. Nat. Acad. Sci.*, *Wash.*, **21**, 137–41.

ABRAMOWITZ, A. A. (1935 c). Colour changes in cancroid crabs of Bermuda. *Proc. Nat. Acad. Sci.*, *Wash.*, **21**, 677–81.

ABRAMOWITZ, A. A. (1936 a). The double innervation of caudal melanophores in *Fundulus*. *Proc. Nat. Acad. Sci.*, *Wash.*, **22**, 233–8.

ABRAMOWITZ, A. A. (1936 b). The action of intermedin on crustacean melanophores and of the crustacean hormone on elasmobranch melanophores. *Proc. Nat. Acad. Sci.*, *Wash.*, **22**, 521–3.

ABRAMOWITZ, A. A. (1936 c). Physiology of the melanophore system in the catfish, *Ameiurus*. *Biol. Bull. Woods Hole*, **71**, 259–81.

ABRAMOWITZ, A. A. (1936 d). The non-identity of the neurohumours for the melanophores and the xanthophores of *Fundulus*. *Amer. Nat.* **70**, 372–8.

ABRAMOWITZ, A. A. (1936 e). Action of crustacean eye-stalk extract on melanophores of hypophysectomized fishes, amphibians, and reptiles. *Proc. Soc. Exp. Biol.*, *N.Y.*, **34**, 714–16.

ABRAMOWITZ, A. A. (1936 f). The chromatophorotropic hormones. *Anat. Rec.* **67**, Suppl. 1, 108.

ABRAMOWITZ, A. A. (1937 a). The opercular approach to the pituitary. *Science*, **85**, 609.

ABRAMOWITZ, A. A. (1937 b). The chromatophorotropic hormone of the Crustacea: standardization, properties and physiology of the eye-stalk glands. *Biol. Bull. Woods Hole*, **72**, 344–65.

ABRAMOWITZ, A. A. (1937 c). The rôle of the hypophysial melanophore hormone in the chromatic physiology of *Fundulus*. *Biol. Bull. Woods Hole*, **73**, 134–42.

ABRAMOWITZ, A. A. (1937 d). The comparative physiology of pigmentary responses in the Crustacea. *J. Exp. Zool.* **76**, 407–22.

ABRAMOWITZ, A. A. (1939 a). The pituitary control of chromatophores in the dogfish. *Amer. Nat.* **73**, 208–18.

ABRAMOWITZ, A. A. (1939 b). A new method for the assay of intermedin. *Biol. Bull. Woods Hole*, **77**, 327.

ABRAMOWITZ, A. A. (1940 a). A new method for the biological assay of intermedin. *J. Pharmacol.* **69**, 156–64.

ABRAMOWITZ, A. A. (1940 b). Purification of the chromatophorotropic hormone of the crustacean eyestalk. *J. Biol. Chem.* **132**, 501–6.

ABRAMOWITZ, A. A. and ABRAMOWITZ, R. K. (1938). On the specificity and related properties of the crustacean chromatophorotropic hormone. *Biol. Bull. Woods Hole*, **74**, 278–96.

ABRAMOWITZ, R. K. and ABRAMOWITZ, A. A. (1940). Moulting, growth, and survival after eyestalk removal in *Uca pugilator*. *Biol. Bull. Woods Hole*, **78**, 179–88.

ADELMANN, H. B. and BUTCHER, E. O. (1937). Experiments on the nervous control of the melanophores in *Fundulus heteroclitus*. *Bull. Mt Desert Isl. Biol. Lab.* pp. 15–16.

ADLER, L. (1914). Metamorphosestudien an Batrachierlarven. *Arch. EntwMech. Org.* **39**, 21–45.

ADRIAN, E. D. (1930). The effect of injury on mammalian nerve fibres. *Proc. Roy. Soc. B,* **106**, 596–618.

ADRIAN, E. D: (1937). The frequency range of neurones in the cerebral cortex. *Melanges Jean Demoor, Paris,* pp. 31–8.

AIDA, T. (1922). On the inheritance of colour in a fresh-water fish, *Aplocheilus latipes* Temmick and Schlegel, with special reference to sex-linked inheritance. *Genetics,* **6**, 554–73.

AIDA, T. (1930). Further genetical studies of *Aplocheilus latipes*. *Genetics,* **15**, 1–16.

ALLEN, B. M. (1916). The results of extirpation of the anterior lobe of the hypophysis and of the thyroid of *Rana pipiens* larvae. *Science,* **44**, 755–7.

ALLEN, B. M. (1917). Effects of the extirpation of the anterior lobe of the hypophysis of *Rana pipiens*. *Biol. Bull. Woods Hole,* **32**, 117–30.

ALLEN, B. M. (1920). Experiments in the transplantation of the hypophysis of adult *Rana pipiens* to tadpoles. *Science,* **52**, 274–6.

ALLEN, B. M. (1921). The effects of transplantation of the several parts of the adult hypophysis into tadpoles in *Rana pipiens*. *Anat. Rec.* **20**, 192–3.

ALLEN, B. M. (1925). Color changes induced in *Rana* larvae by implantation of the intermediate lobe of the hypophysis. *Anat. Rec.* **31**, 302–3.

ALLEN, B. M. (1929*a*). The functional difference between the pars intermedia and pars nervosa of hypophysis of frog. *Proc. Soc. Exp. Biol., N.Y.,* **27**, 11–13.

ALLEN, B. M. (1929*b*). Transplants of the hypophysis anlage into anuran tadpoles. *Anat. Rec.* **44**, 207.

ALLEN, B. M. (1929*c*). The influence of the thyroid gland and hypophysis upon growth and development of amphibian larvae. *Quart. Rev. Biol.* **4**, 325–52.

ALLEN, B. M. (1930). Source of the pigmentary hormone of amphibian hypophysis. *Proc. Soc. Exp. Biol., N.Y.,* **27**, 504–5.

ANDERSON, E. and HAYMAKER, W. (1935). Elaboration of hormones by pituitary cells growing *in vitro*. *Proc. Soc. Exp. Biol., N.Y.,* **33**, 313–16.

ANDRÉ, M. et LAMY, E. (1935). Sur la coloration noire de certaines écrevisses. *Bull. Soc. zool. Fr.* **60**, 40–3.

APGAR, B. D. (1935). A study of the reappearance of melanophores and the formation of melanophore aggregations (spots) in regenerated ventral skin of the common newt, *Triturus viridescens*. *J. Morph.* **58**, 439–61.

ASTWOOD, E. B. and GESCHICKTER, C. F. (1936). The pigmentary response in *Phoxinus laevis*. The effect of blood from a patient with melanosarcoma. *Amer. J. Cancer,* **27**, 493–9.

ASVADOUROVA, N. (1913). Recherches sur la formation de quelques cellules pigmentaires et des pigments. *Arch. Anat. micr.* **15**, 153–314.

ATSATT, S. R. (1939). Color changes as controlled by temperature and light in the lizards of the desert regions of southern California. *Publ. Univ. California Los Angeles Biol. Sci.* **1**, 237–76.

ATWELL, W. J. (1919). On the nature of the pigmentation changes following hypophysectomy in the frog larva. *Science,* **49**, 48–50.

ATWELL, W. J. (1921). Further observations on the pigment changes following removal of the epithelial hypophysis and the pineal gland in the frog tadpole. *Endocrinology,* **5**, 221–32.

ATWELL, W. J. (1934). On the metamorphosis of silvery tadpoles of *Rana sylvatica*. *Anat. Rec.* **58**, Suppl. pp. 48–9.

ATWELL, W. J. (1935). Functioning hypophyseal transplants in Amphibia. *Anat. Rec.* **64**, Suppl. 1, p. 85.

322   BIBLIOGRAPHY OF COLOUR CHANGES

ATWELL, W. J. (1937). Functional transplants of the primordium of the epithelial hypophysis in Amphibia. *Anat. Rec.* **68**, 431–47.
ATWELL, W. J. and HOLLEY, E. (1936). Extirpation of the pars intermedia of the hypophysis in the young amphibian with subsequent silvery condition and metamorphosis. *J. Exp. Zool.* **73**, 23–41.
ATWELL, W. J. and TAFT, J. W. (1940). Functional transplants of epithelial hypophysis in three species of *Amblystoma*. *Proc. Soc. Exp. Biol.*, *N.Y.*, **44**, 53–5.
ATZLER, M. (1930). Untersuchungen über den morphologischen und physiologischen Farbwechsel von *Dixippus* (*Carausius*) *morosus*. *Z. vergl. Physiol.* **13**, 505–33.
AUBRUN, E. A. (1935a). Sinotomas cutaneos de los sapos hipofisoprivos. *Rev. Soc. argent. Biol.* **11**, 371–80.
AUBRUN, E. A. (1935b). Symptomes cutanes de crapaud hypophysoprive. *C.R. Soc. Biol.*, *Paris*, **120**, 734–5.

BABÁK, E. (1910a). Über das Lebensgeschehen in den belichteten und verdunkelten Netzhauten. *Z. Sinnesphysiol.* **44**, 293–315.
BABÁK, E. (1910b). Zur chromatischen Hautfunktion der Amphibien. *Pflüg. Arch. ges. Physiol.* **131**, 87–118.
BABÁK, E. (1912). Über den Einfluss des Nervensystems auf die Pigmentbildung. *Zbl. Physiol.* **25**, 1061–6.
BABÁK, E. (1913). Ueber den Einfluss des Lichtes auf die Vermehrung der Hautchromatophoren. *Pflüg. Arch. ges. Physiol.* **149**, 462–70.
BACQ, Z. M. (1932a). Action des ions potassium sur la musculature des chromatophores des céphalopodes. *C.R. Soc. Biol.*, *Paris*, **111**, 220–2.
BACQ, Z. M. (1932b). Action de l'ergotamine sur les muscles des chromatophores des céphalopodes. *C.R. Soc. Biol.*, *Paris*, **111**, 223–4.
BACQ, Z. M. (1933a). The action of ergotamine on the chromatophores of the catfish (*Ameiurus nebulosus*). *Biol. Bull. Woods Hole*, **65**, 387–8.
BACQ, Z. M. (1933b). Recherches sur la physiologie du système nerveux autonome. III. Les propriétés biologiques et physicochemique de la sympathine comparées à celles de l'adrénaline. *Arch. int. Physiol.* **36**, 167–246.
BACQ, Z. M. (1934a). Réactions de divers tissus isolés du calmar (*Loligo pealii*) à l'adrénaline, à l'acétylcholine, à l'ergotamine et aux ions. *C.R. Soc. Biol.*, *Paris*, **155**, 716–17.
BACQ, Z. M. (1934b). Recherches sur la physiologie du système nerveux autonome. V. Réactions du ventricule median, des chromatophores et de divers organes isolés d'un mollusque céphalopode (*Loligo pealii*) à l'adrénaline, l'acétylcholine, l'ergotamine, l'atropine et aux ions K, Ca et Mg. *Arch. int. Physiol.* **38**, 138–59.
BACQ, Z. M. (1934c). La pharmacologie du système nerveux autonome, et particulièrement de sympathique; d'après la théorie neurohumorale. *Ann. Physiol. Physicochim. biol.* **10**, 467–528.
BACQ, Z. M. et FLORKIN, M. (1935a). Action pharmacologique d'un extrait d'hypophyses et de ganglions nerveux d'une ascidie (*Ciona intestinalis*). *C.R. Soc. Biol.*, *Paris*, **118**, 814–15.
BACQ, Z. M. et FLORKIN, M. (1935b). Mise en évidence, dans le complexe 'ganglion nerveux-glande neuralee' d'une ascidie (*Ciona intestinalis*), de principes pharmacologiquement analogues à ceux du lobe posterieur de l'hypophyse des vertébrés. *Arch. int. Physiol.* **40**, 422–8.
BALLOWITZ, E. (1912). Ueber chromatische Organe in der Haut von Knochenfischen. *Anat. Anz.* **42**, 186–90.
BALLOWITZ, E. (1913a). Ueber schwarz-rote Doppelzellen und andere eigenartige Vereinigungen heterochromer Farbstoffzellen bei Knochenfischen. *Anat. Anz.* **44**, 81–91.

BALLOWITZ, E. (1913*b*). Das Verhalten der Zellkerne bei der Pigmentströmung in den Melanophoren der Knochenfische. *Biol. Zbl.* **33**, 267–72.

BALLOWITZ, E. (1913*c*). Das Verhalten der Kerne bei der Pigmentströmung in der Erythrophoren von Knochenfischen. *Biol. Zbl.* **33**, 490–3.

BALLOWITZ, E. (1913*d*). Ueber chromatische Organe, schwarzrote Doppelzellen und andere eigenartige Chromatophorenvereinigungen, über Chromatophorenfragmentation und über den feineren Bau des Protoplasmas der Farbstoffzellen. *Verh. anat. Ges. Jena*, **27**, 108–16.

BALLOWITZ, E. (1913*e*). Notiz über das Vorkommen alkoholbeständiger karminroter und braunroter Farbstoffe in der Haut von Knochenfischen. *Hoppe-Seyl. Z.* **86**, 215–18.

BALLOWITZ, E. (1913*f*). Die chromatischen Organe in der Haut von *Trachinus vipera* Cuv. *Z. wiss. Zool.* **104**, 471–529.

BALLOWITZ, E. (1913*g*). Über schwarz-rote und sternförmige Farbzellenkombinationen in der Haut der Gobiiden. *Z. wiss. Zool.* **106**, 527–93.

BALLOWITZ, E. (1913*h*). Ueber Erythrophoren besonderer Art in der Haut von Knochenfischen. *Arch. mikr. Anat.* **82**, 205–20, 426.

BALLOWITZ, E. (1913*i*). Ueber die Erythrophoren in der Haut der Seebarbe, *Mullus* L., und ueber das Phänomen der momentanen Ballung und Ausbreitung ihres Pigmentes. *Arch. mikr. Anat.* **83**, 290–403.

BALLOWITZ, E. (1914*a*). Vier Momentaufnahmen der intracellulären Pigmentströmungen in der Chromatophoren erwachsener Knochenfische. *Arch. Zellforsch.* **12**, 553–7.

BALLOWITZ, E. (1914*b*). Zur Kenntnis des feineren Baues des Chromatophoren-Protoplasmas. *Arch. Zellforsch.* **12**, 558–66.

BALLOWITZ, E. (1914*c*). Ueber die Pigmentströmung in den Farbstoffzellen und die Kanälchenstruktur des Chromatophoren-Protoplasmas. *Pflüg. Arch. ges. Physiol.* **157**, 165–210.

BALLOWITZ, E. (1914*d*). Die chromatischen Organe, Melaniridosomen, in der Haut der Barsche (*Perca* und *Acerina*). *Z. wiss. Zool.* **110**, 1–35.

BALLOWITZ, E. (1917*a*). Ueber die Erythrophoren und ihre Vereinigungen mit Iridozyten und Melanophoren bei *Hemichromis bimaculatus* Gill. *Arch. Zellforsch.* **14**, 193–219.

BALLOWITZ, E. (1917*b*). Zur Kenntnis der Gelbzellen, Xanthophoren, in der Haut von *Blennius*. *Arch. Zellforsch.* **14**, 413–16.

BALLOWITZ, E. (1917*c*). Ueber die Vereinigung der Rotzellen mit Guaninzellen in der Haut von *Mullus* und *Crenilabrus*. *Arch. Zellforsch.* **14**, 417–20.

BALLOWITZ, E. (1920*a*). Zur Kenntnis des Peritonäalpigmentes bei Knochenfischen. *Anat. Anz.* **52**, 405–10.

BALLOWITZ, E. (1920*b*). Ueber eigenartige Erscheinungen am Peritonaealpigment bei Knochenfischen. *Arch. mikr. Anat.* **93**, 375–403.

BALLOWITZ, E. (1920*c*). Ueber die Farbzellenvereinigungen bei *Serranus*. *Arch. mikr. Anat.* **93**, 404–13.

BALLOWITZ, E. (1924). Ueber die eigenartiges Verhalten der Melanophoren und Guaninzellen in der Haut des Hornhechtes (*Belone*). *Z. Zell.- u. Gewebelehre*, **1**, 240–51.

BALLOWITZ, E. (1928). Weitere Mitteilungen über die Chromatophorenvereinigungen bei Gobiiden. *Z. mikr.-anat. Forsch.* **14**, 1–12.

BALLOWITZ, E. (1929). Über das Vorkommen alkoholbeständiger Rotzellen ('Allophoren' W. J. Schmidt) in der Haut einheimischer Amphibien. *Z. mikr.-anat. Forsch.* **19**, 277–84.

BALLOWITZ, E. (1930). Weitere Beiträge zur Kenntnis der Farbzellen und Farbzellenvereinigungen der Fische. *Z. mikr.-anat. Forsch.* **21**, 568–78.

BALTZER, F. (1941). Untersuchungen an Chimären von Urodelen und Hyla. I. Die Pigmentierung chimärischer Molch- und Axolotllarven mit Hyla- (Laubfrosch)-Ganglienleiste. *Rev. Suisse Zool.* **48**, 413–82.

BANCROFT, F. W. (1912). Heredity of pigmentation in *Fundulus* hybrids. *J. Exp. Zool.* **12**, 153–78.

BANTA, A. M. (1921). Flat-fish with unsual pigmented area. *Anat. Rec.* **20**, 214–15.

BARBOUR, H. G. and SPAETH, R. A. (1917). Responses of fish melanophores to sympathetic stimulants and depressants. *J. Pharmacol.* **9**, 356–7.

BARDEN, R. B. (1942). The origin and development of the chromatophores of the amphibian eye. *J. Exp. Zool.* **90**, 479–519.

BARDEN, R. B. (1943). Changes in the pigmentation of the iris in metamorphosing amphibian larvae. *J. Exp. Zool.* **92**, 171–97.

BAUER, V. (1910). Ueber die tonische Innervation der Pigmentzellen bei den Plattfischen. *Zbl. Physiol.* **24**, 724–6.

BAUER, V. (1912). Ueber die Ausnutzung strahlender Energie im intermediären Fettstoffwechsel der Garneelen. *Z. allg. Physiol.* **13**, 389–428.

BAUER, V. (1914). Zur Hypothese der physikalischen Wärmeregulierung durch Chromatophoren. *Z. allg. Physiol.* **16**, 191–212.

BAUER, V. und DEGNER, E. (1913). Ueber die allgemein-physiologische Grundlage des Farbenwechsels bei dekapoden Krebsen. *Z. allg. Physiol.* **15**, 363–412.

BAYER, G. (1930). Hypophyse und Chromatophorenreaktion. *Endokrinologie*, **6**, 249–54.

BEALL, D., SHAPIRO, H. A. and ZWARENSTEIN, H. (1937). The melanophore contracting principle of the pineal. *Chem. & Ind.* **56**, 190.

BEAMS, H. W. and KING, R. L. (1938). Pigmentation changes in tadpoles of *Rana pipiens* following centrifugation during the early gastrula. *J. Morph.* **63**, 477–89.

BEAUVALLET, M. (1934). Phénomènes protoplasmiques dans les mélanophores des écailles de poisson en relation avec leur chronaxie, sous l'action de divers poisons. *C.R. Soc. Biol.*, Paris, **115**, 824–6.

BEAUVALLET, M. (1938). Actions comparées de l'acétylcholine et de l'adrénaline sur les mélanophores préalablement atropinés. *C.R. Soc. Biol.*, Paris, **128**, 635–6.

BEAUVALLET, M. et VEIL, C. (1931). Modification de la chronaxie des mélanophores de poisson au cours de leur développement. *C.R. Soc. Biol.*, Paris, **107**, 964–6.

BEAUVALLET, M. et VEIL, C. (1932). Action combinée de courant électrique et de divers poisons sur les mélanophores du poisson. *C.R. Soc. Biol.*, Paris, **110**, 698–700.

BEAUVALLET, M. et VEIL, C. (1933). Réaction de la cellule pigmentaire de l'écaille de poisson selon la nature de milieu ambiant. *C.R. Soc. Biol.*, Paris, **114**, 513–16.

BEAUVALLET, M. et VEIL, C. (1934a). Réaction au courant électrique de la cellule pigmentaire de l'écaille du poisson marin et du poisson d'eau douce. *C.R. Soc. Biol.*, Paris, **116**, 123–5.

BEAUVALLET, M. et VEIL, C. (1934b). Chromatophores de poisson (*Carassius vulgaris*) et chromatophores de crustacés (*Palaemon squilla*). *C.R. Soc. Biol.*, Paris, **117**, 688–90.

BEAUVALLET, M. et VEIL, C. (1935a). Inexcitabilité au courant électrique des cellules pigmentaires à nerfs dégénérés de la carpe. *C.R. Soc. Biol.*, Paris, **120**, 404–7.

BEAUVALLET, M. et VEIL, C. (1935b). Les nerfs pigmento-moteurs de poisson agissent-ils par sécrétion d'adrénaline? *Ann. Physiol. Physicochim. biol.* **11**, 960–2.

BEAUVALLET, M. et VEIL, C. (1936). Action de quelques sympatholytiques sur la cellule pigmentaire de l'écaille détachée du poisson. *C.R. Soc. Biol.*, Paris, **123**, 785–7.

BECHER, H. (1924). Ueber chromatische Organe und andere Chromatophorenzusammenlagerungen in der Haut einheimischen Fische (*Esox luceus, Coregonius fera, Leuciscus rutilus*). *Zool. Jb.* (Abt. Anat.), **45**, 581–618.

BECHER, H. (1929a). Über die Verwendung des Opak-Illuminators zu biologischen Untersuchungen nebst Beobachtungen an den lebenden Chromatophoren der Fischhaut im auffallenden Licht. Z. wiss. Mikr. **46**, 89–124.

BECHER, H. (1929b). Über die Entwicklung der Farbstoffzellen in der Haut der Knochenfische. Verh. Anat. Ges. Jena (Anat. Anz.) Ergänzungsheft 67. Bd., pp. 164–81.

BECHER, H. (1929c). Ueber die Entwicklung der Xanthophoren in der Haut der Knochenfische. Roux Arch. Entw. Mech. Organ. **119**, 143–70.

DE BEER, G. R. (1926). The Comparative Anatomy, Histology, and Development of the Pituitary Body. 108 pp. Edinburgh.

BEHRE, E. H. (1933). Color recognition and color changes in certain species of fishes. Copeia, pp. 49–58.

BEHRE, E. H. (1935). More evidence on the structure of chromatophores. Science, **81**, 292–3.

BELEHRADEK, J. (1937). Étude photoélectrique de fonctionnement des mélanocytes cutanés de grenouille. C.R. Soc. Biol., Paris, **126**, 119–20.

BELLAMY, A. W. (1923). Sex-linked inheritance in the teleost Platypoecilus maculatus Günth. Anat. Rec. **24**, 419–20.

BELLAMY, A. W. (1924). Bionomic studies on certain teleosts (Poeciliinae). I. Statement of problems, description of material, and general notes on life histories and breeding behavior under laboratory conditions. Genetics, **9**, 513–29.

BELLAMY, A. W. (1928). Bionomic studies on certain teleosts (Poeciliinae). II. Color pattern inheritance and sex in Platypoecilus maculatus (Günth.). Genetics, **13**, 226–32.

BELLAMY, A. W. (1936). Interspecific hybrids in Platypoecilus: one species ZZ–WZ; the other XY–XX. Proc. Nat. Acad. Sci., Wash., **22**, 531–6.

BENNITT, R. (1929). Physiological interrelationship in the eyes of decapod Crustacea. Anat. Rec. **44**, 201.

BENNITT, R. (1932). Physiological interrelationship in the eyes of decapod Crustacea. Physiol. Zoöl. **5**, 49–64.

BERGGRÜN, J. (1914). Ueber den Bau der Haut von Hyla arborea L. während ihres Farbenwechsels. Bull. int. Acad. Cracovie, Cl. Sci. Math. Nat. B. Sci. Nat. 1913, pp. 152–9.

BERNDT, W. (1925). Vererbungsstudien an Goldfischrassen. Z. indukt. Abstamm.-u. VererbLehre, **36**, 161–349.

BERNDT, W. (1928). Wildform und Zierrassen bei der Karausche. Zool. Jb. (Abt. allg. Zool.), **45**, 841–972.

BERWEGER, L. (1926). Die Entwicklung der pigmentführenden Zellen in der Haut von Salamandra. Z. mikr.-anat. Forsch. **7**, 231–94.

BIGNEY, A. J. (1919). The effect of adrenin on the pigment migration in the melanophores of the skin and in the pigment cells of the retina of the frog. J. Exp. Zool. **27**, 391–6.

BLACHER, L. J. (1926). The dependence of secondary sex-characters upon testicular hormones in Lebistes reticulatus. Biol. Bull. Woods Hole, **50**, 374–81.

BLACHER, L. J. (1927). The role of the hypophysis and of the thyroid gland in the cutaneous pigmentary function of amphibians and fishes. Trans. Lab. exp. Biol. Zoopark, Moscow, **3**, 79–81.

BLAKE, C. H. (1939). The flight of hummingbirds. New Engl. Nat. no. 3, pp. 1–5.

BLANCHARD, E. W. and McCURDY, H. M. (1934). The origin of melanoblasts in early embryonic stages of Triturus torosus as studied by the Dopa technique. Anat. Rec. **60**, Suppl. p. 62.

BLANCHARD, L., PRUDHOMME, M. et SIMONNET, H. (1932). Action des extraits post-hypophysaires et de l'adrénaline sur les mélanophores d'Acerina cernua L. et de Gobio fluviatilis C. V. C.R. Soc. Biol., Paris, **110**, 760–1.

BLIAKHER, L. I. (1927). The rôle of the hypophysis and of the thyroid gland in the cutaneous pigmentary function of the amphibians and fishes. *Trans. Lab. exp. Biol. Zoopark, Moscow*, 1927, pp. 37–81.

BLOUNT, R. F. (1930). The implantation of additional hypophyseal rudiments in urodele embryos. *Proc. Nat. Acad. Sci., Wash.*, **16**, 218–22.

BLOUNT, R. F. (1932). Transplantation and extirpation of the pituitary rudiment and the effects upon pigmentation in the urodele embryo. *J. Exp. Zool.* **63**, 113–41.

BOCK, F. (1927). Kastration und secondäre Geschlechtsmerkmale bei Teleostiern. *Z. wiss. Zool.* **130**, 455–68.

BOCK, R. (1926). Sur le comportement des greffes de la peau des amphibiens. Greffes homoplastiques de la peau des salamandres adultes. *C.R. Soc. Biol., Paris*, **95**, 506–9.

BOGDANOVITCH, S. B. (1937a). The effects of different drugs on the melanophores of *Fundulus heteroclitus*. *Biol. Bull. Woods Hole*, **73**, 381–2.

BOGDANOVITCH, S. B. (1937b). Further investigations on the effect of tissue on different drugs. *Biol. Bull. Woods Hole*, **73**, 382.

BOGDANOVITCH, S. B. (1938). A pharmacological study of factors influencing the isolated melanophores of *Fundulus heteroclitus*. *Arch. int. Pharmacodyn.* **59**, 227–31.

BOLK, L. (1910). Beobachtungen über Entwicklung und Lagerung von Pigmentzellen bei Knochenfischembryonen. *Arch. mikr. Anat.* **75**, 414–34.

BÖRNSTEIN, W. (1939). Beitrag zur Frage der physiologischen Grundlagen des Wahrnehmens. II. Mitteiling: Über den Einfluß 'heller' und 'dunkler' Reize auf den Melanophoren-Zustand der Amphibien-Haut. *Arch. int. Pharmacodyn.* **61**, 387–417.

BORREL, A. (1913a). Réseau fondamental pigmentaire chez *Alytes obstetricans* et apparition des cellules pigmentaires. *C.R. Soc. Biol., Paris*, **75**, 139–42.

BORREL, A. (1913b). A propos du système pigmentaire chez *Alytes obstetricans*. *C.R. Soc. Biol., Paris*, **75**, 211–14.

BORREL, A. (1914a). Réseau pigmentaire chez *Hemopis sanguisuga*. *C.R. Soc. Biol., Paris*, **76**, 665–6.

BORREL, A. (1914b). Anologie de la formàtion sous-basale de M. Nageotte et du réseau fondamental pigmentaire. *C.R. Soc. Biol., Paris*, **77**, 16–18.

BOSCHMA, H. (1925). On the color changes in the skin of the lizard *Ptychozoön homalocephalum*. *Biol. Bull. Woods Hole*, **48**, 446–54.

BÖTTGER, G. (1935). Ueber einen neuen Intermedintest und die Intermedinreaktion der Elritze. *Z. vergl. Physiol.* **21**, 415–28.

BÖTTGER, G. (1936). Pigmenthormon und antidiuretisches Prinzip der Hypophyse. *Klin. Wschr.* **15**, 73–6.

BÖTTGER, G. (1937a). Ueber das Pigmenthormon. I. Mitteilung. Der Test. *Z. ges. exp. Med.* **101**, 42–7.

BÖTTGER, G. (1937b). Ueber das Pigmenthormon. II. Mitteilung. Zur Darstellung und zur Frage der Diuresewirkung. *Z. ges. exp. med.* **101**, 48–54.

BÖTTGER, G. (1937c). Ueber das Pigmenthormòn. III. Mitteilung. Zur Frage der Einheitlichkeit und über die aktive Substanz alkalischer Extrakte. *Z. ges. exp. Med.* **101**, 55–61.

BOZLER, E. (1928). Über die Tätigkeit der einzelnen glatten Muskelfaser bei der Kontraktion. II. Mitteilung: Die Chromatophorenmuskeln der Cephalopoden. *Z. vergl. Physiol.* **7**, 379–406.

BOZLER, E. (1929). Weitere Untersuchungen zur Frage des Tonussubstrates. *Z. vergl. Physiol.* **8**, 371–90.

BOZLER, E. (1931). Über die Tätigkeit der einzelnen glatten Muskelfaser bei der Kontraktion. 3. Mittheilung: Registrierung der Kontraktionen der Chromatophorenmuskelzellen von Cephalopoden. *Z. vergl. Physiol.* **13**, 762–72.

BRAY, A. W. L. (1918). The reactions of the melanophores of *Ameiurus* to light and to adrenalin. *Proc. Nat. Acad. Sci., Wash.*, **4**, 58–60.

BRECHER, L. (1916). Die Puppenfärbungen des Kohlweisslings *Pieris brassicae* L. (erster bis dritter Teil). *Anz. Akad. Wiss. Wien*, Math.-nat. Kl. **53**, 174–5.

BRECHER, L. (1918). Die Puppenfärbungen des Kohlweisslings, *Pieris brassicae* L. Vierter Teil: Wirkung unsichtbarer und sichtbarer Strahlen. *Anz. Akad. Wiss. Wien*, Math.-nat. Kl. **55**, 249–50.

BRECHER, L. (1919a). Die Puppenfärbungen des Kohlweisslings, *Pieris brassicae* L. Fünfter Teil: Kontrollversuche zur specifischen Wirkung der Spektralbezwirke mit anderen Faktoren. *Anz. Akad. Wiss. Wien*, Math.-nat. Kl. **56**, 244–5.

BRECHER, L. (1919b). Die Puppenfärbungen des Kohlweisslings, *Pieris brassicae* L. Sechster Teil: Chemismus der Farbanpassung. *Anz. Akad. Wiss. Wien*, Math.-nat. Kl. **56**, 246–8.

BRECHER, L. (1920). Die Puppenfärbungen des Kohlweisslings, *Pieris brassicae* L. Siebenter Teil: Wirksamkeit reflektierten und durchgehenden Lichtes. *Anz. Akad. Wiss. Wien*, Math.-nat. Kl. **57**, 157–8.

BRECHER, L. (1921). Die Puppenfärbungen der Vanessiden (*Vanessa Io, V. urticae, Pyrameis cardui, P. atalanta*). *Anz. Akad. Wiss. Wien*, Math.-nat. Kl. **58**, 40–2.

BRECHER, L. (1922a). Die Puppenfärbungen des Kohlweisslings, *Pieris brassicae* L. 8. Teil: Die Farbanpassung der Puppen durch das Raupenauge. *Anz. Akad. Wiss. Wien*, Math.-nat. Kl. **59**, 15–16.

BRECHER, L. (1922b). Die Puppenfärbungen der Vanessiden. II. *Anz. Akad. Wiss. Wien*, Math.-nat. Kl. **59**, 16–18.

BRECHER, L. (1922c). Die Puppenfärbungen des Kohlweisslings, *Pieris brassicae* IX und die Puppenfärbungen der Vanessiden III. *Anz. Akad. Wiss. Wien*, Math.-nat. Kl. **59**, 228–30.

BRECHER, L. (1924a). Die Puppenfärbung des Kohlweisslings, *Pieris brassicae* L. Achter Teil: Die Farbenpassung der Puppen durch das Raupenauge. *Arch. mikr. Anat.* **102**, 501–16.

BRECHER, L. (1924b). Die Puppenfärbung der Vanessiden (*Vanessa Io, V. urticae*). *Arch. mikr. Anat.* **102**, 517–48.

BRECHER, L. (1925). Physico-chemische und chemische Untersuchungen am Raupen- und Puppenblute. *Z. vergl. Physiol.* **2**, 691–713.

BRECHER, L. (1935). Die Puppenfärbungen des Kohlweisslings, *Pieris brassicae* L. X und Puppenfärbungen der Vanessiden (*Vanessa Io* L., *V. urticae* L.) IV: Nervendurchschneidungsversuchs. *Anz. Akad. Wiss. Wien*, Math.-nat. Kl. **72**, 46–7.

BRECHER, L. (1936). Die Puppenfärbungen des Kohlweisslings, *Pieris brassicae* L., XI und Puppenfärbungen der Vanessiden, *Vanessa io* L., *V. urticae* L. V: Der Weg der Farbanpassung vom Rezeptor bis zum Effector. *Anz. Akad. Wiss. Wien*, Math.-nat. Kl. **73**, 236–8.

BREDER, C. M. Jr. and EDGERTON, H. E. (1942). An analysis of the locomotion of the seahorse, *Hippocampus*, by means of high speed cinematography. *Ann. N.Y. Acad. Sci.* **43**, 145–72.

BREIDER, H. (1939a). Über die Vorgänge der Kernvermehrung und -degeneration in sarkomatösen Makromelanophoren. *Z. wiss. Zool.* **152**, 89–106.

BREIDER, H. (1939b). Über die Pigmentbildung in den Zellen von Sarkomen albinotischer und nichtalbinotischer Gattungsbastarde lebendgebärender Zahnkarpfen. *Z. wiss. Zool.* **152**, 107–28.

BREIDER, H. und SEELIGER, R. (1938). Die Farbzellen der Gattungen *Xiphophorus* und *Platypoecilus* und deren Bastarde. *Z. wiss. Zool.* **151**, 243–85.

BRESCA, G. (1910). Experimentelle Untersuchungen über die secondären Sexualcharaktere der Tritonen. *Arch. EntwMech. Org.* **29**, 403–31.

BROWN, F. A., Jr. (1933). The controlling mechanism of chromatophores in *Palaemonetes*. *Proc. Nat. Acad. Sci., Wash.*, **19**, 327–9.

BROWN, F. A., Jr. (1934). The chemical nature of the pigments and the transformations responsible for color changes in *Palaemonetes*. *Biol. Bull. Woods Hole*, **67**, 365–80.

BROWN, F. A., Jr. (1935a). Control of pigment migration within the chromatophores of *Palaemonetes vulgaris*. *J. Exp. Zool.* **71**, 1–15.

BROWN, F. A., Jr. (1935b). Color changes in *Palaemonetes*. *J. Morph.* **57**, 317–34.

BROWN, F. A., Jr. (1936). Light intensity and melanophore response in the minnow, *Ericymba buccata* Cope. *Biol. Bull. Woods Hole*, **70**, 8–15.

BROWN, F. A., Jr. (1939a). The source of chromatophorotropic hormones in crustacean eyestalks. *Biol. Bull. Woods Hole*, **77**, 329.

BROWN, F. A., Jr. (1939b). Humoral control of crustacean chromatophores. *Amer. Nat.* **73**, 247–55.

BROWN, F. A., Jr. (1939c). The coloration and color changes of the gulf-weed shrimp, *Latreutes fucorum*. *Amer. Nat.* **73**, 564–8.

BROWN, F. A., Jr. (1940). The crustacean sinus gland and chromatophore activation. *Physiol. Zoöl.* **13**, 343–55.

BROWN, F. A., Jr. (1941). A summary of our knowledge of endocrine mechanisms in crustaceans. *Trans. Ill. Acad. Sci.* **34**, 24–8.

BROWN, F. A., Jr. and CUNNINGHAM, O. (1941). Upon the presence and distribution of a chromatophorotropic principle in the central nervous system of *Limulus*. *Biol. Bull. Woods Hole*, **81**, 80–95.

BROWN, F. A., Jr. and EDERSTROM, H. E. (1939). On the control of the dark chromatophores of *Orago telson* and uropods. *Biol. Bull. Woods Hole*, **77**, 330.

BROWN, F. A., Jr. and EDERSTROM, H. E. (1940). Dual control of certain black chromatophores of *Orago*. *J. Exp. Zool.* **85**, 53–69.

BROWN, F. A., Jr. and MEGLITSCH, A. (1940). Comparison of the chromatophorotropic activity of insect corpora cardiaca with that of crustacean sinus glands. *Biol. Bull. Woods Hole*, **79**, 409–18.

BROWN, F. A., Jr. and SCUDAMORE, H. H. (1939). Comparative effects of sinus gland extracts of different crustaceans on two chromatophore types. *Biol. Bull. Woods Hole*, **77**, 329–30.

BROWN, F. A., Jr. and SCUDAMORE, H. H. (1940). Differentiation of two principles from the crustacean sinus gland. *J. Cell. Comp. Physiol.* **15**, 103–19.

BROWN, F. A., Jr. and THOMPSON, D. H. (1937). Melanin dispersion and choice of background in fishes, with special reference to *Ericymba buccata*. *Copeia*, pp. 172–81.

BROWN, F. A., Jr. and WULFF, V. J. (1940). Chromatophore types in *Crago* and their endocrine control. *Anat. Rec.* **78**, Suppl. p. 139.

BROWN, F. A., Jr. and WULFF, V. J. (1941a). Different pigmentary types in *Crago* and their humoral control. *Bull. Mt Desert Isl. Biol. Lab.* pp. 27–9.

BROWN, F. A., Jr. and WULFF, V. J. (1941b). Upon the presence of more than one chromatophorotropic principle in both sinus glands and commissural ganglia. *Bull. Mt Desert Isl. Biol. Lab.* p. 29.

BROWN, F. A., Jr. and WULFF, V. J. (1941c). Chromatophore types in *Crago* and their endocrine control. *J. Cell. Comp. Physiol.* **18**, 339–53.

BROWN, G. L. (1937). Transmission at nerve endings by acetylcholine. *Physiol. Rev.* **17**, 485–513.

BUDKER, P. (1936). Sur les changements de coloration d'un ange de mer *Rhine squatina* (L.) et son comportment en aquarium. *Bull. Soc. sci. Arcachon*, **33**, 225–34.

BURCH, A. B. (1938). Suppression of pars intermedia of pituitary body in *Hyla regilla* by operations upon the gastrula. *Proc. Soc. Exp. Biol., N.Y.*, **38**, 608–10.

BURGER, J. W. (1942). Some effects of androgens on the adult male *Fundulus*. *Biol. Bull. Woods Hole*, **82**, 233–42.

BURNS, R. K. (1934). The transplantation of the adult hypophysis into young salamander larvae. *Anat. Rec.* **58**, 415–29.

BUSCHKE, W. (1934). Zur Frage der Einwirkung des Melanophorenhormons auf die Dunkeladaptation des menschlichen Auges. *Klin. Wschr.* **13**, 1785–6.

BUTCHER, E. O. (1929). The pituitary gland of the ascidians. *Anat. Rec.* **44**, 212–13.

BUTCHER, E. O. (1930). The pituitary in the ascidians (*Molgula manhattensis*). *J. Exp. Zool.* **57**, 1–11.

BUTCHER, E. O. (1936). Histology of the pituitaries of several fishes. *Bull. Mt Desert Isl. Biol. Lab.* pp. 18–20.

BUTCHER, E. O. (1937a). The structure and distribution of the rods and cones in the eye of *Fundulus heteroclitus*. *Bull. Mt Desert Isl. Biol. Lab.* pp. 18–19.

BUTCHER, E. O. (1937b). Rods and cones in the retina of *Fundulus heteroclitus*, and the regions of the retina related to the different chromatophoric responses. *Anat. Rec.* **70**, Suppl. 1, p. 56.

BUTCHER, E. O. (1938a). The regions of the retina related to the different chromatophoric responses in *Fundulus heteroclitus*. *Bull. Mt Desert Isl. Biol. Lab.* pp. 18–19.

BUTCHER, E. O. (1938b). The structure of the retina of *Fundulus heteroclitus* and the regions of the retina associated with the different chromatophoric responses. *J. Exp. Zool.* **79**, 275–97.

BUTCHER, E. O. (1939). The illumination of the eye necessary for different melanophoric responses of *Fundulus heteroclitus*. *Biol. Bull. Woods Hole*, **77**, 258–67.

BUTCHER, E. O. and ADELMANN, H. B. (1937). The effects of covering and rotating the eyes on the melanophoric responses in *Fundulus heteroclitus*. *Bull. Mt Desert Isl. Biol. Lab.* pp. 16–18.

BUXTON, C. L. (1936). Transplantation of the hypophysis cerebri to the anterior chamber of the eye in albino rats. *Anat. Rec.* **64**, 277–83.

BUYTENDIJK, F. J. J. (1911). Ueber die Farbe der Tarbutten nach Extirpation der Augen. *Biol. Zbl.* **31**, 593–6.

BYTINSKI-SALZ, H. (1938). Chromatophorenstudien. II. Struktur und Determination des adepidermalen Melanophorennetzes bei *Bombina*. *Arch. exp. Zellforsch.* **22**, 132–70.

BYTINSKI-SALZ, H. e ELIAS, H. (1938). Studi sui chromatophori dei Discoglossidae. 1. Melanophori paraepidermici di *Discoglossus pictus* (Amphibia, Anura). *Arch. ital. Anat. Embriol.* **40**, 1–36.

CANNON, W. B. and ROSENBLUETH, A. (1935a). A comparative study of sympathin and adrenine. *Amer. J. Physiol.* **112**, 268–76.

CANNON, W. B. and ROSENBLUETH, A. (1935b). A comparison of the effects of sympathin and adrenine on the iris. *Amer. J. Physiol.* **113**, 251–8.

CANNON, W. B. and ROSENBLUETH, A. (1937). *Autonomic Neuro-effector Systems*. 229 pp. New York.

CARLSON, S. P. (1935). The color changes in *Uca pugilator*. *Proc. Nat. Acad. Sci., Wash.*, **21**, 549–51.

CARLSON, S. P. (1936). Color changes in Brachyura crustaceans, especially in *Uca pugilator*. *K. fysiogr. Sällsk. Lund Förh.* **6**, 63–80.

CARROLL, P. L. (1931). Chromatophore reaction of an oestrogenic hormone. *Proc. Soc. Exp. Biol., N.Y.*, **28**, 874–6.

CARSTAM, S. P. (1942). Weitere Beiträge zur Farbwechselphysiologie der Crustaceen. *Z. vergl. Physiol.* **29**, 433–72.

TEN CATE, J. (1927). De innervatie van de chromatophoren der huid bij *Octopus vulgaris*. *Bijdr. Biol. physiol. Lab. Univ. Amst.* **17**, no. 13.

TEN CATE, J. (1928). Contribution à la question de l'innervation des chromatophores chez *Octopus vulgaris*. *Arch. néerl. Physiol.* **12**, 568–99.

TEN CATE, J. (1933). L'action de quelques substances pharmacologiques sur le ganglion stellaire des céphalopodes. *Arch. néerl. Physiol.* **18**, 1–14.

CHANG, H. C., HSIEH, W. M. and LU, Y. M. (1939). Light-pituitary reflex and adrenergic-cholinergic sympathetic nerve in a teleost. *Proc. Soc. Exp. Biol.*, *N.Y.*, **40**, 455–6.

CHANG, H. C. and ·LÜ, Y. M. (1939). The light-pituitary reflex. I. Pigmentary response. *Chin. J. Physiol.* **14**, 249–58.

CHARLES, E. (1931). Metabolic changes associated with pigmentary effector activity and pituitary removal in *Xenopus laevis*. II. Calcium and magnesium content of the serum. *Proc. Roy. Soc.* B, **107**, 504–10.

CHEN, G., OLDHAM, F. K. and GEILING, E. M. K. (1940). Appearance of the melanophore-expanding hormone of pituitary gland in developing chick embryo. *Proc. Soc. Exp. Biol.*, *N.Y.*, **45**, 810–13.

CHEN, S. C. (1928). Transparency and mottling, a case of Mendelian inheritance in the goldfish, *Carassius auratus*. *Genetics*, **13**, 434–52.

CHIN, Y. (1939). Does acetylcholine play a part in the mechanism of melanophore expansion? *Proc. Soc. Exp. Biol.*, *N.Y.*, **40**, 454–5.

CHIN, Y. C., LIU, C. C. and LI, J. C. (1938). Melanophoral changes in the wall lizard, *Gekko swinhonis*. *Bull. Peking Soc. Nat. Hist.* **12**, 183–98.

CHUN, C. (1914). Die Cephalopoden. II. Teil: Myopsida, Octopoda. *Wiss. Ergebn. 'Valdivia'*, **18**, 403–552.

CIABATTI, O. (1929). Effetti di sostanze adrenalinsimili sui melanofori di due ciprinodontidi eurialini. *Atti Soc. Sci. med. nat. Cagliari*, **31**, 69–84. (Cited from *Biol. Abstr.* **5**, 20146.)

CLAUSEN, H. J. and MOFSHIN, B. (1936). The pineal of the lizard *Anolis carolinensis*, a photoreceptor as revealed by oxygen consumption studies. *Anat. Rec.* **67**, Suppl. 1, p. 104.

COLE, L. C. (1943). Experiments on toleration of high temperature in lizards with reference to adaptive coloration. *Ecology*, **24**, 94–108.

COLE, W. H. (1922). The transplantation of skin in frog tadpoles, with special reference to the adjustment of grafts over eyes, and to the local specificity of integument. *J. Exp. Zool.* **35**, 353–419.

COLE, W. H. (1939). The effect of temperature on the color changes of *Fundulus* in response to black and to white backgrounds in fresh and in sea water. *J. Exp. Zool.* **80**, 167–72.

COLE, W. H. and SCHAEFFER, K. F. (1936). The rate of adaptation of *Fundulus* in sea water to black and to white backgrounds at different temperatures. *Anat. Rec.* **67**, Suppl. 1, p. 103.

COLE, W. H. and SCHAEFFER, K. F. (1937). The effect of temperature on the adaptation of *Fundulus* to black and white backgrounds. *Bull. Mt Desert Isl. Biol. Lab.* pp. 26–30.

COLLIN, R. (1925). La neurocrinie hypophysaire. *Rev. franç. Endocrin.* **3**, 213–28.

COLLIN, R. (1928). La neurocrine hypophysaire, étude histophysiologique du complexe tubéro-infundibulo-pituitaire. *Arch. Morph. gén. exp.* no. 28, 102 pp.

COLLIN, R. (1933). Sur l'origine histologique des hormones posthypophysaires. L'intermedine. *C.R. Soc. Biol.*, *Paris*, **112**, 1351–3.

COLLIN, R. et DROUET, P. L. (1932a). Le lobe antérieur de la glande pituitaire et la réaction des mélanophores. *C.R. Soc. Biol.*, *Paris*, **110**, 1151–3.

COLLIN, R. et DROUET, P. L. (1932b). Extrait post-hypophysaire et variations pondérales chez la grenouille. *C.R. Soc. Biol.*, *Paris*, **110**, 1153–4.

COLLIN, R. et DROUET, P. L. (1933a). Présence d'un principe mélanophorodilatateur dans le tuber cinereum du cobaye. *C.R. Soc. Biol.*, *Paris*, **112**, 63–5.

COLLIN, R. et DROUET, P. L. (1933b). Dissociation des effets mélanophorique et érythrophorique chez le vairon sous l'influence de produits posthypophysaires. *C.R. Soc. Biol., Paris*, **113**, 1215–17.
COLLIN, R. et DROUET, P. L. (1934). La réaction des érythrophores est-elle liée au fonctionnement hypophysaire? *C.R. Soc. Biol., Paris*, **115**, 1441–3.
COLLIN, R. et DROUET, P. L. (1935). Action de certains extraits hypophysaires sur les mélanophores de la grenouille. *C.R. Soc. Biol., Paris*, **118**, 1008–1114.
COLLIN, R. et FLORENTIN, P. (1934). Ganglions nerveux et réaction des mélanophores. *C.R. Soc. Biol., Paris*, **115**, 162–3.
COLLIN, R., FONTAINE, T. et GROGNOT, P. (1937). Documents histophysiologiques sur la valeur fonctionelle de l'innervation hypothalamique de la glande pituitaire. *C.R. Ass. Anat.* **32**, 88–91.
COLLINS, H. H. and ADOLPH, E. F. (1926). The regulation of skin-pattern in an amphibian, *Diemyctylus*. *J. Morph.* **42**, 473–522.
CONNOLLY, C. J. (1925). Adaptive changes in shade and color of *Fundulus*. *Biol. Bull. Woods Hole*, **48**, 56–77.
CONNOLLY, C. J. (1926). Vasodilatation in *Fundulus* due to a color stimulus. *Biol. Bull. Woods Hole*, **50**, 207–9.
CONSTANTINESCU, G. K. (1928). Kreuzungsversuche mit *Rivulus urophtalmus*. *Z. indukt. Abstamm.- u. VererbLehre*, **47**, 341.
COONFIELD, B. R. (1940). The pigment in the skin' of *Myxine glutinosa* Linn. *Trans. Amer. Micr. Soc.* **59**, 398–403.
COOPER, R. S. (1938). Probable absence of a chromatophore activator in *Limulus polyphemus*. *Biol. Bull. Woods Hole*, **75**, 369.
COSTELLO, D. P. (1942). Neural crest transplantations between *Hyla* and *Triturus* embryos. *Anat. Rec.* **84**, 511.
COURRIER, R. (1921). Glande interstitielle du testicule et caractères sexuels secondaires chez les poissons. *C.R. Acad. Sci., Paris*, **172**, 1316–17.
COURRIER, R. (1922a). Étude préliminaire du déterminisme de caractères sexuel secondaires chez les poissons. *Arch. Anat., Strasbourg*, **1**, 115–44.
COURRIER, R. (1922b). Sur l'indépendance de la glande séminale et des caractères sexuels secondaires chez les poissons. Étude experimentale. *C.R. Acad. Sci., Paris*, **174**, 70–2.
CREED, R. S., DENNY-BROWN, D., ECCLES, J. C., LIDDELL, E. G. T. and SHERRINGTON, C. S. (1932). *Reflex Activity of the Spinal Cord.* 183 pp. Oxford.
CROLL, M. M. (1928). Nerve fibres in the pituitary of a rabbit. *J. Physiol.* **66**, 316–22.
CROWELL, S. (1939). The neurohumoral hypophysis and the color changes in the goldfish, *Carassius auratus*. *Anat. Rec.* **75**, Suppl. p. 62.
CROZIER, W. J. (1918). Note on the coloration of *Planes minutus*. *Amer. Nat.* **52**, 262–3.
CUÉNOT, L. (1927). Recherches sur la valeur protectrice de l'homochromie chez quelques animaux aquatiques. *Ann. Sci. nat.* Ser. 10, Zool. **10**, 123–50.
CUNNINGHAM, J. T. (1921). *Hormones and Heredity.* 246 pp. New York.
CUSHNY, A. R. (1910). The action of atropine, pilocarpine, and physostigmine. *J. Physiol.* **41**, 233–45.

DALE, H. H. (1934). Nomenclature of fibres in the autonomic system and their effects. *J. Physiol.* **80**, 10 P–11 P.
DALE, H. (1935). Pharmacology and nerve-endings. *Proc. Roy. Soc. Med.* **28**, 319–32.
DALE, H. H. and FELDBERG, W. (1934). Chemical transmission at motor nerve endings in voluntary muscle. *J. Physiol.* **81**, 39 P–40 P.

332   BIBLIOGRAPHY OF COLOUR CHANGES

DALE, H. H., FELDBERG, W. and VOGT, M. (1936). Release of acetylcholine at voluntary motor nerve endings. *J. Physiol.* **86**, 353–80.

DALTON, H. C. and GOODRICH, H. B. (1937). Chromatophore reactions in the normal and albino paradise fish. *Biol. Bull. Woods Hole*, **73**, 535–41.

DANIELSON, R. N. (1938). Light intensity and melanophore response in a cyprinid fish. *Physiol. Zoöl.* **11**, 292–8.

DANIELSON, R. N. (1939). Contrast and melanophore response in fishes. *Anat. Rec.* **75**, Suppl. p. 61.

DANIELSON, R. N. (1941). The melanophore response of fishes in relation to contrast in the visual field. *Physiol. Zoöl.* **14**, 96–102.

DAWES, B. (1941a). The melanin content of the skin of *Rana temporaria* under normal conditions and after prolonged light- and dark-adaptation. A photometric study. *J. Exp. Biol.* **18**, 26–49.

DAWES, B. (1941b). Pigmentary changes and the background response in Amphibia. *Nature, Lond.*, **147**, 806–7.

DAWSON, A. B. (1920). The integument of *Necturus maculosus*. *J. Morph.* **34**, 487–580.

DEDERER, P. H. (1921). The behavior of cells in tissue cultures of *Fundulus heteroclitus* with special reference to the ectoderm. *Biol. Bull. Woods Hole*, **41**, 221–41.

DEGNER, E. (1912a). Ueber Bau und Funktion der Krusterchromatophoren. *Z. wiss. Zool.* **102**, 1–78.

DEGNER, E. (1912b). Weitere Beiträge zur Kenntnis der Crustaceen-chromatophoren. *Z. wiss. Zool.* **102**, 701–10.

DE LANNEY, L. E. (1941). The role of the ectoderm in pigment production, studied by transplantation and hybridization. *J. Exp. Zool.* **87**, 323–45.

DEMEREC, M. (1928). A possible explanation of Winge's findings in *Lebistes reticulatus*. *Amer. Nat.* **62**, 90–4.

DENSTEDT, O. F. and COLLIP, J. B. (1939). Studies on the 'melanophore' principle of the pituitary. *Amer. J. Physiol.* **126**, 476.

DESMOND, W. F. (1924). Extirpation and transplantation of the hypophysis of fish and amphibians. *Anat. Rec.* **29**, 103.

DIETEL, F. G. (1932). Untersuchungen über das Melanophorenhormon. I. *Klin. Wschr.* **11**, 2075–8.

DIETEL, F. G. (1933a). Untersuchungen über das Melanophorenhormon. II. Der Einfluss des M.H. auf die Capillaren des Frosches. *Klin. Wschr.* **12**, 1027–8.

DIETEL, F. G. (1933b). Untersuchungen über das Melanophorenhormon. III. *Klin. Wschr.* **12**, 1358–64.

DIETEL, F. G. (1934). Untersuchungen über das Melanophorenhormon. IV. *Klin. Wschr.* **13**, 796–7.

DITTUS, P. (1937). Experimentelle Untersuchungen am Interrenalorgan der Selachier. 1. Atemfrequenz und Melanophoren bei interrenopriven und mit corticotropen Hormon behandelten Selachiern. *Publ. Staz. zool. Napoli*, **16**, 402–35.

VON DOBKIEWICZ, L. (1912). Einfluss der äusseren Umgebung auf die Färbung der indischen Stabheuschrecken—*Dixippus morosus*. *Biol. Zbl.* **32**, 661–3.

DOFLEIN, F. (1910). Lebensgewohnheiten und Anpassungen bei dekapoden Krebsen. *Festschrift R. Hertwig*, **3**, 215–92.

DOUBOVIK, I. A. (1936). Le système nerveux végétatif et l'activité fonctionnelle de l'hypophyse chez les amphibiens. *Rev. Franç. endocrin.* **14**, 151–60.

DRAGER, G. A. and BLOUNT, R. F. (1941). The time of the appearance of melanophore-expanding hormone in the development of *Amblystoma maculatum*. *Anat. Rec.* **81**, Supp. p. 92.

DREYER, N. B. and CLARK, A. J. (1924). Active principles of extracts of the posterior lobe of the pituitary. *J. Physiol.* **58**, xviii–xix.

DuBois-Poulsen, A. (1937). Effets de l'extrait hypophysaire et de l'adrénaline sur les franges de l'épithélium pigmentaire de la rétine de la grenouille. *C.R. Soc. Biol., Paris,* **125**, 248–9.

DuShane, G. P. (1934*a*). The origin of pigment cells in Amphibia. *Science,* **80**, 620–1.

DuShane, G. P. (1934*b*). The source of pigment cells in Amphibia. *Anat. Rec.* **60**, Suppl. pp. 62–3.

DuShane, G. P. (1935). An experimental study of the origin of pigment cells in Amphibia. *J. Exp. Zool.* **72**, 1–31.

DuShane, G. P. (1936). The dopa reaction in Amphibia. *Proc. Soc. Exp. Biol., N.Y.,* **33**, 592–5.

DuShane, G. P. (1938). Neural fold derivatives in the Amphibia. Pigment cells, spinal ganglia and Rohon-Beard cells. *J. Exp. Zool.* **78**, 485–503.

DuShane, G. P. (1939). The role of embryonic ectoderm and mesoderm in pigment production in Amphibia. *J. Exp. Zool.* **82**, 193–215.

DuShane, G. P. (1943). The embryology of vertebrate pigment cells. Part I. Amphibia. *Quart. Rev. Biol.* **18**, 109–27.

Duspiva, F. (1931*a*). Beiträge zur Physiologie der Melanophoren von Fischembryonen. *S.B. Akad. Wiss. Wien,* Math.-nat. Kl. 1, **140**, 553–96.

Duspiva, F. (1931*b*). Der Farbwechsel von Fischembryonen. *Forsch. Forschr. dtsch. Wiss.* **7**, 448–9.

Dutant, J. et Veil, C. (1938). Action de la pyrocatéchine sur les effets de l'adrénaline et l'excitation électrique des cellules pigmentaires de l'écaille de la carpe. *C.R. Soc. Biol., Paris,* **127**, 879–81.

Dye, J. A. (1935). The exhaustibility of the sympathin stores. *Amer. J. Physiol.* **113**, 265–70.

van Dyke, H. B. (1926). Die Verteilung der wirksamen Stoffe der Hypophyse auf die verschiedenen Teile derselben. *Arch. exp. Path. Pharmak.* **114**, 262–74.

van Dyke, H. B. (1943). Introduction to the conference on protein hormones of the pituitary body. *Ann. N.Y. Acad. Sci.* **43**, 255–8.

Eakin, R. M. (1939*a*). Developmental relationship between epithelial hypophysis and infundibulum in *Triturus torosus. Proc. Soc. Exp. Biol., N.Y.,* **41**, 308–10.

Eakin, R. M. (1939*b*). Correlative differentiation of the intermediate lobe of the pituitary in *Triturus torosus. Growth,* **3**, 373–80.

Eales, N. B. (1933). Albinism in the common frog. *Nature, Lond.,* **132**, 278–9.

Eccles, J. C. (1936). Synaptic and neuro-muscular transmission. *Ergebn. Physiol.* **38**, 339–444.

Eccles, J. C. (1937). Synaptic and neuro-muscular transmission. *Physiol. Rev.* **17**, 538–55.

Ehrhardt, K. (1927). Die Hypophysen-Melanophore-Reaktion und ihre klinische Auswertung. *Münch. med. Wschr.* **74**, 1879–81.

Elias, H. (1932). Die Entwicklung des Farbkleides des Wasserfrosches (*Rana esculenta*). *Z. Zellforsch.* **14**, 55–72.

Elias, H. (1934). Ueber die Entwicklung der Chromatophoren und anderer Zellen in der Haut von *Bufo viridis. Z. Zellforsch.* **21**, 529–44.

Elias, H. (1936). Die Hautchromatophoren von *Bombinator pachypus* und ihre Entwicklung. *Z. Zellenforsch.* **24**, 622–40.

Elias, H. (1937). Zur vergleichenden Histologie und Entwicklungsgeschichte der Haut der Anuren. *Z. mikr.-anat. Forsch.* **41**, 359–416.

Elias, H. (1939*a*). Die adepidermalen Melanophoren der Discoglossiden, ein Beispiel für den phylogenetischen Funktionswechsel eines Organs, seinen Ersatz in der früheren Funktion durch ein neues Organ und schliessliches Verschwinden. *Z. Zellforsch.* **29**, 448–61.

ELIAS, H. (1939b). Development of the network of the adepidermal melanophores in the tadpoles of *Discoglossus pictus*. *Anat. Rec.* **75**, Suppl. pp. 39–40.

ELIAS, H. (1940). The adepidermal melanophores of the Discoglossidae (Amphibia, Anura). *Anat. Rec.* **77**, Suppl. p. 117.

ELIAS, H. (1941a). The development of xanthophores in *Discoglossus pictus* (Amphibia Anura). *Anat. Rec.* **81**, Suppl. p. 80.

ELIAS, H. (1941b). Growth of the adepidermal melanophore network of *Discoglossus pictus*, studied in living tadpoles. *J. Morph.* **69**, 127–40.

ELIAS, H. (1942). Chromatophores as evidence of phylogenetic evolution. *Amer. Nat.* **76**, 405–14.

ELIAS, H., COHEN, D. and LIEBERMAN, I. (1942). The chromatophores of *Rana pipiens*. *Anat. Rec.* **84**, 476.

ELIAS, H., QUIGG, E. and SAURIS, E. (1942). The origin of subcutaneous melanophores. *Anat. Rec.* **84**, 464–5.

ELLINGER, F. (1940). Roentgen-pigmentation in the gold fish. *Proc. Soc. Exp. Biol.*, *N.Y.*, **45**, 148–50.

ELZE, C. (1923). Untersuchungen am sympathischen Nervensystem des Frosches, besonders uber seinen Einfluss auf die Skelettmuskelkontraktion. *Pflug. Arch. ges. Physiol.* **198**, 348–58.

ENAMI, S. (1939). Rôle de la sécrétion hypophysaire sur le changement de coloration chez un poisson-chat, *Parasilurus asotus* (L.). *C.R. Soc. Biol.*, *Paris*, **130**, 1498–1501.

ETKIN, W. (1935). Hyperactivity of the pars intermedia as a graft in the tadpole. *Anat. Rec.* **64**, Suppl. 1, p. 75.

ETKIN, W. (1940). Developmental relationship between pars intermedia of pituitary and brain in tadpoles. *Proc. Soc. Exp. Biol.*, *N.Y.*, **44**, 471–3.

ETKIN, W. (1941a). The first appearance of functional activity in the pars intermedia in the frog, *Xenopus*. *Proc. Soc. Exp. Biol.*, *N.Y.*, **47**, 425–8.

ETKIN, W. (1941b). On the control of growth and activity of the pars intermedia of the pituitary by the hypothalamus in the tadpole. *J. Exp. Zool.* **86**, 113–39.

ETKIN, W. and ROSENBERG, L. (1938). Infundibular lesion and pars intermedia activity in the tadpole. *Proc. Soc. Exp. Biol.*, *N.Y.*, **39**, 332–4.

EVANS, L. T. (1935). Winter mating and fighting behavior of *Anolis carolinensis* as induced by pituitary injections. *Copeia*, pp. 3–6.

FARIS, H. S. (1924). A study of pigment in embryos of *Amblystoma*. *Anat. Rec.* **27**, 63–76.

FARIS, H. S. (1926). The nature, origin and significance of pigment in embryos of *Amblystoma*. *Univ. Kansas Sci. Bull.* **16**, 181–227.

FAWCETT, D. W. (1939). Absence of the epithelial hypophysis in a fetal dogfish associated with abnormalities of the head and of pigmentation. *Biol. Bull. Woods Hole*, **77**, 174–83.

FEDERIGHI, H. (1934). Color changes in transplanted tadpole skin. *Physiol. Zoöl.* **7**, 271–8.

FENN, W. O. (1924). Active principles of the pituitary posterior lobe. *J. Physiol.* **59**, xxxv–xxxvi.

FIELD, E. H. (1931). Reactions of dermal melanophores in *Necturus* to heat and cold. *Proc. Nat. Acad. Sci.*, *Wash.*, **17**, 137–40.

FIESER, L. F. (1936). *The Chemistry of Natural Products related to Phenanthrene*. 358 pp. New York.

FIGGE, F. H. J. (1938a). Indophenol depigmentation of normal and hypophysectomized Amphibia. *J. Exp. Zool.* **78**, 471–83.

FIGGE, F. H. J. (1938b). Inhibition of tyrosinase melanin formation by sodium benzenone-indophenol. *Proc. Soc. Exp. Biol.*, *N.Y.*, **39**, 569–71.

FIGGE, F. H. J. (1939). Melanin: a natural reversible oxidation-reduction system and indicator. *Proc. Soc. Exp. Biol., N.Y.*, **41**, 127.

FIGGE, F. H. J. (1940). Squid melanin: a naturally occurring reversible oxidizable pigment. *Proc. Soc. Exp. Biol., N.Y.*, **44**, 293–4.

FISCHEL, A. (1920). Beiträge zur Biologie der Pigmentzelle. *Anat. Hefte* (Abt. Arb.), **58**, 1–136.

FLEISCHMANN, W. and KANN, S. (1937). Wirkung von Hypophysenhormonen auf den Farbwechsel einiger Adriafische. *Z. vergl. Physiol.* **25**, 251–5.

FOLIN, O. (1932). The micro method for the determination of blood sugar. *New Engl. J. Med.* **206**, 727–9.

FONTAINE, M. et BUSNEL, R. G. (1938). Sur la localization et le rôle de la flavine ou d'un corps voisin de la flavine dans la peau des poissons. *C.R. Acad. Sci., Paris*, **206**, 372–4.

FORD, E. (1921). A contribution to our knowledge of the life-histories of the dogfishes landed at Plymouth. *J. Mar. Biol. Ass.* **12**, 468–505.

FORD, W. E. and STEGGERDA, F. R. (1940). The effects of pituitrin and adrenalin on the skin potentials and melanophore size of the frog. *Anat. Rec.* **78**, Suppl. p. 156.

FOSI, V. (1934). Le modificazioni istologichi della pelle delle anguille nell' acquisizione dell' abito di nozzi. *Arch. zool. (ital.)*, *Napoli*, **20**, 285–302.

FOSTER, K. W. (1933). Color changes in *Fundulus* with special references to the color changes of the iridosomes. *Proc. Nat. Acad. Sci., Wash.*, **19**, 535–40.

FOSTER, K. W. (1937). The blue phase in the color changes of fish with special reference to the role of the guanin deposits in the skin of *Fundulus heteroclitus*. *J. Exp. Zool.* **77**, 169–213.

FOSTER, P. C. (1935). Effect of infra-red on tissue temperature gradient as influenced by pigment. *Proc. Soc. Exp. Biol., N.Y.*, **33**, 62–4.

FRANKENBERGER, Z. (1924). Sur le developpement du pigment chez les larves de *Hyla arborea*. *Biol. Listy*, **10**, 293–300.

FRANZ, S. and GRAY, S. (1941). The response of melanophores in normal and hypophysectomized frogs to varying concentrations of pituitrin. *Anat. Rec.* **81**, Suppl. p. 92.

FRANZ, V. (1910a). Zur Physiologie und Pathologie der Chromatophoren. *Biol. Zbl.* **30**, 150–8.

FRANZ, V. (1910b). Zur Struktur der Chromatophoren bei Crustaceen. *Biol. Zbl.* **30**, 424–30.

FRANZ, V. (1935a). Struktur und Mechanismus der Melanophoren im Farbenkleid der Teleostier. *Z. Zellforsch.* **23**, 150–97.

FRANZ, V. (1935b). Nachtrag zu vorstehender Arbeit. *Z. Zellforsch.* **23**, 198–200.

FRANZ, V. (1939). Struktur und Mechanismus der Melanophoren. Teil II: Das Endoskelett. *Z. Zellforsch.* Abt. A, **30**, 194–234.

FREDERICQ, H. (1927). La transmission humorale des excitations nerveuses. *C.R. Soc. Biol., Paris*, **96**, Suppl. pp. 3–38.

FRIES, E. F. B. (1927). Nervous control of xanthophore changes in *Fundulus*. *Proc. Nat. Acad. Sci., Wash.*, **13**, 567–9.

FRIES, E. F. B. (1931). Color changes in *Fundulus*, with special consideration of the xanthophores. *J. Exp. Zool.* **60**, 384–426.

FRIES, E. F. B. (1942a). Pituitary and nervous control of chromatic responses, especially of xanthophores, in killifish (*Fundulus*). *Proc. Soc. Exp. Biol., N.Y.*, **51**, 170–1.

FRIES, E. F. B. (1942b). Some neurohumoral evidence for double innervation of xanthophores in killifish (*Fundulus*). *Biol. Bull. Woods Hole*, **82**, 261–72.

FRIES, E. F. B. (1942c). Notes on color change and pigmentary innervation in a goby, a wrasse, and the plaice. *Biol. Bull. Woods Hole*, **82**, 273–83.

FRIES, E. F. B. (1942*d*). White pigmentary effectors (leucophores) in killifishes. *Proc. Nat. Acad. Sci., Wash.*, **28**, 396–401.

FRIES, E. F. B. (1943). Pituitary and nervous control of pigmentary effectors, especially xanthophores, in killifish (*Fundulus*). *Physiol. Zoöl.* **16**, 199–212.

VON FRISCH, K. (1910). Ueber die Beziehungen der Pigmentzellen in der Fischhaut zum sympathischen Nervensystem. *Festschrift R. Hertwig*, **3**, 15–28.

VON FRISCH, K. (1911*a*). Ueber den Einfluss der Temperatur auf die schwarzen Pigmentzellen der Fischhaut. *Biol. Zbl.* **31**, 236–48.

VON FRISCH, K. (1911*b*). Die Pigmentzellen der Fischhaut. *Biol. Zbl.* **31**, 412–15.

VON FRISCH, K. (1911*c*). Beiträge zur Physiologie der Pigmentzellen in der Fischhaut. *Pflüg. Arch. ges. Physiol.* **138**, 319–87.

VON FRISCH, K. (1912*a*). Ueber das Parietalorgan der Fische als functionierendes Organ. *S.B. Ges. Morph. Physiol. München*, **27**, 16–18.

VON FRISCH, K. (1912*b*). Ueber farbige Anpassung bei Fischen. *Zool. Jb.* (Abt. allg. Zool. Physiol.), **32**, 171–230.

VON FRISCH, K. (1912*c*). Ueber die Farbenanpassung des Crenilabrus. *Zool. Jb.* (Abt. allg. Zool. Physiol.), **33**, 151–64.

VON FRISCH, K. (1913). Ueber Färbung und Farbensinn der Tiere. *S.B. Ges. Morph. Physiol. München*, **28**, 30–8.

VON FRISCH, K. (1920). Ueber den Einfluss der Bodenfarbe auf die Fleckenzeichnung des Feuersalamanders. *Biol. Zbl.* **40**, 390–414.

FRÖHLICH, A. (1910). Farbwechselreaktionen bei *Palaemon*. *Arch. EntwMech. Org.* **29**, 432–8.

FRÖLICH, F. W. (1910). Experimentelle am Nervensystem der Mollusken. 7. Ueber den peripheren Tonus der Cephalopodenchromatophoren und seine Hemmung. *Z. allg. Physiol.* **11**, 99–106.

FROMMEL, E. et ZIMMET, D. (1937). L'action de tartrate d'ergotamine sur les chromatophores de la grenouille. *Arch. int. Pharmacodyn.* **55**, 175–83.

FUCHS, R. F. (1910*a*). Die elektrischen Erscheinungen am glatten Muskel. *Pflüg. Arch. ges. Physiol.* **136**, 65–100.

FUCHS, R. F. (1910*b*). Zur Physiologie der Pigmentzellen, zugleich ein Beitrag zur Funktion des Stellarganglions der Cephalopoden. *Arch. EntwMech. Org.* **30**, 389–410.

FUCHS, R. F. (1912). Die physiologische Funkton des Chromatophorsystems als Organ der physikalischen Wärmeregulierung der Poikilothermen. *S.B. phys.-med. Soz. Erlangen*, **44**, 134–77.

FUCHS, R. F. (1913). Die physiologische Funktion der Pigmentzellen. *Naturwissenschaften*, **1**, 903–6, 927–31.

FUJITA, H. (1911). Pigmentbewegung und Zapfenkontraktion im Dunkelauge des Frosches bei Einwirkung verschiedener Reize. *Arch. vergl. Ophthal.* **2**, 164–79.

FUKUI, K. (1923). Notes on experiments on the color change of the file-fish, *Monacanthus cirrhifer*. *Folia anat. japon.* **1**, 345–8.

FUKUI, K. (1927). On the color pattern produced by various agents in the goldfish. *Folia anat. japon.* **5**, 257–302.

FUKUI, K. (1930). The definite localization of the color pattern in the goldfish. *Folia anat. japon.* **8**, 283–312.

GAMBLE, F. W. (1910). The relation between light and pigment-formation in *Crenilabrus* and *Hippolyte*. *Quart. J. Micr. Sci.* **55**, 541–83.

GAUPP, R., Jr und SCHARRER, E. (1935). Die Zwischenhirnsekretion bei Mensch und Tier. *Z. ges. Neurol. Psychiat.* **153**, 326–55.

GEILING, E. M. K. (1926). The pituitary body. *Physiol. Rev.* **6**, 62–123.

GEILING, E. M. K. and LE MESSURIER, D. H. (1936). The pressor antidiuretic and oxytocic hormones of the hypophysis cerebri of certain selachians and teleosts *Bull. Mt Desert Isl. Biol. Lab.* pp. 21–2.

GEILING, E. M. K. and LEWIS, M. E. (1935). Further information regarding the melanophore hormone of the hypophysis cerebri. *Amer. J. Physiol.* **113**, 534–7.

GEIRINGER, M. (1935). Die Rolle von Hypophysenstoffen für die Farbkleidung von Amphibien (*Salamandra maculosa* Laur. und *Hyla arborea* L.). *Anz. Akad. Wiss. Wien*, Math.-nat. Kl. pp. 235–7.

GEIRINGER, M. (1937). Einfluss des Zentralnervensystems auf die Farbanpassung der Frösche (*Hyla arborea* L.). *Anz. Akad. Wiss. Wien*, Math.-nat. Kl. **74**, 44–5.

GEIRINGER, M. (1938). Die Beziehung der basalen Optikuswurzel zur Hypophyse und ihre Bedeutung für den Farbwechsel der Amphibien. *Anat. Anz.* **86**, 202–7.

VON GELDERN, C. E. (1921). Color changes and structure of the skin of *Anolis carolinensis*. *Proc. Calif. Acad. Sci.* Ser. 4, **10**, 77–117.

VON GELEI, G. (1942). Zur Frage der Doppelinnervation der Chromatophoren. *Z. vergl. Physiol.* **29**, 532–40.

GERSCHLER, M. W. (1914). Über alternative Vererbung bei Kreuzung von Cyprinodontiden-Gattungen. *Z. indukt. Abstamm.- u. VererbLehre*, **12**, 73–96.

GERSH, I. and TARR, A. D. (1935). The so-called hyaline bodies of herring in the posterior lobe of the hypophysis. *Anat. Rec.* **63**, 231–8.

GIANFERRARI, L. (1922). Influenza dell' alimentazione con capsule surrenali, ipofisi ed su la pigmentazione cutanea ed il ritmo respiratirio di *Salmo fario*. *Arch. Sci. biol.*, *Napoli*, **3**, 39–52.

GIERSBERG, H. (1928). Ueber den morphologischen und physiologischen Farbwechsel der Stabheuschrecke *Dixippus* (*Carausius*) *morosus*. *Z. vergl. Physiol.* **7**, 657–95.

GIERSBERG, H. (1930). Der Farbwechsel der Fische. *Z. vergl. Physiol.* **13**, 258–79.

GIERSBERG, H. (1931). Ueber den Zusammenhang von morphologischem und physiologischem Farbwechsel. Nach Untersuchungen an Insekten und Fischen. *Arch. Zool.* (*ital.*), *Napoli*, **16**, 363–70.

GIERSBERG, H. (1932). Der Einfluss der Hypophyse auf die farbigen Chromatophoren der Elritze. *Z. vergl. Physiol.* **18**, 369–77.

GILSON, A. S., Jr. (1922). The diverse effects of adrenalin upon the migration of the scale pigment and the retinal pigment in the fish, *Fundulus heteroclitus* Linn. *Proc. Nat. Acad. Sci.*, *Wash.*, **8**, 130–3.

GILSON, A. S., Jr. (1926a). Melanophores in developing and adult *Fundulus*. *J. Exp. Zool.* **45**, 415–55.

GILSON, A. S., Jr. (1926b). The control of melanophore activity in *Fundulus*. *J. Exp. Zool.* **45**, 457–68.

GINSBERG, J. (1929). Beiträge zur Kenntnis der Guaninophoren und Melanophoren. *Zool. J.* (Abt. Anat.), **51**, 227–60.

GIUSTI, L. et HOUSSAY, B. A. (1921). Altérations cutanées chez les crapauds hypophysectomisés. *C.R. Soc. Biol.*, *Paris*, **85**, 597–8.

GIUSTI, L. et HOUSSAY, B. A. (1924). Modifications cutanées et génitales produites chez le crapaud par l'extirpation de l'hypophyse ou par lésion du cerveau. *C.R. Soc. Biol.*, *Paris*, **91**, 313–17.

GLASER, E. und HAEMPEL, O. (1932). Das experimentell hervorgerufene Hochzeitskleid des kastrierten Fisches als Stigma einer Test- und Standardisierungsmethode des männlichen Sexualhormons. *Pflüg. Arch. ges Physiol.* **229**, 1–14.

GLICK, D. (1938). Cholin esterase and the theory of chemical mediation of nerve impulses. *J. Gen. Physiol.* **21**, 431–8.

GODA, T. (1929). Cytoplasmic inclusions of amphibian cells with special reference to melanin. *J. Fac. Sci. Tokyo Univ.* Sect IV. Zool. **2**, 51–122.

GOODRICH, H. B. (1927). A study of the development of Mendelian characters in *Oryzias latipes*. *J. Exp. Zool.* **49**, 261–87.

GOODRICH, H. B. (1929). Mendelian inheritance in fish. *Quart. Rev. Biol.* **4**, 83–99.

GOODRICH, H. B. (1933). One step in the development of hereditary pigmentation in the fish *Oryzias latipes*. *Biol. Bull. Woods Hole*, **65**, 249–52.

GOODRICH, H. B. (1935 a). Studies by means of tissue transplants on the color patterns of tropical marine fish. *Anat. Rec.* **64**, Suppl. 1, p. 95.

GOODRICH, H. B. (1935 b). The development of hereditary color patterns in fish. *Amer. Nat.* **69**, 267–77.

GOODRICH, H. B. (1935 c). Studies on color patterns of Tortugas fish. *Ann. Rep. Tortugas Lab. Wash.* 1934–5, pp. 81–2.

GOODRICH, H. B. (1939). Chromatophores in relation to genetic and specific distinctions. *Amer. Nat.* **73**, 198–207.

GOODRICH, H. B. and ANDERSON, P. L. (1939). Variations of color pattern in hybrids of the goldfish, *Carassius auratus*. *Biol. Bull. Woods Hole*, **77**, 184–91.

GOODRICH, H. B. and HANSEN, I. B. (1931). The postembryonic development of Mendelian characters in the goldfish, *Carassius auratus*. *J. Exp. Zool.* **59**, 337–58.

GOODRICH, H. B. and HEDENBURG, M. (1941). The cellular basis of colors in some Bermuda parrot fish with special reference to blue pigment. *J. Morph.* **68**, 493–505.

GOODRICH, H. B., HILL, G. A. and ARRICK, M. S. (1941). The chemical identification of gene-controlled pigments in *Playtpoecilus* and *Xiphophorus* and comparisons with other tropical fish. *Genetics*, **26**, 573–86.

GOODRICH, H. B. and MERCER, R. N. (1934). Genetics and colors of the Siamese fighting fish, *Betta splendens*. *Science*, **79**, 318–19.

GOODRICH, H. B. and NICHOLS, R. (1931). The development and the regeneration of the color pattern in *Brachydanio rerio*. *J. Morph.* **52**, 513–23.

GOODRICH, H. B. and NICHOLS, R. (1933). Scale transplantation in the goldfish *Carassius auratus*. *Biol. Bull. Woods Hole*, **65**, 253–65.

GOODRICH, H. B. and SMITH, M. A. (1937). Genetics and histology of the color pattern in the normal and albino paradise fish, *Macropodus opercularis* L. *Biol. Bull. Woods Hole*, **73**, 527–34.

GOODRICH, H. B. and TRINKAUS, J. P. (1939). The differential effect of radiations on mendelian phenotypes of the goldfish, *Carassius auratus*. *Biol. Bull. Woods Hole*, **77**, 192–9.

GORDON, M. (1926 a). Melanophores of *Platypoecilus*, the top-minnow of geneticists. *Anat. Rec.* **34**, 138.

GORDON, M. (1926 b). Variation in the tropical, viviparous killifish, *Platypoecilus*. *Anat. Rec.* **34**, 166.

GORDON, M. (1926 c). Inheritance in fishes. *Anat. Rec.* **34**, 172–3.

GORDON, M. (1927). The genetics of a viviparous top-minnow *Platypoecilus*; the inheritance of two kinds of melanophores. *Genetics*, **12**, 253–83.

GORDON, M. (1928 a). Morphological studies of genetically different color patterns in *Platypoecilus*. *Anat. Rec.* **41**, 55.

GORDON, M. (1928 b). Pigment inheritance in the Mexican killifish. *J. Hered.* **19**, 551–6.

GORDON, M. (1931 a). Morphology of the heritable color patterns in the Mexican killifish, *Platypoecilus*. *Amer. J. Cancer*, **15**, 732–87.

GORDON, M. (1931 b). Hereditary basis of melanosis in hybrid fishes. *Amer. J. Cancer*, **15**, 1495–1523.

GORDON, M. (1931 c). The hereditary basis for melanosis in hybrids of Mexican killifishes. *Proc. Nat. Acad. Sci., Wash.*, **17**, 276–80.

GORDON, M. (1933 a). Studies on pigmentation of fishes. *Yearb. Carneg. Instn*, **32**, 268.

GORDON, M. (1933 b). Sex-linked factors in crosses between varieties, species, and genera of Mexican fishes. *Amer. Nat.* **67**, 75–6.

GORDON, M. (1934). Wild types in the platyfish. *Amer. Nat.* **68**, 174–5.

GORDON, M. (1935a). Genetics of *Xiphophorus*, the swordtail killifish. I. The inheritance of micromelanophores. *Amer. Nat.* **69**, 64.

GORDON, M. (1936a). The production of spontaneous melanotic neoplasms with a single type of melanophore in hybrid fishes, *Amer. Nat.* **70**, 50–1.

GORDON, M. (1936b). The production of spontaneous melanotic neoplasms at birth in hybrid fishes. *Amer. Nat.* **70**, 51.

GORDON, M. (1937a). The production of spontaneous melanotic neoplasms in fishes by selective matings. *Amer. J. Cancer*, **30**, 362–75.

GORDON, M. (1937b). Heritable color variations in the Mexican swordtailfish. *J. Hered.* **28**, 223–30.

GORDON, M. (1938). The genetics of *Xiphophorus helleri*: heredity in montezuma, a Mexican swordtail fish. *Copeia*, pp. 19–29.

GORDON, M. and FLATHMAN, F. (1943). The genetics of melanoma in fishes. VI. Mendelian segregation of melanophore reaction types in embryos of a melanomatous mother. *Zoologica, N.Y.*, **28**, 9–12.

GORDON, M. and FRASER, A. C. (1931). Pattern genes in the platyfish. *J. Hered.* **22**, 169–85.

GORDON, M. and LANSING, W. (1943). Cutaneous melanophore eruptions in young fishes during stages preceding melanic tumor formation. *J. Morph.* **73**, 231–45.

GORDON, M. and SMITH, G. M. (1938a). Progressive growth stages of a heritable melanotic neoplastic disease in fishes from the day of birth. *Amer. J. Cancer*, **34**, 255–72.

GORDON, M. and SMITH, G. M. (1938b). The production of a melanotic neoplastic disease in fishes by selective matings. IV. Genetics of geographical species hybrids. *Amer. J. Cancer*, **34**, 543–65.

GOUBEAUD, W. (1931). Die histologischen Grundlagen von Farbkleid und Farbwechsel bei *Bufo viridis*. *Z. Morph. Ökol. Tiere*, **21**, 702–39.

GRAND, C. G., GORDON, M. and CAMERON, G. (1941). Neoplasm studies. VIII. Cell types in tissue culture of fish melanomas compared with mammalian melanomas. *Cancer Res.* **1**, 660–6.

GRAUPNER, H. (1933). Ueber die Entstehung des schwarzen Farbstoffes in der Fischhaut. *Verh. dtsch. Zool. Ges.* **35**, 203–9.

GRAUPNER, H. and FISCHER, I. (1933). Beiträge zur Kenntnis der Goldfischhaut. I. *Z. mikr.-anat. Forsch.* **33**, 91–142.

GRAUPNER, H. and FISCHER, I. (1935). Die Entwicklung und Degeneration der Melanophoren von *Atherina mocho*. *Z. Zellforsch.* **22**, 434–44.

GRAY, S. W. and FORD, W. (1940). The effect of the crustacean eye-stalk hormone upon water metabolism and melanophore expansion in the frog. *Endocrinology*, **26**, 160–2.

GRAY, S. W. and LITTLE, W. M. (1939). Humoral effects of the eyes in amphibian pigment responses. *Anat. Rec.* **75**, Suppl. p. 62.

GREVING, R. (1928). Das Zwischenhirnhypophysensystem. *Klin. Wschr.* **7**, 734–7.

GRIFFITHS, M. (1936). The colour-changes of batoid fishes. *Proc. Linn. Soc. N.S.W.* **61**, 318–21.

GUDGER, E. W. (1934). Ambicoloration in the winter flounder, *Pseudopleuronectes americanus*. *Amer. Mus. Novit.* no. 717, 8 pp.

GUDGER, E. W. (1935a). Two partially ambicolorate flatfishes (Heterosomata). *Amer. Mus. Novit.* no. 768, 8 pp.

GUDGER, E. W. (1935b). Abnormalities in flatfishes (Heterosomata). *J. Morph.* **58**, 1–39.

GUDGER, E. W. (1936). A reversed almost wholly ambicolorate summer flounder, *Paralichthys dentatus*. *Amer. Mus. Novit.* no. 896, 5 pp.

GUDGER, E. W. (1941). A totally ambicolorate flounder, *Platichthys stellatus*, from Alaskan waters. *Copeia*, pp. 28–30.

GUDGER, E. W. and FIRTH, F. E. (1935). An almost totally ambicolorate halibut, *Hippoglossus hippoglossus*, with partially rotated eye and hooked dorsal fin—the only recorded specimen. *Amer. Mus. Novit.* no. 811, 7 pp.

GUDGER, E. W. and FIRTH, F. E. (1936). Three partially ambicolorate four-spotted flounders, *Paralichthys oblongus*, two each with a hooked dorsal fin and a partially rotated eye. *Amer. Mus. Novit.* no. 885, 9 pp.

GUDGER, E. W. and FIRTH, F. E. (1937). Two reversed partially ambicolorate halibuts: *Hippoglossus hippoglossus*. *Amer. Mus. Novit.* no. 925, 1–10.

HACHLOV, L. (1910). Die Körperwand von *Hirudo medicinalis*, nebst einigen Bemerkungen über die Bayer'schen Organe von *Clepsine sexoculata*. *Zool. Jb.* (Abt. Anat.), **29**, 449–84.

HADJIOLOFF, A. (1929a). Les chromatophores bleus dans la peau de la grenouille. *C.R. Soc. Biol.*, Paris, **100**, 669–71.

HADJIOLOFF, A. (1929b). La coloration de la peau chez la grenouille et ses mécanismes histologiques. *Bull. Histol. Tech. micr.* **6**, 286–305.

HADLEY, C. E. (1928). Color changes in excised pieces of the integument of *Anolis equestris* under the influence of light. *Proc. Nat. Acad. Sci., Wash.*, **14**, 822–4.

HADLEY, C. E. (1929). Color changes in two Cuban lizards. *Bull. Mus. Comp. Zool. Harv.* **69**, 105–14.

HADLEY, C. E. (1931). Color changes in excised and intact reptilian skin. *J. Exp. Zool.* **58**, 321–31.

HAFTER, E. (1932). Untersuchungen über den Mechanismus der retinalen Umstimmung hinsichtlich einer Abhängigkeit vom vegetativen Nervensystem. *Pflüg. Arch. ges. Physiol.* **229**, 447–65.

HAIR, G. W. (1938). The nerve supply of the hypophysis of the cat. *Anat. Rec.* **71**, 141–60.

HANAOKA, K. I. (1936). Experiments on the nuptial coloration of the common Japanese newt, *Triturus pyrrhogaster* (Boie). *J. Fac. Sci. Hokkaido Univ.* Ser. VI, Zool. **5**, 113–19.

HANCE, R. T. (1924). Heredity in goldfish. *J. Hered.* **15**, 177–82.

HANSEN, I. B. (1931). Gonadectomy in the goldfish, *Carassius auratus*. *Science*, **73**, 293–5.

HANSTRÖM, B. (1931). Neue Untersuchungen über Sinnesorgane und Nervensystem der Crustaceen. I. *Z. Morph. Ökol. Tiere*, **23**, 80–236.

HANSTRÖM, B. (1933). Neue Untersuchungen über Sinnesorgane und Nervensystem der Crustaceen. II. *Zool. Jb.* (Abt. Anat.), **56**, 367–520.

HANSTRÖM, B. (1934a). Neue Untersuchungen über Sinnesorgane und Nervensystem der Crustaceen. III. *Zool. Jb.* (Abt. Anat.), **58**, 101–44.

HANSTRÖM, B. (1934b). Über das Organ X, ein inkretorische Gehirndrüse der Crustaceen. *Psychiat. neurol. Bl., Amst.*, pp. 1–14.

HANSTRÖM, B. (1935). Preliminary report on the probable connection between the blood gland and the chromatophore activator in decapod crustaceans. *Proc. Nat. Acad. Sci., Wash.*, **21**, 584–5.

HANSTRÖM, B. (1936). Ueber eine Substanz im Insektenkopf, die zusammenballend auf das Pigment der Garneelenkromatophoren wirkt. *Kl. fysiogr. Sällsk. Lund Förh.* **6**, 58–62.

HANSTRÖM, B. (1937a). Die Sinusdrüse und der hormonal bedingte Farbwechsel der Crustaceen. *Kl. svenska VetensAkad. Handl.* 3 ser. **16** (99 pp.).

HANSTRÖM, B. (1937b). Vermischte Beobachtungen über die chromatophoraktivierenden Substanzen der Augenstiele der Crustaceen und des Kopfes der Insekten. *Kl. fysiogr. Sällsk. Handl.* **47** (11 pp.).

HANSTRÖM, B. (1937c). Inkretorische Organe und Hormonfunktionen bei den Wirbellosen. *Ergebn. Biol.* **14**, 143–224.

HANSTRÖM, B. (1938a). Der Einfluss der Lackierung der Augen auf die Expansion der Chromatophoren bei *Leander adspersus*. *Acta Univ. Lund.* N.F. Adv. 234, nr. 11, 10 pp.

HANSTRÖM, B. (1938b). Der Einfluss der Lackierung der Augen auf die Expansion der Chromatophoren bei *Leander adspersus. Kl. fysiogr. Sällsk. Handl.* 49 (10 pp.).

HANSTRÖM, B. (1939). *Hormones in Invertebrates.* 198 pp. Oxford.

HANSTRÖM, B. (1940a). Inkretorische Organe, Sinnesorgane und Nervensystem des Kopfes einiger niederer Insektenordnungen. *Kl. svenska VetenskAkad. Handl.* 3 Ser. **18**, no. 8 (265 pp.).

HANSTRÖM, B. (1940b). Die chromatophoraktivierends Substanz des Insektenkopfes. *Acta Univ. Lund,* N.F. Adv. 2, **36**, nr. 12, 20 pp.

HANSTRÖM, B. (1941). Einige Parallelen im Bau und in der Herkunft der inkretorischen Organe der Arthropoden und Vertebraten. *Acta Univ. Lund,* N.F. Adv. 2, **37**, nr. 4, 19 pp.

HARGITT, C. W. (1912). Behavior and color changes of tree frogs. *J. Anim. Behav.* **2**, 51–78.

HASAMA, B. (1931). On the effect of the pituitary body upon the epidermal melanophores of the toad. *J. Pharmacol.* **41**, 179–94.

HEALEY, E. G. (1940). Ueber den Farbwechsel der Elritze (*Phoxinus laevis* Ag.). *Z. vergl. Physiol.* **27**, 545–86.

HENZE, M. (1910). Ueber das Vorkommen des Betains bei Cephalopoden. *Hoppe-Seyl. Z.* **70**, 253–5.

HENZE, M. (1913). *p*-Oxyphenyläthylamin, das Speicheldrüsengift der Cephalopoden. *Hoppe-Seyl. Z.* **87**, 51–8.

HERBST, K. (1919). Beiträge zur Entwicklungsphysiologie der Färbung und Zeichnung der Tiere. 1. Der Einfluss gelber, weisser und schwarzer Umgebung auf die Zeichnung von *Salamandra maculosa. Abh. heidelberg. Akad. Wiss.* Math.-nat. Kl. 7 Abh. 64 pp.

HERBST, C. (1924). Beiträge zur Entwicklungsphysiologie der Färbung und Zeichnung der Tiere. 2. Die Weiterzucht der Tiere in gelber und schwartzer Umgebung. *Arch. mikr. Anat.* **102**, 130–67.

HERBST, C. (1927). Beiträge zur Entwicklungsphysiologie der Färbung und Zeichnung der Tiere. IV. Kritische Bemerkungen zu der Arbeit von MacBride 'Influence of the colour of the background on the colour of the skin of *Salamandra maculosa'. Roux Arch. Entw. Mech. Organ.* **112**, 59–60.

HERBST, C. und ASCHER, F. (1927). Beiträge zur Entwicklungsphysiologie der Färbung und Zeichnung der Tiere. III. Der Einfluss der Beleuchtung von unten auf das Farbkleid des Feuersalamanders. *Roux Arch. Entw. Mech. Organ.* **112**, 1–59.

HERRICK, E. H. (1932). Mechanism of movement of epidermis, especially its melanophores in wound healing, and behavior of skin grafts in frog tadpoles. *Biol. Bull. Woods Hole,* **63**, 271–86.

HERRICK, E. H. (1933). The structure of epidermal melanophores in frog tadpoles. *Biol. Bull. Woods Hole,* **64**, 304–8.

HERRICK, E. H. (1934). Do chromatophore walls cause movement of pigment granules? *Science,* **80**, 96–7.

HESS, C. (1914). Untersuchungen zur Physiologie des Gesichtssinnes der Fische. *Z. Biol.* **63**, 245–74.

HEWER, H. R. (1923). Studies in amphibian colour change. *Proc. Roy. Soc. B,* **95**, 31–41.

HEWER, H. R. (1926). Studies in colour changes of fish. I. The action of certain endocrine secretions in the minnow. *Brit. J. Exp. Biol.* **3**, 123–40.

HEWER, H. R. (1927). Studies in colour changes of fish. II. An analysis of the colour patterns of the dab. III. The action of nicotine and adrenalin in the dab. IV. The action of caffeine in the dab, and a theory of the control of colour changes in fish. *Philos. Trans.* B, **215**, 177-200.

HEWER, H. R. (1931). Studies in colour-changes in fish. V. The colour-patterns in certain flat-fish and their relation to the environment. *J. Linn. Soc.* (Zool.), **37**, 493-513.

HILL, A. V., PARKINSON, J. L. and SOLANDT, D. Y. (1935). Photoelectric records of the colour change in *Fundulus heteroclitus*. *J. Exp. Biol.* **12**, 397-9.

HIMMER, A. (1923). Untersuchungen über den physiologischen und morphologischen Farbwechsel bei Amphibien. *Arch. mikr. Anat.* **100**, 110-63.

HINSBERG, K. und RODEWALD, W. (1939). Über die Wirkung von Porphyrin auf die Hypophyse, besonders auf die Melanophorenhormonausschüttung. *Arch. exp. Path. Pharmak.* **191**, 1-11.

HITCHCOCK, H. B. (1941). The coloration and color changes of the gulf-weed crab, *Planes minutus*. *Biol. Bull. Woods Hole*, **80**, 26-30.

HJORT, A. M. (1935). Some observations on the relationship between chemical constitution and physiological action: the comparative effects of benzyl-, phenylethyl- and di-(phenylethyl) amines and some of their derivatives. *J. Pharmacol.* **50**, 131-50.

HLOBIL, J. (1924). Wirkung der Augen und der Beleuchtung auf Farbänderung bei *Dixippus morosus*. *Biol. Listy*, **10**, 65-74.

HOFMANN, F. B. (1910a). Gibt es in der Muskulatur der Mollusken periphere, kintinuierlich leitende Nervennetze bei Abwesenheit von Ganglienzellen? II. Weitere Untersuchungen an den Chromatophoren der Kephalopoden. Innervation der Mantellappen von *Aplysia. Pflüg. Arch. ges. Physiol.* **132**, 43-81.

HOFMANN, F. B. (1910b). Chemische Reizung und Lähmung markloser Nerven und glatter Muskeln wirbelloser Tiere. Untersuchungen an den Chromatophoren der Kephalopoden. *Pflüg. Arch. ges. Physiol.* **132**, 82-130.

HOGBEN, L. T. (1923). A method of hypophysectomy in adult frogs and toads. *Quart. J. Exp. Physiol.* **13**, 177-9.

HOGBEN, L. T. (1924). The pigmentary effector system. IV. A further contribution to the role of pituitary secretion in amphibian colour response. *Brit. J. Exp. Biol.* **1**, 249-70.

HOGBEN, L. T. (1936). The pigmentary effector system. VII. The chromatic function in elasmobranch fishes. *Proc. Roy. Soc.* B, **120**, 142-58.

HOGBEN, L. T. and DE BEER, G. R. (1925). Studies on the pituitary. VI. Localization and phyletic distribution of active materials. *Quart. J. Exp. Physiol.* **15**, 163-76.

HOGBEN, L., CHARLES, E. and SLOME, D. (1931). Studies on the pituitary. VIII. The relation of the pituitary gland to calcium metabolism and ovarian function in *Xenopus. J. Exp. Biol.* **8**, 345-54.

HOGBEN, L. T. and CREW, F. A. E. (1923). Studies on internal secretion. II. Endocrine activity in foetal and embryonic life. *Brit. J. Exp. Biol.* **1**, 1-13.

HOGBEN, L. and GORDON, C. (1930). Studies on the pituitary. VII. The separate identity of the pressor and melanophore principles. *J. Exp. Biol.* **7**, 286-92.

HOGBEN, L. and LANDGREBE, F. (1940). The pigmentary effector system. IX. The receptor fields of the teleostean visual response. *Proc. Roy. Soc.* B, **128**, 317-42.

HOGBEN, L. and MIRVISH, L. (1928a). Some observations on the production of excitement pallor in reptiles. *Trans. Roy. Soc. S. Afr.* **16**, 45-52.

HOGBEN, L. T. and MIRVISH, L. (1928b). The pigmentary effector system. V. The nervous control of excitement pallor in reptiles. *Brit. J. Exp. Biol.* **5**, 295-308.

HOGBEN, L. T. and SCHLAPP, W. (1924). Studies on the pituitary. III. The vasomotor activity of pituitary extracts throughout the vertebrate series. *Quart. J. Exp. Physiol.* **14**, 229-58.

HOGBEN, L. T., SCHLAPP, W. and MACDONALD, A. D. (1924). Studies on the pituitary. IV. Quantitative comparison of pressor activity. *Quart. J. Exp. Physiol.* **14**, 301-18.

HOGBEN, L. and SLOME, D. (1931). The pigmentary effector system. VI. The dual character of endocrine coordination in amphibian colour change. *Proc. Roy. Soc. B*, **108**, 10–53.

HOGBEN, L. and SLOME, D. (1936). The pigmentary effector system. VIII. The dual receptive mechanism of the amphibian background response. *Proc. Roy. Soc. B*, **120**, 158–73.

HOGBEN, L. T. and WINTON, F. R. (1922a). The pigmentary effector system. I. Reaction of frog's melanophores to pituitary extracts. *Proc. Roy. Soc. B*, **93**, 318–29.

HOGBEN, L. T. and WINTON, F. R. (1922b). Studies on the pituitary. I. The melanophore stimulant in posterior lobe extracts. *Biochem. J.* **16**, 619–30.

HOGBEN, L. T. and WINTON, F. R. (1922c). The pigmentary effector system. II. *Proc. Roy. Soc. B*, **94**, 151–62.

HOGBEN, L. T. and WINTON, F. R. (1923). The pigmentary effector system. III. Colour response in the hypophysectomized frog. *Proc. Roy. Soc. B*, **95**, 15–31.

HOLMES, S. J. (1913). Observations on isolated living pigment cells from the larvae of amphibians. *Univ. Calif. Publ. Zool.* **11**, 143–54.

HOLMES, S. J. (1914). The movements and reactions of the isolated melanophores of the frog. *Univ. Calif. Publ. Zool.* **13**, 167–74.

HOLMES, W. (1940). The colour changes and colour patterns of *Sepia officinalis* L. *Proc. Zool. Soc. Lond.* A, **110**, 17–35.

HOLTFRETER, J. (1933). Der Einfluss von Wirtsalter und verschiedenen Organbezirken auf die Differenzierung von angelagertem Gastrulaektoderm. *Roux Arch. Entw. Mech. Organ.* **127**, 619–775.

HOOKER, D. (1912). The reactions of the melanophores of *Rana fusca* in the absence of nervous control. *Z. allg. Physiol.* **14**, 93–104.

HOOKER, D. (1914a). The development of stellate pigment cells in plasma cultures of frog epidermis. *Anat. Rec.* **8**, 103–4.

HOOKER, D. (1914b). Amoeboid movement in the corial melanophores of *Rana. Amer. J. Anat.* **16**, 237–50.

HOOKER, D. (1914c). The reactions to light and darkness of the melanophores of frog tadpoles. *Science*, **39**, 473.

HOSOI, T. (1934). Chromatophore-activating substance in the shrimps. *J. Fac. Sci. Tokyo Univ.* **3**, 265–70.

HOU, H. C. (1930). The action of ephedrine on melanophores. *Proc. Soc. Exp. Biol., N.Y.*, **28**, 221–2.

HOUSSAY, B. A. (1936). Funciones de la hipofisis del sapo. *Bol. Acad. Med. B. Aires*, pp. 133–52.

HOUSSAY, B. A. et GIUSTI, L. (1929). Les fonctions de l'hypophyse et de la région infundibulo-tubérienne chez le crapaud. *C.R. Soc. Biol., Paris*, **101**, 935–8.

HOUSSAY, B. A. et GIUSTI, L. (1930). Les fonctions de l'hypophyse et de la région infundibulo-tubérienne chez le crapaud *Bufo arenarum* (Hens.). *C.R. Soc. Biol., Paris*, **104**, 1105–12.

HOUSSAY, B. A. et UNGAR, I. (1924a). Modifications produites chez la grenouille par l'hypophysectomie ou par les lésions cérébrales. *C.R. Soc. Biol., Paris*, **91**, 317–18.

HOUSSAY, B. A. et UNGAR, I. (1924b). Action de l'hypophyse sur la coloration des batraciens. *C.R. Soc. Biol., Paris*, **91**, 318–20.

HOUSSAY, B. A. et UNGAR, I. (1925a). Influence de divers agents pharmacodynamiques sur la couleur de *Leptodactylus ocellatus* (L). Gir. *C.R. Soc. Biol., Paris*, **93**, 253–5.

HOUSSAY, B. A. and UNGAR, I. (1925b). Factors that govern pigmentation of the frog *Leptodactylus ocellatus. Rev. Soc. argent. Biol.* pp. 1–65. (Cited from McLean, 1928, p. 318.)

344   BIBLIOGRAPHY OF COLOUR CHANGES

HSIAO, S. C. T. (1941). Melanosis in the common cod, *Gadus callarias* L., associated with trematode infection. *Biol. Bull. Woods Hole*, **80**, 37–44.
HUNTER, G. W. III and WASSERMAN, E. (1941). Observations on the melanophore control of the cunner *Tautogolabrus adspersus* (Walbaum). *Biol. Bull. Woods Hole*, **81**, 300.
HUSSAKOF, L. (1914). On two ambicolorate specimens of the summer flounder, *Paralichthys dentatus*, with an explanation of ambicoloration. *Bull. Amer. Mus. Nat. Hist.* **33**, 95–100.
HUXLEY, J. S. (1935). Chemical regulation and the hormone concept. *Biol. Rev.* **10**, 427–41.
HUXLEY, J. S. and HOGBEN, L. T. (1922). Experiments on amphibian metamorphosis and pigment responses in relation to internal secretions. *Proc. Roy. Soc. B*, **93**, 36–53.

IKEDA, K. (1933). Effect of castration on the secondary sexual characters of anadromous three-spined stickleback, *Gasterosteus aculeatus aculeatus* (L.). *Japan. J. Zool.* **5**, 135–57.
IPPISCH, G. (1928). Beobachtungen über Albinismus usw. bei *Bombinator pachypus* Laur. *Bl. Aquar.- u. Terrarienk.* **39**, 380–3.
ISHIWARA, M. (1917). On the inheritance of body color in *Oryzias latipes*. *Mitt. med. Fac. Kyushu*, **4**, 43–51. (Cited from Goodrich, 1929, p. 98.)
IUGA, V. G. (1931). Les pigments de la *Glossosiphonia paludosa*. *Arch. Zool. exp. gén.* **71**, 1–97.

JANDA, V. (1935). Contribution à l'étude des changements périodiques de la coloration chez *Dixippus morosus* Br. et Redt. *Mem. Soc. Sci. Bohème*, Cl. Sci. 1934, Art. XII, 30 pp.
JANDA, V. (1936). Ueber den Farbwechsel transplantierter Hautstücke und künstlich verbundener Körperfragmente bei *Dixippus morosus* (Br. et Redt.). *Zool. Anz.* **115**, 177–85.
JANZEN, R. (1932a). Ueber das Verkommen eines physiologischen Farbwechsels bei einigen einheimischen Hirudineen. *Zool. Anz.* **101**, 35–40.
JANZEN, R. (1932b). Der Farbwechsel von *Piscicola geometra* L. I. Mitteilung: Beschreibung des Farbwechsels und seiner Elemente. *Z. Morph. Ökol. Tiere*, **24**, 327–41.
JELIASKOWA-PASPALEWA, A. (1930). Cytologische Untersuchungen über die Entstehung des melanotischen Pigments. *Z. wiss. Zool.* **137**, 365–402.
JOHNSON, M. L. (1935). Visual cells of the amphibian retina in the absence of the epithelial pigment layer. *Anat. Rec.* **63**, 53–75.
JONES, M. E. and STEGGERDA, F. R. (1935a). Effects of light and dark environment on weight changes in normal and hypophysectomized frogs. *Proc. Soc. Exp. Biol., N.Y.*, **32**, 1369–71.
JONES, M. E. and STEGGERDA, F. R. (1935b). Studies on water metabolism in normal and hypophysectomized frogs. *Amer. J. Physiol.* **112**, 397–400.
JORDAN, H. (1917). Integumentary photosensitivity in a marine fish, *Epinephelus striatus* Bloch. *Amer. J. Physiol.* **44**, 259–74.
JORES, A. (1932). Ueber das Melanophorenhormon. *Klin. Wschr.* **11**, 2116.
JORES, A. (1933a). Das Farbwechselhormon beim Menschen. *Umschau*, **37**, 1015–17.
JORES, A. (1933b). Das interrenotrope Hormon der Hypophyse. *Klin. Wschr.* **12**, 1989–90.
JORES, A. (1933c). Melanophorenhormon und Auge. *Klin. Wschr.* **12**, 1599–1601.
JORES, A. (1933d). Untersuchungen über das Melanophorenhormon und seinen Nachweis im menschlichen Blut. *Z. ges. exp. Med.* **87**, 266–82.

JORES, A. (1933*d*). Ueber die Funktionen des Pigmenthormons im menschlichen Organismus. *Verh. dtsch. Ges. inn. Med.* **45**, 166–8.

JORES, A. (1934*a*). Einige prinzipielle Bemerkungen zur Hypophysenhormonforschung. *Klin. Wschr.* **13**, 1269–70.

JORES, A. (1934*b*). Ueber das Melanophorenhormon. *S.B. naturf. Ges. Rostock,* **4**, 10–17.

JORES, A. (1935*a*). Zur Frage des Identität zwischen Pigmenthormon und corticotropem Hormon. *Klin. Wschr.* **14**, 132–3.

JORES, A. (1935*b*). Änderungen des Hormongehaltes der Hypophyse mit dem Wechsel von Licht und Dunkelheit. *Klin. Wschr.* **14**, 1713–16.

JORES, A. (1935*c*). Untersuchungen über die Funktion des Pigmenthormons im Warmblüterorganismus. I. Mittheilung. Die Wirkungen des Hormons auf Temperatur und Blutzucker bei intraventriculärer Injektion beim Kaninchen. *Z. ges. exp. Med.* **97**, 207–13.

JORES, A. (1936*a*). Welche Schlüsse lassen sich aus einer mit menschlichem Harn positiven Melanophorenreaktion ziehen? *Klin. Wschr.* **15**, 1433–4.

JORES, A. (1936*b*). Licht und Hypophysentätigkeit. *Verh.* 3 *Intern. Kongr. Lichtforsch.* pp. 570–3.

JORES, A. (1937*a*). Endokrine Korrelationen. *Klin. Wschr.* **16**, 1777–9.

JORES, A. (1937*b*). Experimentelle Untersuchungen über die Wirkungen der Nebennieren auf die Hypophyse. III. Mittheilung. Über die histologischen Änderungen des Hypophysenvorderlappens nach Zufuhr von Adrenalin und Cortidyn. *Z. ges. exp. Med.* **102**, 289–91.

JORES, A. (1938). Über die Funktionen der Hypophyse. *Klin. Wschr.* **17**, 689–93.

JORES, A. und BECK, H. (1934). Melanophoren und Nebennieren. *Z. ges. exp. Med.* **94**, 293–9.

JORES, A. und CAESAR, K. G. (1935). Ueber die Wirkung des Melanophorenhormons auf Pigmentwanderung und Pupillenweite des Froschauges. *Pflüg. Arch. ges. Physiol.* **235**, 724–32.

JORES, A. und GLOGNER, O. (1933). Gibt es einen funktionstüchtigen Zwischenlappen der meschlichen Hypophyse? Untersuchungen über Gehalt und Bildungsstätte des Melanophorenhormons der menschlichen Hypophyse. *Z. ges. exp. Med.* **91**, 91–9.

JORES, A. und HELBRON, O. (1933). Ueber das Verhalten des Melanophorenhormons im menschlichen Blut während der Gestationsphasen. *Arch. Gynaek.* **154**, 243–50.

JORES, A. und HOELTJE, K, (1936). Untersuchungen über die das Melanophorenhormon bindende Substanz im Blut von Tieren (Fröschen und Kaninchen) und des Menschen nach Dunkelaufenthalt. *Z. vergl. Physiol.* **23**, 571–7.

JORES, A. und HOTOP, H. (1934). Vergleichende Untersuchungen über den Gehalt verschiedener Tierhypophysen an Melanophoren- und Erythrophorenhormon. *Z. vergl. Physiol.* **20**, 699–701.

JORES, A. und LENSSEN, E. W. (1933). Sind die Erythrophorenreaktion der Ellritze und die Melanophorenreaktion des Frosches identisch? *Endokrinologie,* **12**, 90–101.

JORES, A. und WILL, G. (1934). Erythrophoren- und Melanophorenhormon. *Z. ges. exp. Med.* **94**, 389–93.

JOST, F. (1926). Die Farbzellen und Farbzellvereinigungen in der Haut der Nordseefisches *Callionymus lyra* L. *Z. mikr.-anat. Forsch.* **7**, 461–502.

JUSZCYK, W. (1937). Die Verteilung der Chromatophoren in der Haut eines normalen und flavistischen *Pelobates fuscus* Laur. *Bull. int. Acad. Cracovie,* Cl. Sci. Math. Nat. B, Sci. Nat. pp. 215–25.

KAHN, R. H. (1922). Studien über die Innervation der Chromatophoren auf Grund gegensätzlicher Giftwirkungen. *Pflüg. Arch. ges. Physiol.* **195**, 337–60.

KALMUS, H. (1938). Tagesperiodisch verlaufende Vorgänge an der Stabheuschrecke (*Dixippus morosus*) und ihre experimentelle Beeinflussung. *Z. vergl. Physiol.* **25**, 496–508.

KAMADA, T. (1937). Parker's effect in melanophore reactions of *Macropodus opercularis*. *Proc. Imp. Acad. Japan*, **13**, 217–19.

KAMMERER, P. (1910). Vererbung erzwungener Farbveränderungen. I. und II. Mitteilung: Induktion von weiblichem Dimorphismus bei *Lacerta muralis*, von männlichem Dimorphismus bei *Lacerta fiumana*. *Arch. EntwMech. Org.* **29**, 456–98.

KAMMERER, P. (1912). Experimente über Fortpflanzung, Farbe, Augen und Körperreduction bei *Proteus anguinus* Laur. *Arch. EntwMech. Org.* **33**, 349–461.

KAMMERER, P. (1913). Vererbung erzwungener Farbveränderungen. IV. Mitteilung: Das Farbkleid des Feuersalamanders (*Salamandra maculosa* Laurenti) in seiner Abhängigkeit von der Umwelt. *Arch. EntwMech. Org.* **36**, 4–193.

KARÁSEK, F. (1933). Die Innervation der Melanophoren beim Frosch. *Biol. gen.* **9**, 403–16.

KARPLUS, I. P. und PECZENIK, O. (1930). Über die Beeinflussung der Hypophysentätigkeit durch die Erregung des *Hypothalamus*. *Pflüg. Arch. ges. Physiol.* **225**, 654–68.

KAWAMURA, T. (1934). Über die Melanophoren in der embryonalen Epidermis von *Coturnix coturnix japonica* Temminck et Schlegel. *J. Sci. Hiroshima Univ.* Ser. B, Div. 1, **3**, 87–97.

KEATY, C. and STANLEY, A. J. (1941). A new method of assay of chromatophorotropic hormone by means of excised lizard skin. *Proc. Soc. Exp. Biol., N.Y.*, **47**, 403–4.

KINOSHITA, Y. (1934). On the differentiation of the male colour-pattern, and the sex ratio in *Halichoeres poecilopterus* (Temminck and Schlegel). *J. Sci. Hiroshima Univ.* Ser. B, Div. 2, Zool. **3**, 65–76.

KINOSHITA, Y. (1935). Effects of gonadectomies on the secondary sexual characters in *Helichoeres poecilopterus* (Temminck and Schlegel). *J. Sci. Hiroshima Univ.* Ser. B, Div. 1, Zool. **4**, 1–14.

KINOSHITA, Y. (1938). On the secondary sexual characters, with special remarks on the influence of hormone preparations upon the nuptial coloration in *Chloea sarchynnis* Jordan & Snyder. *J. Sci. Hiroshima Univ.* Ser. B, Div. 1, Zool. **6**, 5–22.

KLATT, B. (1931). Hypophysenexstirpationen und -implantationen an Tritonlarven. *Roux Arch. Entw. Mech. Organ.* **123**, 747–91.

KLAUBER, L. M. (1939). Studies of reptile life in the arid southwest. Part II. Speculations on protective coloration and protective reflectivity. *Bull. Zool. Soc. S. Diego*, **14**, 65–79.

KLEINHOLZ, L. H. (1934). Eye-stalk hormone and the movement of distal retinal pigment in *Palaemonetes*. *Proc. Nat. Acad. Sci., Wash.*, **20**, 659–61.

KLEINHOLZ, L. H. (1935). The melanophore-dispersing principle in the hypophysis of *Fundulus heteroclitus*. *Biol. Bull. Woods Hole*, **69**, 379–90.

KLEINHOLZ, L. H. (1936a). Crustacean eye-stalk hormone and retinal pigment migration. *Biol. Bull. Woods Hole*, **70**, 159–84.

KLEINHOLZ, L. H. (1936b). Studies in reptilian color changes. I. A preliminary report. *Proc. Nat. Acad. Sci., Wash.*, **22**, 454–6.

KLEINHOLZ, L. H. (1937a). Studies in the pigmentary system of Crustacea. I. Color changes and diurnal rhythm in *Ligia bandiniana*. *Biol. Bull. Woods Hole*, **72**, 24–36.

KLEINHOLZ, L. H. (1937b). Studies in the pigmentary system of Crustacea. II. Diurnal movements of the retinal pigments of Bermudan decapods. *Biol. Bull. Woods Hole*, **72**, 176–89.

KLEINHOLZ, L. H. (1938a). Studies in reptilian colour changes. II. The pituitary and adrenal glands in the regulation of the melanophores of *Anolis carolinensis*. *J. Exp. Biol.* **15**, 474–91.

KLEINHOLZ, L. H. (1938b). Studies in reptilian colour changes. III. Control of the light phase and behaviour of isolated skin. *J. Exp. Biol.* **15**, 492–9.

KLEINHOLZ, L. H. (1938c). Studies in the pigmentary system of Crustacea. IV. The unitary versus the multiple hormone hypothesis of control. *Biol. Bull. Woods Hole*, **75**, 510–32.

KLEINHOLZ, L. H. (1938d). Color changes in echinoderms. *Pubbl. Staz. zool. Napoli*, **17**, 53–7.

KLEINHOLZ, L. H. (1940a). The distribution of intermedin: first appearance of the hormone in the early ontogeny of *Rana pipiens*. *Anat. Rec.* **78**, Suppl. p. 140.

KLEINHOLZ, L. H. (1940b). The distribution of intermedin: first appearance of the hormone in the early ontogeny of *Rana pipiens*. *Biol. Bull. Woods Hole*, **79**, 432–8.

KLEINHOLZ, L. H. (1941). Behavior of melanophores in the alligator. *Anat. Rec.* **81**, Suppl. p. 121.

KLEINHOLZ, L. H. and KNOWLES, F. G. W. (1938). Studies in the pigmentary system of Crustacea. III. Light-intensity and the position of the distal retinal pigment in *Leander adspersus*. *Biol. Bull. Woods Hole*, **75**, 266–73.

KLEINHOLZ, L. H. and RAHN, H. (1940). The distribution of intermedin: a new biological method of assay and results of tests under normal and experimental conditions. *Anat. Rec.* **76**, 157–72.

KLEINHOLZ, L. H. and WELSH, J. H. (1937). Colour changes in *Hippolyte varians*. *Nature, Lond.*, **140**, 851.

KLEINSCHMIDT, A. (1938). Das Verhalten der Melanophoren bei hypophysektomierten Urodelen (*Amblystoma mexicanum*, Shaw. und *Triturus vulgaris* L.) und parellele Befunde an einem anormal neotenen *Triturus vulgaris* L. *Verh. anat. Ges. Jena*, **45**, 262–6.

KLEITMAN, N. (1940). The modifiability of the diurnal pigmentary rhythm in isopods. *Biol. Bull. Woods Hole*, **78**, 403–6.

KNAUS, H. H., DREYER, N. B. and CLARK, A. J. (1925). A note on the melanophore dilator action of the pituitary. *J. Physiol.* **60**, xviii–xix.

KNOWLES, F. G. W. (1939). The control of the white reflecting chromatophores in Crustacea. *Pubbl. Staz. zool. Napoli*, **17**, 174–82.

KNOWLES, F. G. W. and CALLAN, H. G. (1940). A change in the chromatophore pattern of Crustacea at sexual maturity. *J. Exp. Biol.* **17**, 262–6.

KOBAYASHI, E. (1928). Versuche an den Chromatophoren des Frosches. I. Theil: Über die Wirkung des Lichts und den Einfluss von Giften auf dieselben. *Folia pharm. japon.* **6**, 271–9.

KOKETSU, R. (1915). The color changes of the young goldfish. *Zool. Mag., Tokyo*, **27**, 125–30, 183–90, 251–7 (Japanese).

KOLLER, G. (1925). Farbwechsel bei *Crangon vulgaris*. *Verh. dtsch. zool. Ges.* **30**, 128–32.

KOLLER, G. (1927). Über Chromatophorensystem, Farbensinn und Farbwechsel bei *Crangon vulgaris*. *Z. vergl. Physiol.* **5**, 191–246.

KOLLER, G. (1928). Versuche über die inkretorischen Vorgänge beim Garneelenfarbwechsel. *Z. vergl. Physiol.* **8**, 601–12.

KOLLER, G. (1929). Die innere Sekretion bei wirbellosen Tieren. *Biol. Rev.* **4**, 269–306.

KOLLER, G. (1930). Weitere Untersuchungen über Farbwechsel und Farbwechselhormone bei *Crangon vulgaris*. *Z. vergl. Physiol.* **12**, 632–67.

KOLLER, G. (1934). Über den Farbwechsel von Coregonenlarven. *Biol. Zbl.* **54**, 419–36.

348 BIBLIOGRAPHY OF COLOUR CHANGES

KOLLER, G. und MEYER, E. (1930). Versuche uber den Wirkungsbereich von Farbwechselhormonen. *Biol. Zbl.* **50**, 759–68.

KOLLER, G. and RODEWALD, W. (1933). Über den Einfluss des Lichtes auf die Hypophysentätigkeit des Frosches. *Pflüg. Arch. ges. Physiol.* **232**, 637–42.

KOLLER, P. C. (1929). Experimental studies in pigment-formation. 1. The development *in vitro* of the mesodermal pigment cells of the fowl. *Arch. Zellforsch.* **8**, 490–8.

KOLTZOFF, M. (1928). Physiologie der Pigmentzellen. *Arch. Zellforsch.* **6**, 107–8.

KOLTZOFF, N. K. (1929). Les principes physico-chimiques de l'irritabilité des cellules pigmentaires, musculaires et glandulaires. *Rev. gén. Sci. pur. appl.* **40**, 165–71.

KOPEĆ, S. (1918). Contribution to the study of the development of the nuptial colours of fishes. *C.R. Soc. Sci. Varsovie,* **11**, 88–114.

KOPEĆ, S. (1927). Experiments on the dependence of the nuptial hue on the gonads in fish. *Biol. gen.* **3**, 259–80.

KORNFELD, W. (1919). Über die Beziehungen der Pigmentzellen im Corium und in der Epidermis bei Anuren. *Verh. zool.-bot. Ges. Wien,* **69**, 158–60.

KOSSWIG, C. (1929a). Melanotische Geschwulstbildungen bei Fischbastarden. *Verh. dtsch. zool. Ges.* **33**, 90–8.

KOSSWIG, C. (1929b). Über die veränderte Wirkung von Farbgenen des *Platypoecilus* in der Gattungskreuzung mit *Xiphophorus*. *Z. indukt. Abstamm.- u. VererbLehre,* **50**, 63–73.

KOSSWIG, C. (1929c). Zur Frage der Geschwulstbildung bei Gattungsbastarden der Zahnkarpfen *Xiphophorus* und *Platypoecilus*. *Z. indukt. Abstamm.- u. VererbLehre,* **52**, 114–20.

KOSSWIG, C. (1931). Zur Frage der Geschwulstbildungen bei Gattungbastarden der Zahnkarpfen *Xiphophorus* und *Platypoecilus*. *Z. indukt. Abstamm.- u. VererbLehre,* **59**, 61–76.

KOSSWIG, C. (1935). Über Albinismus bei Fischen. *Zool. Anz.* **110**, 41–7.

KROGH, A. (1922). *The Anatomy and Physiology of Capillaries.* 276 pp. New Haven.

KROGH, A. (1926). The pituitary (posterior lobe) principle in circulating blood. *J. Pharmacol.* **29**, 177–89.

KROGH, A. and REHBERG, P. B. (1923). Kinematographic methods in the study of capillary circulation. *Amer. J. Physiol.* **68**, 153–60.

KROPP, B. (1927). The control of the melanophores in the frog. *J. Exp. Zool.* **49**, 289–318.

KROPP, B. (1929). The melanophore activator of the eye. *Proc. Nat. Acad. Sci., Wash.,* **15**, 693–4.

KROPP, B. (1932). The crustacean chromatophore activator and the gonads of the rat. *Proc. Nat. Acad. Sci., Wash.,* **18**, 690.

KROPP, B. (1936). Some physiological effects of the crustacean eye stalk hormone. *Bull. Mt Desert Isl. Biol. Lab.* pp. 15–16.

KROPP, B. and CROZIER, W. J. (1934). The production of the crustacean chromatophore activator. *Proc. Nat. Acad. Sci., Wash.,* **20**, 453–6.

KROPP, B. and PERKINS, E. B. (1933a). The occurrence of the humoral chromatophore activator among marine crustaceans. *Biol. Bull. Woods Hole,* **64**, 28–32.

KROPP, B. and PERKINS, E. B. (1933b). The action of the crustacean chromatophore activator on the melanophores of fishes and amphibians. *Biol. Bull. Woods Hole,* **64**, 226–32.

KRÜGER, P. (1929). Über die Bedeutung der ultraroten Strahlen für den Warmehaushalt der Poikilothermen. *Biol. Zbl.* **49**, 65–82.

KRÜGER, P. und KERN, H. (1924). Die physikalische und physiologische Bedeutung des Pigmentes bei Amphibien und Reptilien. *Pflug. Arch. ges. Physiol.* **202**, 119–38.

KUDÔ, T. (1922). Veränderungen der Melaninmenge bei Farbwechsel der Fische *Esox, Carassius, Phoxinus, Gobius, Nemachilus. Arch. EntwMech. Org.* **50**, 309–25.

KÜHN, A. (1930). Über Farbensinn und Anpassung der Körperfarbe an die Umgebung bei Tintenfischen. *Nach. Ges. Wiss. Göttingen,* Math.-nat. Kl. pp. 10–16.

KÜHN, A. und HEBERDEY, R. F. (1929). Über die Anpassung von *Sepia officinalis* L. an Helligkeit und Farbton der Umgebung. *Verh. dtsch. zool. Ges.* pp. 231–7.

KUMAGAI, K. (1923). The blinded fish. *Tokyo Igakukai Zasshi,* **37**. (Cited from Fukui, 1927, p. 300.)

KUNTZ, A. (1917). The histological basis of adaptive shades and colors in the flounder *Paralichthys albiguttus. Bull. U.S. Bur. Fish.* **35**, 1–29.

KURZ, F. (1920). Versuch über den Einfluss farbigen Lichtes auf die Entwicklung und Veränderung der Pigmente bei den Fischen. *Zool. Jb.* (Abt. Physiol.), **37**, 239–78.

LANDGREBE, F. W., REID, E. and WARING, H. (1943). Further observations on the intermediate lobe pituitary hormone. *Quart. J. Exp. Physiol.* **32**, 121–41.

LANDGREBE, F. W. and WARING, H. (1941). Intermediate lobe pituitary hormone. *Quart. J. Exp. Physiol.* **31**, 31–62.

LANGE, B. (1931). Integument der Sauropsiden. In L. Bolk, E. Göppert, E. Kallius, und W. Lubosch, *Handb. vergl. Anat. Wirbelt.* **1**, 375–448.

LAURENS, H. (1914a). The reactions of normal and eyeless amphibian larvae to light. *Science,* **39**, 471.

LAURENS, H. (1914b). The reactions of the melanophores of *Amblystoma* larvae. *Proc. Soc. Exp. Biol., N.Y.,* **12**, 31–2.

LAURENS, H. (1914c). The reactions of normal and eyeless amphibian larvae to light. *J. Exp. Zool.* **16**, 195–210.

LAURENS, H. (1915). The reactions of the melanophores of *Amblystoma* larvae. *J. Exp. Zool.* **18**, 577–638.

LAURENS, H. (1916). The reactions of the melanophores of *Amblystoma* larvae. The supposed influence of the pineal organ. *J. Exp. Zool.* **20**, 237–61.

LAURENS, H. (1917). The reactions of the melanophores of *Amblystoma tigrinum* larvae to light and darkness. *J. Exp. Zool.* **23**, 195–205.

LEDERER, E. (1935). Sur le pigment rouge de la peau de deux poissons: dorade (*Beryx decadactylus* C.V.) et poisson rouge (*Carassius auratus*). *C.R. Soc. Biol., Paris,* **118**, 542–4.

LEE, R. E. (1942a). Pituitary function in the chromatic physiology of *Opsanus tau. Biol. Bull. Woods Hole,* **83**, 299–300.

LEE, R. E. (1942b). Notes on the color changes of the sea robin (*Prionotus strigatus* Cuvier) with special reference to the erythrophores. *J. Exp. Zool.* **91**, 131–53.

LEHMANN, C. (1923). Farbwechsel bei *Hyperia galba. Biol. Zbl.* **43**, 173–5.

LEINER, M. (1930). Fortsetzung der ökologischen Studien an *Gasterosteus aculeatus. Z. Morph. Ökol. Tiere,* **16**, 499–540.

DE LERMA, B. (1936). L' attivita endocrina negli invertebrati. *Attual. Zool.* **2**, 83–135. (*Arch. zool.* (ital.), Napoli, Suppl. al vol. **23**.)

LESZCZYNSKI, R. J. (1933). Über die Wirkung von Giften von verschiedenem Typus auf die chromatische Funktion der Froschhaut. *Arch. int. Pharmacodyn.* **45**, 89–112.

LEWIS, D., LEE, F. G. and ASTWOOD, E. B. (1937). Some observations on intermedin. *Bull. Johns Hopk. Hosp.* **61**, 198–209.

LEWIS, M. R. (1932). The disappearance of the pigmentation of the eye and the skin of tadpoles (*Rana sylvatica*) that develop in solutions of indophenol dyes. *J. Exp. Zool.* **64**, 57–69.

LEWIS, M. R. (1936). Further studies on the hypophysis cerebri of certain selachians by means of tissue cultures. *Bull. Mt Desert Isl. Biol. Lab.* p. 22.

LEWIS, M. R. and BUTCHER, E. O. (1936*a*). The separation of the hypophysis cerebri of certain selachians (*Squalus acanthias* and *Raja strabuliforis*) into six distinct lobes. *Bull. Mt Desert Isl. Biol. Lab.* pp. 16–18.

LEWIS, M. R. and BUTCHER, E. O. (1936*b*). The melanophore hormone of the hypophysis cerebri of certain selachians. *Bull. Mt Desert Isl. Biol. Lab.* pp. 20–1.

LINDEMAN, V. F. (1928). Integumentary pigmentation in the frog (*R. pipiens*) during metamorphosis, with especial reference to tail-skin histolysis. *Anat. Rec.* **41**, 39.

LISSÁK, K. (1939). Liberation of acetylcholine and adrenaline by stimulating isolated nerves. *Amer. J. Physiol.* **127**, 263–71.

LOEWI, O. (1935). The Ferrier lecture on problems connected with the principle of humoral transmission of nervous impulses. *Proc. Roy. Soc.* B, **118**, 299–316.

LOEWI, O. (1936). Quantitative und qualitative Untersuchungen über den Sympathicusstoff. *Pflüg. Arch. ges. Physiol.* **237**, 504–14.

LOEWI, O. (1937*a*). Über eine wasserunlösliche Zustandsform des Acetylcholins im Zentralnervensystem des Frosches. *Pflüg. Arch. ges. Physiol.* **239**, 430–9.

LOEWI, O. (1937*b*). *Die chemische Übertragung der Nervenwirkung.* Les Prix Nobel en 1937. 14 pp. Stockholm.

LOEWI, O. (1937*c*). *Über portganglionäre Gefässwirkungen von Nikotin und Acetylcholin.* pp. 349–56. Volume Jubilaire J. Demoor.

LOISON, C. (1923). Modifications des chromatophores de la grenouille sous l'influence de diverses substances. *C.R. Soc. Biol., Paris,* **88**, 799–800.

LONGLEY, W. H. (1924). A new index to the function of coloration in fishes. *Anat. Rec.* **29**, 93.

LÖNNBERG, E. (1930*a*). Einige Studien über die Lipochrome der Fische. *Ark. Zool.* **21** A, no. 10, 29 pp.

LÖNNBERG, E. (1930*b*). Zur Kenntnis des gelben Farbstoffes in der Haut einiger Batrachier und einer Eidechse. *Ark. Zool.* **21** B, no. 3, 4 pp.

LÖNNBERG, E. (1932*a*). Zur Kenntnis der Carotinoide bei marinen Evertebraten. *Ark. Zool.* **23** A, no. 15, 74 pp.

LÖNNBERG, E. (1932*b*). Some observations on carotinoid colour substances of fishes. *Ark. Zool.* **23** A, no. 16, 11 pp.

LÖNNBERG, E. (1933*a*). Zur Kenntnis der Carotinoide bei marinen Evertebraten. II. *Ark. Zool.* **25** A, no. 1, 17 pp.

LÖNNBERG, E. (1933*b*). Some observations on carotinoid colour substances of fishes. II. *Ark. Zool.* **25** A, no. 2, 8 pp.

LÖNNBERG, E. (1934*a*). Note on the carotinoids of the hagfish, *Myxine glutinosa. Ark. Zool.* **26** A, no. 3, 3 pp.

LÖNNBERG, E. (1934*b*). Weitere Beiträge zur Kenntnis der Carotinoide der marinen Evertebraten. *Ark. Zool.* **26** A, no. 7, 36 pp.

LÖNNBERG, E. (1936). On the occurrence of carotinoid substances in cephalopods. *Ark. Zool.* **28** B, no. 8, 4 pp.

LORENTE DE NO, R. (1935). The electrical excitability of the motoneurones. *J. Cell. Comp. Physiol.* **7**, 47–71.

LORENTE DE NO, R. (1938). Liberation of acetylcholine by the superior cervical sympathetic ganglion and the nodosum ganglion of the vagus. *Amer. J. Physiol.* **121**, 331–49.

LOWE, J. N. (1917). The action of various pharmacological and other chemical agents on the chromatophores of the brook trout *Salvelinus fontinalis* Mitchill. *J. Exp. Zool.* **23**, 147–93.

LÜ, Y. M. (1939). On the question of a myelencephalic sympathetic centre. VIII. Further studies in the fish. *Chin. J. Physiol.* **14**, 225–30.

LUNDSTROM, H. M. and BARD, P. (1932). Hypophysial control of cutaneous pigmentation in an elasmobranch fish. *Biol. Bull. Woods Hole*, **62**, 1–9.

LWOFF, A. (1927). Le cycle du pigment carotinoïde chez *Idya furcata* (Baird). *Bull. Biol.* **61**, 193–240.

LYNN, W. G. (1940). Effects of implantation of extra pituitaries in *Rana pipiens*. *Anat. Rec.* **78**, Suppl. p. 139.

MACBRIDE, E. W. (1925). The influence of the colour of the background on the colour of the skin of *Salamandra maculosa*. *Proc. Zool. Soc. Lond.* pp. 983–93.

McCORD, C. P. and ALLEN, F. P. (1917). Evidences associating pineal gland function with alteration in pigmentation. *J. Exp. Zool.* **23**, 207–24.

McLEAN, A. J. (1928). The anuran in bio-titration of pituitrin. *J. Pharmacol.* **33**, 301–19.

MARSLAND, D. A. (1940). The effect of high hydrostatic pressure on the melanophores of the isolated scales of *Fundulus heteroclitus*. *Anat. Rec.* **78**, Suppl. p. 168.

MARSLAND, D. A. (1942a). Contractile mechanism in unicellular pigmentary effectors. *Collect. Net*, **17**, 81–3.

MARSLAND, D. A. (1942b). The contractile mechanism in unicellular chromatophores (melanophores of *Fundulus*). *Biol. Bull. Woods Hole*, **83**, 292.

MARTINI, E. und ACHUNDOW, I. (1929). Versuche über Farbenanpassung bei Culiciden. *Zool. Anz.* **81**, 25–44.

MARX, L. (1929). Entwicklung und Ausbildung des Farbkleides beim Feuersalamander nach Verlust der Hypophyse. *Roux Arch. Entw. Mech. Organ.* **114**, 512–48.

MAST, S. O. (1916). Changes in shade, color, and pattern in fishes and their bearing on the problems of adaptation and behavior, with especial reference to the flounders *Paralichthys* and *Ancylopsetta*. *Bull. U.S. Bur. Fish.* **34**, 173–238.

MAST, S. O. (1933). 'Expansion and contraction' of chromatophores. *Science*, **78**, 435–6.

MAST, S. O. (1934). Movement of pigment granules in chromatophores. *Science*, **79**, 249.

MATSUSHITA, K. (1938). Studies on the color changes of the catfish, *Parasilurus asotus* (L.). *Sci. Rep. Tôhoku Univ. Sendai*, 4 Ser. Biol. **13**, 171–200.

MATTHEWS, S. A. (1931). Observations on pigment migration within the fish melanophore. *J. Exp. Zool.* **58**, 471–86.

MATTHEWS, S. A. (1933). Color changes in *Fundulus* after hypophysectomy. *Biol. Bull. Woods Hole*, **64**, 315–20.

MATTHEWS, S. A. (1936). The pituitary gland of *Fundulus*. *Anat. Rec.* **65**, 357–67.

MATTHEWS, S. A. (1937). The development of the pituitary gland in *Fundulus*. *Biol. Bull. Woods Hole*, **73**, 93–8.

MATTHEWS, S. A. (1940). The effects of implanting adult hypophyses into sexually immature *Fundulus*. *Biol. Bull. Woods Hole*, **79**, 207–14.

MAY, R. M. (1924). Skin grafts in the lizard, *Anolis carolinensis* Cuv. *Brit. J. Exp. Biol.* **1**, 539–55.

MEGUŠAR, F. (1912). Experiments über den Farbwechsel der Crustaceen (I. *Gelasimus*. II. *Potamobius*. III. *Palaemonetes*. IV. *Palaemon*). *Arch. EntwMech. Org.* **33**, 462–665.

MENDES, E. G. (1942). Respostas dos melanóforos de traíra (*Hoplias malabaricus*) a vários excitantes. *Bol. Fac. Filos. Cien. Letr. Univers. Sao Paulo*, **15**, Zool. 6, 285–99.

MENKE, H. (1911). Periodische Bewegungen und ihr Zusammenhang mit Licht und Stoffwechsel. *Pflüg. Arch. ges. Physiol.* **140**, 37–91.

MERIAN, L. (1913). In welchem Sinne vermag Licht von verschiedenen Wellenlängen die Pigmentbildung im Froschlarvenschwanz zu beeinflussen? *Arch. Anat. Physiol., Lpz.* (Abt. Physiol.), pp. 57–76.

MEYER, E. (1930). Ueber die Mitwirkung von Hormonen beim Farbwechsel der Fische. *Forsch. Fortschr. dtsch. Wiss.* Jahrg. 6, 379–80.

MEYER, E. (1931). Versuche über den Farbwechsel von *Gobius* und *Pleuronectes*. *Zool. Jb.* (Abt. allg. Zool. Physiol.), **49**, 231–70.

MEYER, H. H. (1939). Über ein Hormon der Fischretina. *Endokrinologie*, **22**, 261–79.

MILLOT, J. (1922a). Signification biologique de l'argenture des poissons. *Bull. Soc. Zool. Fr.* **47**, 194–8.

MILLOT, J. (1922b). Formation des iridocytes chez les batraciens. *C.R. Soc. Biol.*, *Paris*, **87**, 26–8.

MILLOT, J. (1922c). Contribution à la physiologie du pigment purique chez les vertébrés inférieurs. *C.R. Soc. Biol.*, *Paris*, **87**, 63–5.

MILLOT, J. (1923a). Le pigment purique chez les vertébrés inférieurs. *Bull. Biol.* **57**, 261–363.

MILLOT, J. (1923b). Influence de l'alimentation sur la pigmentation des vertébrés inférieurs. *C.R. Ass. Anat.* **18**, 361–5.

MILLOT, J. (1929). Les cellules pigmentaires de la peau chez la grenouille. *Bull. Histol. Tech. micr.* **6**, 412–15.

MILLS, S. M. (1932a). Double innervation of melanophores. *Proc. Nat. Acad. Sci.*, *Wash.*, **18**, 538–40.

MILLS, S. M. (1932b). Neuro-humoral control of fish melanophores. *Proc. Nat. Acad. Sci.*, *Wash.*, **18**, 540–3.

MILLS, S. M. (1932c). The double innervation of fish melanophores. *J. Exp. Zool.* **64**, 231–44.

MILLS, S. M. (1932d). Evidence for a neurohumoral control of fish melanophores. *J. Exp. Zool.* **64**, 245–55.

MINKIEWICZ, R. (1933). Rôle des facteurs optiques dans les changements de livrée, chez les grenouilles adultes. *Acta Biol. exp.*, *Varsovie*, **8**, 102–77.

DE MIOMANDRE, F. (1938). *Mon caméléon.* 219 pp. Paris.

MOSAUER, W. (1936). The toleration of solar heat in desert reptiles. *Ecology*, **17**, 56–66.

MOSAUER, W. and LAZIER, E. L. (1933). Death from isolation in desert snakes. *Copeia*, p. 149.

MURISIER, P. (1913). Note sur les chromatocytes intraépidermiques des amphibiens. *C.R. Ass. Anat.* **15**, 232–9.

MURISIER, P. (1920–1). Le pigment mélanique de la truite (*Salmo lucustris* L.) et le mécanisme de sa variation quantitative sous l'influence de la lumière. *Rev. suisse Zool.* **28**, 45–97, 149–95, 243–99 (= 343–99).

MUZLERA, J. M. (1934). Acción de la temperatura sobre la pigmentación de *Jenynsia lineata* (Jenyns) Günther. *Rev. Soc. argent. Biol.* **10**, 369–70.

NADLER, J. E. (1927). A quantitative comparison and toxicological study of ephedrine and epinephrin. *Chin. J. Physiol.* **1**, 271–6.

NAGEOTTE, J. (1914). Histologie comparée de la peau des têtards d'anoures. *C.R. Soc. Biol.*, *Paris*, **77**, 323–8.

NAVEZ, A. E. and KROPP, B. (1934). The growth-promoting action of crustacean eye-stalk extract. *Biol. Bull. Woods Hole*, **67**, 250–8.

NEILL, R. M. (1940). On the existence of two types of chromatic behaviour in teleostean fishes. *J. Exp. Biol.* **17**, 74–98.

NEWMANN, H. H. (1918). Hybrids between *Fundulus* and mackerel. *J. Exp. Zool.* **26**, 391–421.

NIGRELLI, R. F. (1934). Pseudo-melanosis in the tail of trout and salmon. *Copeia*, pp. 61–6.

NOBLE, G. K. and BRADLEY, H. T. (1933). The relation of the thyroid and the hypophysis to the molting process in the lizard, *Hemidactylus Brookii*. *Biol. Bull. Woods Hole*, **64**, 289–98.

OCHOA, S. (1928). The action of guanidins on the melanophores of the skin of the frog (*Rana temporaria*). *Proc. Roy. Soc.* B, **102**, 256–63.

ODIORNE, J. M. (1933*a*). Degeneration of melanophores in *Fundulus*. *Proc. Nat. Acad. Sci., Wash.*, **19**, 329–32.

ODIORNE, J. M. (1933*b*). The effects of the pituitary hormones on the melanophores of fishes. *Proc. Nat. Acad. Sci., Wash.*, **19**, 745–9.

ODIORNE, J. M. (1933*c*). The occurrence of guanophores in *Fundulus*. *Proc. Nat. Acad. Sci., Wash.*, **19**, 750–4.

ODIORNE, J. M. (1936). The degeneration of melanophores in *Fundulus*. *J. Exp. Zool.* **74**, 7–39.

ODIORNE, J. M. (1937). Morphological color changes in fishes. *J. Exp. Zool.* **76**, 441–65.

O'DONNELL, A. M. (1927). A study of the effect of measured doses of ultra-violet radiation on the irritability of melanophores. *Amer. J. Physiol.* **83**, 254–68.

OGNEFF, J. F. (1911). Über die Aenderungen in den Organen der Goldfische nach dreijährigem Verbleiben im Finsternis. *Anat. Anz.* **40**, 81–7.

OHASHI, Y. (1921*a*). Der Farbensinn bei Fischen (Karpfen und Goldfische). *J. Coll. Agric. Tokyo*, **6**, 403–41.

OHASHI, Y. (1921*b*). Einige Bemerkungen über den Farbenwechsel bei Karpfen durch physikalische Einflüsse. *J. Coll. Agric. Tokyo*, **6**, 443–57.

OKAMOTO, T. (1937). Hat die Hypophysenexstirpation bei der Kröte auf die Verschiebung der Netzhautpigmente irgendeiner Einfluss? *Z. ges. exp. Med.* **101**, 155–65.

OLDHAM, F. K., LAST, J. H. and GEILING, E. M. K. (1940). Distribution of melanophore-dispersing hormone in anterior lobe of cetaceans and armadillo. *Proc. Soc. Exp. Biol., N.Y.*, **43**, 407–10.

OLDHAM, F. K., MCCLEERY, D. P. and GEILING, E. M. K. (1938). A note on the histology and pharmacology of the hypophysis of the manatee (*Trichechus inunguis*). *Anat. Rec.* **71**, 27–32.

OMURA, K. (1930). Über die melanophoreausbreitende Wirkung des Pituitrins und einiger anderer Organpräparate. *Jap. J. Med. Sci.* IV, Pharmacol. **4**, 84. (Cited from *Ber. ges. Physiol.* **62**, 438, 1931.)

OPPENHEIMER, J. M. (1937). The normal stages of *Fundulus heteroclitus*. *Anat. Rec.* **68**, 1–15.

OSBORN, C. M. (1936). The inhibition of molting in Urodeles following thyroidectomy or hypophysectomy. *Anat. Rec.* **66**, 257–69.

OSBORN, C. M. (1938*a*). The role of the melanophore-dispersing hormone of the pituitary in the color changes of the catfish. *Proc. Nat. Acad. Sci., Wash.*, **24**, 121–5.

OSBORN, C. M. (1938*b*). The role of the melanophore-dispersing principle of the pituitary in the color change of the catfish. *J. Exp. Zool.* **79**, 309–30.

OSBORN, C. M. (1939*a*). The experimental production of melanin pigment on the ventral surface of the summer flounder (*Paralichthys dentatus*). *Anat. Rec.* **75**, Suppl. p. 60.

OSBORN, C. M. (1939*b*). The physiology of color changes in flatfishes. *J. Exp. Zool.* **81**, 479–515.

OSBORN, C. M. (1939*c*). The effects of partial and total blinding on the color changes of the summer flounder (*Paralichthys dentatus*). *Anat. Rec.* **75**, Suppl. p. 136.

OSBORN, C. M. (1940*a*). The experimental production of melanin pigment on the lower surface of summer flounders (*Paralichthys dentatus*). *Proc. Nat. Acad. Sci., Wash.*, **26**, 155–61.

OSBORN, C. M. (1940*b*). The growth of melanophores on the normally unpigmented surface of the black catfish, *Ameiurus melas*. *Anat. Rec.* **78**, Suppl. p. 167.

Osborn, C. M. (1941 a). Factors influencing the pigmentation of regenerating scales on the ventral surface of the summer flounder. *Biol. Bull. Woods Hole*, **81**, 301–2.

Osborn, C. M. (1941 b). Studies on the growth of integumentary pigment in the lower vertebrates. I. The origin of artificially developed melanophores on the normally unpigmented ventral surface of the summer flounder (*Paralichthys dentatus*). *Biol. Bull. Woods Hole*, **81**, 341–51.

Osborn, C. M. (1941 c). Studies on the growth of integumentary pigment in the lower vertebrates. II. The rôle of the hypophysis in melanogenesis in the common catfish (*Ameiurus melas*). *Biol. Bull. Woods Hole*, **81**, 352–63.

Osborn, C. M. (1941 d). The effect of hypophysectomy upon the growth of melanophores in the catfish (*Ameiurus melas*). *Anat. Rec.* **81**, Suppl. pp. 59–60.

Osterhage, K. H. (1932). Morphologische und physiologische Studien an Pigmentzellen der Fische. *Z. mikr.-anat. Forsch.* **30**, 551–98.

Parhon, C. I. and Cahane, T. (1935). Recherches sur l'antagonisme intermédine-adrénaline et sur les rapports de ces substances ou des substances qui dilatent les mélanophores, avec les glandes génitales et la glande thyroïde. *Bull. Sect.*

Parhon, C. I. et Caraman, Z. (1930). Sur les cellules mélanophores du lobe intermédiaire et les cellules lipoïdophores du lobe postérieur de l'hypophyse du rat pie. *C.R. Soc. Biol., Paris*, **103**, 283–4. *Endocrin. Soc. Roum. Neur.* **1**, 59–61.

Parker, G. H. (1930). The color changes of the tree toad in relation to nervous and humoral control. *Proc. Nat. Acad. Sci., Wash.*, **16**, 395–6.

Parker, G. H. (1931 a). The color changes in the sea-urchin *Arbacia*. *Proc. Nat. Acad. Sci., Wash.*, **17**, 594–6.

Parker, G. H. (1931 b). Effects of acetyl choline on chromatophores. *Proc. Nat. Acad. Sci., Wash.*, **17**, 596–7.

Parker, G. H. (1932 a). *Humoral Agents in Nervous Activity with special reference to Chromatophores.* 79 pp. Cambridge.

Parker, G. H. (1932 b). Some aspects of neurohumoralism. *Science*, **76**, 543.

Parker, G. H. (1932 c). Transfusion of neurohumoral substances to chromatophores by other means than blood and lymph. *Anat. Rec.* **54**, Suppl. p. 34.

Parker, G. H. (1933 a). The cellular transmission of neurohumoral substances in melanophore reactions. *Proc. Nat. Acad. Sci., Wash.*, **19**, 175–7.

Parker, G. H. (1933 b). Transmission of neurohumors in animals by other means than blood and lymph. *Proc. Soc. Exp. Biol., N.Y.*, **30**, 555–8.

Parker, G. H. (1933 c). The color changes of elasmobranch fishes. *Proc. Nat. Acad. Sci., Wash.*, **19**, 1038–9.

Parker, G. H. (1934 a). Cellular transfer of substances, especially neurohumors. *J. Exp. Biol.* **11**, 81–8.

Parker, G. H. (1934 b). What part of the melanophore system in *Fundulus* is acted upon by adrenalin? *J. Cell. Comp. Physiol.* **5**, 311–18.

Parker, G. H. (1934 c). The expansion and contraction of chromatophores. *Science*, **79**, 428–9.

Parker, G. H. (1934 d). Neurohumors as activating agents for fish melanophores. *Proc. Amer. Phil. Soc.* **74**, 177–84.

Parker, G. H. (1934 e). Color changes of the catfish *Ameiurus* in relation to neurohumors. *J. Exp. Zool.* **69**, 199–223.

Parker, G. H. (1934 f). The prolonged activity of momentarily stimulated nerves. *Proc. Nat. Acad. Sci., Wash.*, **20**, 306–10.

Parker, G. H. (1934 g). Acetyl choline and chromatophores. *Proc. Nat. Acad. Sci., Wash.*, **20**, 596–9.

Parker, G. H. (1934 h). Oil-soluble and water-soluble neurohumors. *Anat. Rec.* **60**, Suppl. p. 30.

PARKER, G. H. (1935a). The electric stimulation of the chromatophoral nerve-fibers in the dogfish. *Biol. Bull. Woods Hole*, **68**, 1-3.
PARKER, G. H. (1935b). The chromatophoral neurohumors of the dogfish. *J. Gen. Physiol.* **18**, 837-46.
PARKER, G. H. (1935c). The disappearance of primary caudal bands in the tail of *Fundulus* and its relation to the neurohumoral hypothesis. *Proc. Amer. Phil. Soc.* **75**, 1-10.
PARKER, G. H. (1935d). Neurohumors: novel agents in the action of the nervous system. *Science*, **81**, 279-83.
PARKER, G. H. (1935e). An oil-soluble neurohumour in the catfish *Ameiurus*. *J. Exp. Biol.* **12**, 239-45.
PARKER, G. H. (1935f). What are the resting and the active states of chromatophores? *Proc. Nat. Acad. Sci., Wash.*, **21**, 286-92.
PARKER, G. H. (1935g). The cellular transmission of substances, especially neurohumors. *Quart. Rev. Biol.* **10**, 251-71.
PARKER, G. H. (1936a). *Color Changes of Animals in relation to Nervous Activity.* 74 pp. Philadelphia.
PARKER, G. H. (1936b). Integumentary color changes in the newly born dogfish, *Mustelus canis. Biol. Bull. Woods Hole*, **70**, 1-7.
PARKER, G. H. (1936c). The reactions of chromatophores as evidence for neurohumors. *Cold Spr. Harb. Symp. Quant. Biol.* **4**, 358-70.
PARKER, G. H. (1936d). Are there antidromic responses in the melanophore system? *Anat. Rec.* **67**, Suppl. 1, p. 37.
PARKER, G. H. (1936e). Color changes in elasmobranchs. *Proc. Nat. Acad. Sci., Wash.*, **22**, 55-60.
PARKER, G. H. (1936f). The reactivation by cutting of severed melanophore nerves in the dogfish *Mustelus. Biol. Bull. Woods Hole*, **71**, 255-8.
PARKER, G. H. (1937a). The control of color changes in the dogfish *Mustelus* by neurohumors. *Anat. Rec.* **70**, Suppl. 1, pp. 56-7.
PARKER, G. H. (1937b). Integumentary color changes of elasmobranch fishes especially of *Mustelus. Proc. Amer. Phil. Soc.* **77**, 223-47.
PARKER, G. H. (1937c). Color changes due to erythrophores in the squirrel fish *Holocentrus. Proc. Nat. Acad. Sci., Wash.*, **23**, 206-11.
PARKER, G. H. (1937d). The relation of melanophore responses to vascular disturbances. *Biol. Bull. Woods Hole*, **73**, 374.
PARKER, G. H. (1937e). A catalogue of neurohumors. *Science*, **85**, 436-7.
PARKER, G. H. (1937f). Antidromic responses from the melanophore nerves of the catfish *Ameiurus. Proc. Nat. Acad. Sci., Wash.*, **23**, 595-6.
PARKER, G. H. (1937g). Antagonism in neurohumors as seen in the pectoral bands of *Mustelus. Proc. Nat. Acad. Sci., Wash.*, **23**, 596-600.
PARKER, G. H. (1937h). Do melanophore nerves show antidromic responses? *J. Gen. Physiol.* **20**, 851-8.
PARKER, G. H. (1938a). The colour changes in lizards, particularly in *Phrynosoma. J. Exp. Biol.* **15**, 48-73.
PARKER, G. H. (1938b). Melanophore responses in the young of *Mustelus canis. Proc. Amer. Acad. Arts Sci.* **72**, 269-82.
PARKER, G. H. (1938c). Melanophore responses and blood supply (vasomotor changes). *Proc. Amer. Phil. Soc.* **78**, 513-27.
PARKER, G. H. (1939a). The eye in relation to chromatophoral color changes in animals. *Science*, **89**, 411.
PARKER, G. H. (1939b). The relation of the eyes to the integumentary color changes in the catfish *Ameiurus. Proc. Nat. Acad. Sci., Wash.*, **25**, 499-502.
PARKER, G. H. (1939c). Color responses of catfishes with single eyes. *Biol. Bull. Woods Hole*, **77**, 312-13.
PARKER, G. H. (1939d). Color changes in animals, their significance and activation. Introductory remarks. *Amer. Nat.* **73**, 193-7.

PARKER, G. H. (1939e). The active and the resting states of melanophores tested experimentally. *Anat. Rec.* **75**, Suppl. p. 61.

PARKER, G. H. (1940a). Types of animal reflexes. *Science*, **91**, 216.

PARKER, G. H. (1940b). Novel types of nerve reflexes. *Proc. Nat. Acad. Sci., Wash.*, **26**, 246–9.

PARKER, G. H. (1940c). On the neurohumors of the color changes in catfishes and on fats and oils as protective agents for such substances. *Proc. Amer. Phil. Soc.* **83**, 379–409.

PARKER, G. H. (1940d). Neurohumors as chromatophore activators. *Proc. Amer. Acad. Arts Sci.* **73**, 165–95.

PARKER, G. H. (1940e). The chromatophore system in the catfish *Ameiurus*. *Biol. Bull. Woods Hole*, **79**, 237–51.

PARKER, G. H. (1940f). Lipoids and their probable relation to melanophore activity. *Biol. Bull. Woods Hole*, **79**, 355–6.

PARKER, G. H. (1940g). The active and the resting states of catfish melanophores tested experimentally. *J. Cell. Comp. Physiol.* **15**, 137–46.

PARKER, G. H. (1941a). Melanophore bands and areas due to nerve cutting, in relation to the protracted activity of nerves. *J. Gen. Physiol.* **24**, 483–504.

PARKER, G. H. (1941b). The activity of peripherally stored neurohumors in catfishes. *J. Gen. Physiol.* **25**, 177–84.

PARKER, G. H. (1941c). The method of activation of melanophores and the limitations of melanophore responses in the catfish *Ameiurus*. *Proc. Amer. Phil. Soc.* **85**, 18–24.

PARKER, G. H. (1941d). The responses of catfish melanophores to ergotamine. *Biol. Bull. Woods Hole*, **81**, 163–7.

PARKER, G. H. (1941e). The organization of the melanophore system in bony fishes. *Biol. Bull. Woods Hole*, **81**, 280.

PARKER, G. H. (1941f). Hypersensitization of catfish melanophores to adrenaline by denervation. *Biol. Bull. Woods Hole*, **81**, 302.

PARKER, G. H. (1942a). Color changes in *Mustelus* and other elasmobranch fishes. *J. Exp. Zool.* **89**, 451–73.

PARKER, G. H. (1942b). Sensitization of melanophores by nerve cutting. *Proc. Nat. Acad. Sci., Wash.*, **28**, 164–70.

PARKER, G. H. (1943a). Methods of estimating the effects of melanophore changes on animal coloration. *Biol. Bull. Woods Hole*, **84**, 273–84.

PARKER, G. H. (1943b). Animal color changes and their neurohumors. *Quart. Rev. Biol.* **18**, 205–27.

PARKER, G. H. (1943c). Color changes in the American eel *Anguilla rostrata*. *Anat. Rec.* **87**, 463–4.

PARKER, G. H. (1943d). The time factor in chromatophore responses. *Proc. Amer. Phil. Soc.* **87**, 429–34.

PARKER, G. H. and BROWER, H. P. (1935). A nuptial secondary sex-character in *Fundulus heteroclitus*. *Biol. Bull. Woods Hole*, **68**, 4–6.

PARKER, G. H. and BROWER, H. P. (1937). An attempt to fatigue the melanophore system in *Fundulus* and a consideration of lag in melanophore responses. *J. Cell. Comp. Physiol.* **9**, 315–29.

PARKER, G. H., BROWN, F. A., Jr. and ODIORNE, J. M. (1935). The relation of the eyes to chromatophoral activities. *Proc. Amer. Acad. Arts Sci.* **69**, 439–62.

PARKER, G. H. and LANCHNER, A. J. (1922). The responses of *Fundulus* to white, black and darkness. *Amer. J. Physiol.* **61**, 548–50.

PARKER, G. H. and PORTER, H. (1933). Regeneration of chromatophore nerves. *J. Exp. Zool.* **66**, 303–9.

PARKER, G. H. and PORTER, H. (1934). The control of the dermal melanophores in elasmobranch fishes. *Biol. Bull. Woods Hole*, **66**, 30–7.

PARKER, G. H. and PUMPHREY, S. M. (1936). The relation of nerves to chromatophore pulsations. *J. Cell. Comp. Physiol.* **7**, 325–31.

PARKER, G. H. and ROSENBLUETH, A. (1941). The electric stimulation of the concentrating (adrenergic) and the dispersing (cholinergic) nerve-fibres of the melanophores in the catfish. *Proc. Nat. Acad. Sci., Wash.*, **27**, 198–204.

PARKER, G. H. and SCATTERTY, L. E. (1937). The number of neurohumors in the control of frog melanophores. *J. Cell. Comp. Physiol.* **9**, 297–314.

PARKER, H. W. (1935). A new melanic lizard from Transjordania, and some speculations concerning melanism. *Proc. Zool. Soc. Lond.* pp. 137–42.

PAULI, W. F. (1926). Versuche über den physiologischen Farbenwechsel der Salamanderlarve und der Pfrille. *Z. wiss. Zool.* **128**, 421–508.

PEABODY, E. B. (1939). Pigmentary responses in the isopod, *Idothea*. *J. Exp. Zool.* **82**, 47–83.

PEARSON, J. F. W. (1930). Changes in pigmentation exhibited by the fresh-water catfish, *Ameiurus melas*, in response to differences in illumination. *Ecology*, **11**, 703–12.

PECZENIK, O. (1933). Über den Mechanismus der Intermedinreaktion. *Z. vergl. Physiol.* **19**, 84–93.

PERKINS, E. B. (1928). Color changes in crustaceans, especially in *Palaemonetes*. *J. Exp. Zool.* **50**, 71–105.

PERKINS, E. B. and COLE, W. H. (1938). A cinematographic method of studying dispersion and concentration of chromatophoric pigment. *Anat. Rec.* **72**, 115.

PERKINS, E. B. and KROPP, B. (1932a). The crustacean eye hormone and its effect upon the chromatophores of crustaceans, fishes and amphibians. *Anat. Rec.* **54**, 34.

PERKINS, E. B. and KROPP, B. (1932b). The crustacean eye hormone as a vertebrate melanophore activator. *Biol. Bull. Woods Hole*, **63**, 108–12.

PERKINS, E. B. and KROPP, B. (1933). The occurrence of the humoral chromatophore activator among marine crustaceans, and its effect upon the chromatophores of crustaceans, fishes, and amphibia. *Rep. Mt Desert Isl. Biol. Lab.* pp. 24–6.

PERKINS, E. B. and SNOOK, T. (1931). Control of pigment migration in the chromatophores of crustaceans. *Proc. Nat. Acad. Sci., Wash.*, **17**, 282–5.

PERKINS, E. B. and SNOOK, T. (1932). The movement of pigment within the chromatophores of *Palaemonetes*. *J. Exp. Zool.* **61**, 115–28.

PERNITZSCH, F. (1913). Zur Analyse der Rassenmerkmale der Axolotl. I. Die Pigmentierung junger Larven. *Arch. mikr. Anat.* **82**, 148–205.

PEROTTI, P. (1928). Innervazione nella cute dei batraci. *Atti Soc. ital. Sci. nat.* **67**, 193–207.

PIERCE, M. E. (1939a). Activity of melanophores in response to injections of adrenalin. *Anat. Rec.* **75**, Suppl. p. 61.

PIERCE, M. E. (1939b). Activity of melanophores in two lizards, *Phrynosoma douglassii* and *P. modestum*. *Anat. Rec.* **75**, Suppl. p. 137.

PIERCE, M. E. (1939c). Activity of melanophores in a teleost, *Mollienesia latipinna*. *Anat. Rec.* **75**, Suppl. p. 137.

PIERCE, M. E. (1941a). Response of melanophores of the skin to injections of adrenalin, with special reference to body weight of the animal. *J. Exp. Zool.* **86**, 189–203.

PIERCE, M. E. (1941b). The activity of the melanophores of a teleost *Mollienisia latipinna*, to light, heat, and anesthetics. *J. Exp. Zool.* **87**, 1–15.

PIERCE, M. E. (1941c). Activity of melanophores in an amphibian, *Rana clamitans*, with special reference to injections of adrenalin. *Anat. Rec.* **81**, Suppl. pp. 92–3.

PIERCE, M. E. (1942). The activity of the melanophores of an amphibian, *Rana clamitans*, with special reference to the effect of injection of adrenalin in relation to body weight. *J. Exp. Zool.* **89**, 283–95.

PIÉRON, H. (1913). Le mécanisme de l'adaptation chromatique et la livrée nocturne de l'*Idotea tricuspidata* Desm. *C.R. Acad. Sci., Paris*, **157**, 951–3.

PIÉRON, H. (1914). Recherches sur le comportement chromatique des invertébrés et en particulier des isopodes. *Bull. Sci. Fr. Belg.* **48**, 30–79.

PLATTNER, F. und KRANNICH, E. (1932). Über das Vorkommen eines acetyl-cholinartigen Körpers in den Skeletmuskeln. *Pflüg. Arch. ges. Physiol.* **229**, 730–7.

POLIMANTI, O. (1912). Einfluss der Augen und der Bodenbeschaffenheit auf die Farbe der Pleuronectiden. *Biol. Zbl.* **32**, 296–307.

POMINI, F. P. (1938). Primi risultati di incroci fra due razze di trota iridea gialla e bruna. I. Comportamento della pigmentazione melanica cutanea. *Boll. Soc. ital. Biol. Sper.* **13**, 812–13.

POPA, G. T. et FIELDING, U. (1933). L'expansion des mélanophores sous l'action de l'extrait du lobe antérieur de l'hypophyse mélangé avec des globules rouges. *C.R. Soc. Biol., Paris*, **114**, 1139–40.

POTTER, G. E. and BROWN, S. O. (1941). Effect of sex and gonadotropic hormones on the development of the gonads in *Phrynosoma cornutum* during reproductive and non-reproductive phases. *Anat. Rec.* **81**, Suppl. p. 59.

PRENANT, A. (1914). Développement du 'réseau d'Asvadourova' chez les têtards d'alyte. *C.R. Soc. Biol., Paris*, **77**, 236–8.

PRENANT, A. (1920). Sur les phénomènes de la pigmentation chez les larves d'anoures. *C.R. Soc. Biol., Paris*, **83**, 839–42.

PRENANT, A. (1923). Recherches sur le développement du réseau pigmentaire et des autres chromatocytes chez les larves de batraciens anoures. *Arch. Anat., Strasbourg*, **2**, 461–504.

PRIEBATSCH, I. (1933). Der Einfluss des Lichtes auf Farbwechsel und Phototaxis von *Dixippus (Carausius) morosus*. *Z. vergl. Physiol.* **19**, 453–88.

PRZIBRAM, H. (1922). Verpuppung kopfloser Raupen von Tagfaltern. *Arch. EntwMech. Org.* **50**, 203–8.

PRZIBRAM, H. (1932a). Einfluss von Infundin- und Adrenalindosierung auf die Färbung unserer Frösche, *Rana esculenta* L., *R. fusca* Rösel, *Hyla arborea* L. *Z. vergl. Physiol.* **17**, 565–73.

PRZIBRAM, H. (1932b). Rolle der Gesichtswahrnehmungen für den Farbwechsel der Frösche *Rana esculenta* L., *R. fusca* Rösel und *Hyla arborea* L. *Z. vergl. Physiol.* **17**, 574–90.

PRZIBRAM, H. und BRECHER, L. (1920). Die Farbmodifikationen der Stabheu-schrecke *Dixippus morosus* Br. et Redt. *Anz. Akad. Wiss. Wien*, Math.-nat. Kl. **57**, 164–5.

PRZIBRAM, H. und BRECHER, L. (1922). Die Farbmodifikationen der Stabheu-schrecke *Dixippus morosus* Br. et Redt. *Arch. EntwMech. Org.* **50**, 147–85.

PRZIBRAM, H. und SUSTER, P. (1931). Fühler- und Beinregeneration bei Phasmiden. VII. Einflusslosigkeit des Fühlerganglions nach Versuchen an *Dixippus* (=*Carausius*) *morosus* Br. et Redt. *Anz. Akad. Wiss. Wien*, Math.-nat. Kl. **68**, 268–9.

PYLE, R. W. (1943). The histogenesis and cyclic phenomena of the sinus gland and X-organ in Crustacea. *Biol. Bull. Woods Hole*, **85**, 87–102.

QUINTO, C. (1933a). Osservazioni sulla regolazione neurohumorale cardiaca. *Boll. Accad. Med.-chir. Perugia*, 1932–3, pp. 24–8.

QUINTO, C. (1933b). Rapporto tra i prodotti di nervi misti eccitati e la soglia di eccitazione cerebrale. *Riv. Biol.* **15**, 502–20.

QUINTO, C. (1933c). Sulla regolazione neuro-umorale cardiaca. *Riv. Biol.* **15**, 299–320.

QUINTO, C. (1935). Liquidi da eccitazione nervosa e loro attività sperimentale. *Riv. Biol.* **19**, 82–103.

RABL, H. (1931). Integument der Anamnier. In Bolk, L., Goppert, E., Kallius, E. und Lubosch, W., *Handb. vergl. Anat. Wirbelt.* **1**, 271–374.

RAHN, H. (1940). The pituitary regulation of melanophores in the rattlesnake. *Anat. Rec.* **78**, Suppl. pp. 138–49.

RAHN, H. (1941 a). The distribution and development of the melanophore hormone in the pituitary of the chick. *Coll. Net*, **16**, 87–9.

RAHN, H. (1941 b). The pituitary regulation of melanophores in the rattlesnake. *Biol. Bull. Woods Hole*, **80**, 228–37.

RAHN, H. and DRAGER, G. A. (1941). Quantitative assay of the melanophore-dispersing hormone during the development of the chick pituitary. *Endocrinology*, **29**, 725–30.

RAHN, H. and ROSENDALE, F. (1941). Diurnal rhythm of melanophore hormone secretion in the *Anolis* pituitary. *Proc. Soc. Exp. Biol., N.Y.*, **48**, 100–2.

RAMON Y CAJAL, S. (1928). *Degeneration and Regeneration of the Nervous System.* Trans. by R. M. May. 2 vols., 792 pp. Oxford Univ. Press.

RAND, H. W. and PIERCE, M. E. '(1932). Skin grafting in frog tadpoles: local specificity of skin and behavior of epidermis. *J. Exp. Zool.* **62**, 125–70.

RASMUSSEN, A. T. (1938). Innervation of the hypophysis. *Endocrinology*, **23**, 263–78.

RASS, T. S. (1937). Pigmentation of embryos and larvae in the herrings (family Clupeidae) as an adaptation to a pelagic mode of life. *Bull. Soc. Nat. Moscou*, **46**, 163–4.

REDFIELD, A. C. (1916). The coordination of chromatophores by hormones. *Science*, **43**, 580–1.

REDFIELD, A. C. (1918). The physiology of the melanophores of the horned toad, *Phrynosoma. J. Exp. Zool.* **26**, 275–333.

REED, H. D. and GORDON, M. (1931). The morphology of melanotic overgrowths in hybrids of Mexican killifishes. *Amer. J. Cancer*, **15**, 1524–46.

REED, S. C., WILLIAMS, C. M. and CHADWICK, L. E. (1942). Frequency of wing-beat as a character for separating species, races and geographic varieties of *Drosophila. Genetics*, **27**, 349–61.

REGAMEY, J. (1935). Les caractères sexuels du lézard (*Lacerta agilis* L.). *Rev. Suisse Zool.* **42**, 87–168.

RFIS, K. (1926). Sur le comportement des greffes de la peau des amphibiens. Métamorphose des greffes de la peau larvaire sur les salamandres adultes. *C.R. Soc. Biol., Paris*, **94**, 349–51.

REMANE, A. (1931). Farbwechsel, Farbrassen und Farbanpassung bei der Meeresassel *Idothea tricuspidata. Verh. dtsch. zool. Ges.* **34**, 109–14.

RODEWALD, W. (1935). Die Wirkung des Lichtes auf die Hypophvse von *Rana temporaria* L. *Z. vergl. Physiol.* **21**, 767–800.

RODEWALD, W. (1939). Über die oestrogene Wirkung von Porphyrinen. *Arch. Exp. Path. Pharmak.* **194**, 76–7.

RODEWALD, W. (1940). Über das Auftreten eines gegen das Melanophorenhormon gerichteten Hemmstoffes während des Sexualzyklus bei der weiblichen weissen Ratte. *Dtsch. med. Wschr.* **66**, 238–40.

ROSENBLUETH, A. (1937). The transmission of sympathetic nerve impulses. *Physiol. Rev.* **17**, 514–37.

ROSENBLUETH, A. and LUCO, J. V. (1937). A study of denervated mammalian skeletal muscle. *Amer. J. Physiol.* **120**, 781–97.

ROSIN, S. (1940). Zur Frage der Pigmentmusterbildung bei Urodelen. *Rev. Suisse Zool.* **47**, 203–8.

ROTH, A. (1932). Über die Melanophorenwirksamkeit des menschlichen Hypophysenvorderlappens. *Zbl. allg. Path. path. Anat.* **54**, 234–42.

ROWE, L. W. (1928). Studies on oxytocin and vasopressin: the effect on frog melanophores. *Endocrinology*, **12**, 663–70.

360   BIBLIOGRAPHY OF COLOUR CHANGES

RUSSELL, A. M. (1939). Pigment inheritance in the *Fundulus-Scomber* hybrid. *Biol. Bull. Woods Hole*, **77**, 316–17.

RUTH, E. S. and GIBSON, R. B. (1917). Disappearance of the pigment in the melanophore of Philippine house lizards. *Philippine J. Sci.* B, **12**, 181–8.

VAN RYNBERK, G. (1911). Kleinere Beiträge zur vergleichenden Physiologie. *Zbl. Physiol.* **24**, 1161–3.

SAITO, T. (1936). Note on the influence of the sex hormone preparations upon the nuptial coloration of fin and the protrusion of oviduct in *Archeilognathus intermedium. Zool. Mag., Japan,* **48**, 503–5.

SAND, A. (1935). The comparative physiology of colour response in reptiles and fishes. *Biol. Rev.* **10**, 361–82.

SANTA, N. et VEIL, C. (1939). Action de la cortine sur la cellule pigmentaire. Possibilité d'utilisation de cette réaction. *C.R. Soc. Biol., Paris,* **131**, 1172–6.

SAPHIR, W. (1934). Artificial production of the 'wedding dress' in *Chrosomus erythrogaster. Proc. Soc. Exp. Biol., N.Y.,* **31**, 864–6.

SASAKI, M. and NAKAMURA, H. (1937). Relation of endocrine system to neoteny and skin pigmentation in a salamander *Hynobius lichenatus* Boulenger. *Annot. zool. jap.* **16**, 81–97.

SASYBIN, N. (1934). Über die Innervation der Pigmentzellen bei Säugetieren. *Z. Zellforsch.* **20**, 476–88.

SATO, M. (1935). Note on the nuptial coloration and pearl organs of *Tribolodon hakonensis* (Günther). *Sci. Rep. Tôhoku Univ.* Ser. 4, Biol. **10**, 499–514.

SAWAYA, P. (1939). Sobra a mudanca da côr nos crustaceos. *Bol. Fac. Filos. Ciê. Let. Univ. S. Paulo,* **13**, 1–109.

SCHAEFER, J. G. (1921). Beiträge zur Physiologie des Farbenwechsels der Fische. I. Untersuchungen an Pleuronectiden. II. Weitere Untersuchungen. *Pflüg. Arch. ges. Physiol.* **188**, 25–48.

SCHÄFER, W. (1937). Bau, Entwicklung und Farbenentstehung bei den Flitterzellen von *Sepia officinalis. Z. Zellforsch.* **27**, 222–45.

SCHARRER, B. (1941a). Neurosecretion. II. Neurosecretory cells in the central nervous system of cockroaches. *J. Comp. Neurol.* **74**, 93–108.

SCHARRER, B. (1941b). Neurosecretion. III. The cerebral organ of the nemerteans. *J. Comp. Neurol.* **74**, 109–30.

SCHARRER, B. (1941c). Neurosecretion. IV. Localization of neurosecretory cells in the central nervous system of *Limulus. Biol. Bull. Woods Hole,* **81**, 96–104.

SCHARRER, E. (1928). Die Lichtempfindlichkeit blinder Elritzen (Untersuchungen über das Zwischenhirn der Fische. I.) *Z. vergl. Physiol.* **7**, 1–38.

SCHARRER, E. (1930). Über sekretorisch tätige Zellen im Thalamus von *Fundulus heteroclitus* L. *Z. vergl. Physiol.* **11**, 767–73.

SCHARRER, E. (1932a). Die Secretproduction im Zwischenhirn einiger Fische. *Z. vergl. Physiol.* **17**, 491–509.

SCHARRER, E. (1932b). Secretory cells in the midbrain of the European minnow (*Phoxinus laevis* L.). *J. Comp. Neurol.* **55**, 573–6.

SCHARRER, E. (1933a). Über neurokrine Organe der Wirbeltiere. *Verh. dtsch. zool. Ges.* **35**, 217–20.

SCHARRER, E. (1933b). Ein inkretorisches Organ im Hypothalamus der Erdkrote, *Bufo vulgaris* Laur. *Z. wiss. Zool.* **144**, 1–11.

SCHARRER, E. (1941). Neurosecretion. I. The nucleus preopticus of *Fundulus heteroclitus* L. *J. Comp. Neurol.* **74**, 81–92.

SCHARRER, E. und SCHARRER, B. (1937). Über Drüsen-nervenzellen und neurosekretorische Organe bei Wirbellosen und Wirbeltieren. *Biol. Rev.* **12**, 185–216.

SCHARRER, E. and SCHARRER, B. (1940). Secretory cells within the Hypothalamus *Res. Publ. Ass. Nerv. Ment. Dis.* **20**, 170–94.

SCHLEIP, W. (1910). Der Farbenwechsel von *Dixippus morosus* (Phasmidae). *Zool. Jb.* (Abt. allg. Zool. Physiol.), **30**, 45–132.

SCHLEIP, W. (1915). Über die Frage nach der Beteiligung des Nervensystems beim Farbenwechsel von *Dixippus*. *Zool. Jb.* (Abt. allg. Zool. Physiol.), **34**, 225–32.

SCHLEIP, W. (1921). Über den Einfluss des Lichtes auf die Färbung von *Dixippus* und die Frage der Erblichkeit des erworbenen Farbkleides. *Zool. Anz.* **52**, 151–60.

SCHLIEPER, C. (1926). Der Farbwechsel von *Hyperia galba*. *Z. vergl. Physiol.* **3**, 547–57.

SCHMIDT, W. J. (1910). Das Integument von *Voeltzkowia mira* Bttgr. Ein Beitrag zur Morphologie und Histologie der Eidechsenhaut. *Z. wiss. Zool.* **94**, 605–720.

SCHMIDT, W. J. (1911). Beobachtungen an der Haut von *Gecholepsis* und einigen anderen Geckoniden. Völtzkow, *Reise in Ostafrika in* 1903–1905. Stuttgart.

SCHMIDT, W. J. (1912). Studien am Integument der Reptilien. I. Die Haut der Geckoniden. *Z. wiss. Zool.* **101**, 139–258.

SCHMIDT, W. J. (1913a). Studien am Integument der Reptilien. III. Ueber die Haut der Gerrhosauriden. *Zool. Jb.* (Abt. Anat.), **35**, 75–104.

SCHMIDT, W. J. (1913b). Studien am Integument der Reptilien. IV. *Uroplatus fimbriatus* (Schneid.) und die Geckoniden. *Zool. Jb.* (Abt. Anat.), **36**, 377–464.

SCHMIDT, W. J. (1917). Die Chromatophoren der Reptilienhaut. *Arch. mikr. Anat.* Abt. 1, **90**, 98–259.

SCHMIDT, W. J. (1918a). Ueber Chromatophorenvereinigungen bei Amphibien, insbesondere bei Froschlarven. *Anat. Anz.* **51**, 493–501.

SCHMIDT, W. J. (1918b). Die Chromatophoren der Reptilienhaut. *Arch. mikr. Anat.* Abt. 1, **90**, 98–259.

SCHMIDT, W. J. (1918c). Zur Kenntnis der lipochromführenden Farbzellen in der Haut nach Untersuchungen an *Salamandra maculosa*. *Derm. Z.* **25**, 324–8.

SCHMIDT, W. J. (1918d). Ueber Riesenepithel- und -drüsenzellen in der Epidermis des Laubfrosches. *Anat. Anz.* **51**, 535–47.

SCHMIDT, W. J. (1918e). Ueber die Methoden zur mikroskopischen Untersuchung der Farbzellen und Pigmente in der Haut der Wirbeltiere. *Z. wiss. Mikr* **35**, 1–43.

SCHMIDT, W. J. (1919a). Vollzieht sich Ballung und Expansion des Pigmentes in den Melanophoren von *Rana* nach Art amoeboider Bewegungen oder durch intrazelluläre Körnchenströmung. *Biol. Zbl.* **39**, 140–4.

SCHMIDT, W. J. (1919b). Einige Versuche mit Bruno Blochs 'Dopa' on Amphibienhaut. *Derm. Z.* **27**, 284–94.

SCHMIDT, W. J. (1919c). Über Chromatophoren bei Insekten. *Arch. mikr. Anat.* Abt. 1, **93**, 118–36.

SCHMIDT, W. J. (1920a). Einiges über die Entwicklung der Guanophoren bei den Amphibien. *Anat. Hefte*, Arb. **59**, pp. 293–319.

SCHMIDT, W. J. (1920b). Einige Bemerkungen über 'Doppelsternchromatophoren' bei Urodelenlarven. *Anat. Anz.* **53**, 230–9.

SCHMIDT, W. J. (1920c). Über pigmentfreie Ausläufer, Kerne und Centren der Melanophoren bei den Fröschen. *Arch. Zellforsch.* **15**, 269–82.

SCHMIDT, W. J. (1920d). Beobachtungen an den roten Chromatophoren in der Haut von *Rana fusca* nebst Bemerkungen über die anderen hier vorkommenden Farbzellen. *Anat. Hefte*, Arb. 58, pp. 641–71.

SCHMIDT, W. J. (1920e). Ueber das Verhalten der verschiedenartigen Chromatophoren beim Farbenwechsel des Laubfrosches. *Arch. mikr. Anat.* Abt. 1, **93**, 414–55.

SCHMIDT, W. J. (1920f). Ueber Chromatophoren bei Insekten. *Arch. mikr. Anat.* Abt. 1, **93**, 118–36.

SCHMIDT, W. J. (1920g). Ueber die sog. Xantholeukophoren beim Laubfrosch. *Arch. mikr. Anat.* **93**, 93–117.

SCHMIDT, W. J. (1920h). Einige Beobachtungen an (melaninhaltigen) Zellformen des Froschlarvenschwanzes. *Zool. Anz.* **51**, 49–63.

SCHMIDT, W. J. (1921). Ueber die Xantholeukosomen von *Rana esculenta. Jena. Z. Naturwiss.* **50**, 219–28.

SCHMIDT, W. J. (1926). Das Glanzepithel und die Schillerfarben der Sapphirinen nebst Bemerkungen über die Erzeugung von Strukturfarben durch Guanin bei anderen Tieren. *Verh. naturh. Ver. preuss. Rheinl.* **82**, 227–300.

SCHMITT-AURACHER, A. (1921). Die 3 Arten von Farbenänderungen bei *Carausus morosus*, ihre Resultate und Ursachen. *Zool. Anz.* **53**, 108–110.

SCHMITT-AURACHER, A. (1925). Physiologisch-biologische Beobachtungen an *Carausus morosus. Pflüg. Arch. ges. Physiol.* **210**, 149–86.

SCHNAKENBECK, W. (1921). Zur Analyse der Rassenmerkmale der Axolotl. II. Die Entstehung und das Schicksal der epidermalen Pigmentträger. *Z. indukt. Abstamm.- u. VererbLehre,* **27**, 178–226.

SCHNAKENBECK, W. (1925). Vergleichende Untersuchungen über die Pigmentierung mariner Fische. *Z. mikr.-anat. Forsch.* **4**, 203–89.

SCHNURMANN, F. (1920). Untersuchungen an Elritzen über Farbenwechsel und Lichtsinn der Fische. *Z. Biol.* **71**, 69–98.

SCHÜRMEYER, A. (1926). Über die Innervation der Pars intermedia der Hypophyse der Amphibien. *Klin. Wschr.* **5**, 2311–12.

SCUDAMORE, H. H. (1941). A correlation between the rate of heart beat and the state of certain chromatophores in the shrimp, *Palaemonetes. Trans. Ill. Acad. Sci.* **34**, 238–40.

SEARS, M. (1931). The responses of the deep-seated melanophores in the frog to adrenalin and pituitrin. *Proc. Nat. Acad. Sci., Wash.,* **17**, 280–2.

SEARS, M. (1935). Responses of deep-seated melanophores in fishes and amphibians. *Biol. Bull. Woods Hole,* **68**, 7–24.

SECEROV, S. (1912). Weitere Farbwechsel- und Hauttransplantationsversuche an der Bartgrundel (*Nemachilus barbatula* L.). *Arch. EntwMech. Org.* **33**, 716–22.

SERENI, E. (1927a). Ricerche sui cromatofori der cefalopodi. *Boll. Soc. ital. Biol. sper.* **2**, 377–81.

SERENI, E. (1927b). Ulteriori ricerche sui cromatofori dei cefalopodi. *Boll. Soc. ital. Biol. sper.* **2**, 667–70.

SERENI, E. (1927c). Ricerche sui cromatofori der cefalopodi. *Atti Acad. Nat. Lincei Rend.* **6**, 74–7.

SERENI, E. (1928a). Sulla innervazione dei cefalopodi. *Boll. Soc. ital. Biol. sper.* **3**, 707–11.

SERENI, E. (1928b). Sui cromatofori dei cefalopodi. I. Azione di alcuni veleni *in vivo. Z. vergl. Physiol.* **8**, 488–600.

SERENI, E. (1929a). Fenomeni fisiologici consecutivi alla sezione dei nervi nei cefalopodi. *Boll. Soc. ital. Biol. sper.* **4**, 736–40.

SERENI, E. (1929b). Sulla funzione delle ghiandole salivari posteriori dei cefalopodi. *Boll. Soc. ital. Biol. sper.* **4**, 749–53.

SERENI, E. (1929c). Correlazioni umorali nei cefalopodi. *Amer. J. Physiol.* **90**, 512.

SERENI, E. (1929d). Sul meccanismo d' azione della veratrina. *Boll. Soc. ital. Biol. sper.* **4**, 1211–15.

SERENI, E. (1930a). Sulla funzione dei corpi branchiali dei cefalopodi. *Boll. Soc. ital. Biol. sper.* **5**, 1156–61.

SERENI, E. (1930b). Sui cromatofori dei cefalopodi. II. Azione della betaine e della arecolina. *Arch. Farmacol. Sper.* **48**, 223–4.

SERENI, E. (1930c). Sui cromatofori dei cefalopodi. III. Azione di alcuni veleni *in vitro. Z. vergl. Physiol.* **12**, 329–503.

SERENI, E. (1930d). The chromatophores of the Cephalopods. *Biol. Bull. Woods Hole,* **59**, 247–68.

SERENI, E. (1932). Sulla funzione dei corpi branchiali dei cefalopodi. *Arch. zool. (ital.), Napoli,* **16**, 941–7.

SHANES, A. M. and NIGRELLI, R. F. (1941 a). Chromatophores of fishes in polarized light. *Anat. Rec.* **81**, Suppl. p. 76.

SHANES, A. M. and NIGRELLI, R. F. (1941 b). The chromatophores of *Fundulus heteroclitus* in polarized light. *Zoologica,* **26**, 237–40.

SHAPIRO, S. (1936). The presence of an oxytocic substance (posterior hypophysis extract) in cerebrospinal fluid. *Arch. Neurol. Psychiat., Chicago,* **15**, 331–40.

SHEN, T. C. R. (1937 a). The role of hypophysis in the pharmacological actions upon the melanophores of frogs (*Rana temporaria*). *J. Physiol.* **90**, 51 P–53 P.

SHEN, T. C. R. (1937 b). The pharmacology of melanophores in frogs (*Rana temporaria*) and the rôle of the hypophysis. *Arch. int. Pharmacodyn.* **57**, 289–334.

SHEN, T. C. R. (1937 c). Influence du pipéridinométhyl-x-benzodioxane (F. 933) sur les mélanophores de la grenouille. Rôle de l'hypophyse. *C.R. Soc. Biol., Paris,* **124**, 991–3.

SHEN, T. C. R. (1937 d). Sur la pharmacologie des mélanophores de la grenouille. Rôle de l'hypophyse. *C.R. Soc. Biol., Paris,* **126**, 433–4.

SHEN, T. C. R. (1939 a). The site of the stimulating action of several drugs upon the melanophore hormone secretion from the hypophysis of frogs. *J. Physiol.* **95**, 47 P–48 P.

SHEN, T. C. R. (1939 b). The mechanism of the melanophore-expanding action of several drugs and its relationship to the internal secretion of the hypophysis in frogs. *Arch. int. Pharmacodyn.* **62**, 295–329.

SJÖGREN, S. (1934). Die Blutdrüse und ihre Ausbildung bei den Decapoden. *Zool. Jb.* (Abt. Anat.), **58**, 145–70.

SLOME, D. and HOGBEN, L. (1928). The chromatic function in *Xenopus laevis. S. Afr. J. Sci.* **25**, 329–35.

SLOME, D. and HOGBEN, L. (1929). The time factor in the chromatic responses of *Xenopus laevis. Trans. Roy. Soc. S. Afr.* **17**, 141–50.

SMITH, D. C. (1928). The effect of temperature on the melanophores of fishes. *J. Exp. Zool.* **52**, 183–234.

SMITH, D. C. (1929). The direct effect of temperature changes upon the melanophores of the lizard *Anolis equestris. Proc. Nat. Acad. Sci., Wash.,* **15**, 48–56.

SMITH, D. C. (1930 a). The effects of temperature changes upon the chromatophores of crustaceans. *Biol. Bull. Woods Hole,* **58**, 193–202.

SMITH, D. C. (1930 b). Melanophore pulsations in the isolated scales of *Fundulus heteroclitus. Proc. Nat. Acad. Sci., Wash.,* **16**, 381–5.

SMITH, D. C. (1931 a). The influence of humoral factors upon the melanophores of fishes, especially *Phoxinus. Z. vergl. Physiol.* **15**, 613–36.

SMITH, D. C. (1931 b). The action of certain autonomic drugs upon the pigmentary responses of *Fundulus. J. Exp. Zool.* **58**, 423–53.

SMITH, D. C. (1931 c). The effect of temperature changes upon the pulsations of isolated scale melanophores of *Fundulus heteroclitus. Biol. Bull. Woods Hole,* **60**, 269–87.

SMITH, D. C. (1933). Color changes in the isolated scale iridocytes of the squirrel fish, *Holocentrus ascensionis* (Osbeck). *Proc. Nat. Acad. Aci., Wash.,* **19**, 885–92.

SMITH, D. C. (1936 a). A method for recording chromatophore pulsations in isolated fish scales by means of a photo-electric cell. *J. Cell. Comp. Physiol.* **8**, 83–7.

SMITH, D. C. (1936 b). An independent response to light on the part of the scale melanophores of *Tautoga onitis* (L.). *Anat. Rec.* **67**, Suppl. 1, pp. 103–4.

SMITH, D. C. (1939). The responses of melanophores in isolated fish scales. *Amer. Nat.* **73**, 247–55.

SMITH, D. C. (1941). The effect of denervation upon the response to adrenanin in the isolated fish scale melanophore. *Amer. J. Physiol.* **132**, 245–8.

SMITH, D. C. and SMITH, M. T. (1934). Observations on the erythrophores of *Scorpaena ustulata*. *Biol. Bull. Woods Hole*, **67**, 45–58.

SMITH, D. C. and SMITH, M. T. (1935). Observations on the color changes and isolated scale erythrophores of the squirrel fish, *Holocentrus ascensionis* (Osbeck). *Biol. Bull. Woods Hole*, **68**, 131–9.

SMITH, D. E. and HARTMAN, F. A. (1943). Influence of adrenal preparations on fish melanophores. *Endocrinology*, **32**, 145–8.

SMITH, G. M. (1931). Occurrence of melanophores in certain experimental wounds of the goldfish (*Carassius auratus*). *Biol. Bull. Woods Hole*, **61**, 73–84.

SMITH, G. M. (1932a). Melanophores induced by X-ray compared with those existing in patterns as seen in *Carassius auratus*. *Biol. Bull. Woods Hole*, **63**, 484–91.

SMITH, G. M. (1932b). Eruption of corial melanophores and general cutaneous melanosis in the goldfish (*Carassius auratus*) following exposure to X-ray. *Amer. J. Cancer*, **16**, 863.

SMITH, G. M. (1934). The formation of melaniridosomes in healing wounds of Haemulidae. *Quart. J. Micr. Sci.* **76**, 647–54.

SMITH, H. G. (1938). The receptive mechanism of the background response in chromatic behaviour of Crustacea. *Proc. Roy. Soc.* B, **125**, 250–63.

SMITH, P. E. (1916a). Experimental ablation of the hypophysis in the frog embryo. *Science*, **44**, 280–2.

SMITH, P. E. (1916b). The effect of hypophysectomy in the early embryo upon the growth and development of the frog. *Anat. Rec.* **11**, 57–64.

SMITH, P. E. (1919). The pigment changes in frog larvae deprived of the epithelial hypophysis. *Proc. Soc. Exp. Biol., N.Y.*, **16**, 74–78.

SMITH, P. E. (1920). The pigmentary, growth and endocrine disturbances induced in the anuran tadpole by the early ablation of the pars buccalis of the hypophysis. *Amer. Anat. Mem.* **11**, 151 pp.

SMITH, P. E. and GRAESER, J. B. (1924). A differential response of the melanophore stimulant and oxytocic autocoid of the posterior hypophysis. *Anat. Rec.* **27**, 187.

SMITH, P. E. and SMITH, I. P. (1923). The function of the lobes of the hypophysis as indicated by replacement therapy with different portions of the ox gland. *Endocrinology*, **7**, 579–91.

SMITH, R. I. (1942). Nervous control of chromatophores in the leech *Placobdella parasitica*. *Physiol. Zoöl.* **15**, 410–17.

SNYDER, F. F. (1928). The presence of melanophore-expanding and uterus-stimulating substance in the pituitary body of early pig embryos. *Amer. J. Anat.* **41**, 399–409.

SODERWALL, A. L. and STEGGERDA, F. R. (1937). The relationship of the pars tuberalis to melanophore response in frogs (*Rana pipiens*). *Anat. Rec.* **70**, Suppl. 1, p. 57.

SOLLAUD, E. (1908). Rôle du système nerveux dans les changements de coloration chez la grenouille. *C.R. Acad. Sci., Paris*, **147**, 536–8.

SPAETH, R. A. (1913a). The mechanism of the contraction in the melanophores of fishes. *Anat. Anz.* **44**, 520–4.

SPAETH, R. A. (1913b). The physiology of the chromatophores of fishes. *J. Exp. Zool.* **15**, 527–85.

SPAETH, R. A. (1916a). Evidence proving the melanophore to be disguised type of smooth muscle cell. *J. Exp. Zool.* **20**, 193–215.

SPAETH, R. A. (1916b). The responses of single melanophores to electrical stimulation. *Amer. J. Physiol.* **41**, 577–96.

SPAETH, R. A. (1916c). A device for recording the physiological responses of single melanophores. *Amer. J. Physiol.* **41**, 597–602.

SPAETH, R. A. (1917). The physiology of the chromatophores of fishes. II. Responses to alkaline earths and to certain neutral combinations of electrolytes. *Amer. J. Physiol.* **42**, 595–6.

SPAETH, R. A. (1918). Concerning a new method for the biological standardization of pituitary extract and other drugs. *J. Pharmacol.* **11**, 209–19.

SPAETH, R. A. and BARBOUR, H. G. (1917). The action of epinephrin and ergotoxin upon single physiologically isolated cells. *J. Pharmacol.* **9**, 431–40.

SPAUL, E. A. (1925). Experiments on the localization of the substances in pituitary extracts responsible for metamorphic and pigmentary changes in Amphibia. *Brit. J. Exp. Biol.* **2**, 427–37.

SPEIDEL, C. C. (1919). Gland-cells of internal secretion in the spinal cord of the skates. *Publ. Carneg. Instn*, no. 281, pp. 1–31.

SPEIDEL, C. C. (1922). Further comparative studies in other fishes of cells that are homologous to the large irregular glandular cells in the spinal cord of the skate. *J. Comp. Neurol.* **34**, 303–17.

SPEIDEL, C. C. (1926). Studies of hyperthyroidism. IV. The behavior of the epidermal mitochondria and the pigment in frog tadpoles under conditions of thyroid-accelerated metamorphosis and of regeneration following wound infliction. *J. Morph.* **43**, 57–79.

SPEIDEL, C. C. (1929). Studies in hyperthyroidism. VI. Regenerative phenomena in thyroid-treated amphibian larvae. *Amer. J. Anat.* **43**, 103–65.

SPERANSKAJA-STEPANOVA, E. (1930). Einfluss des sympathischen Nervensystems und des Adrenalins auf die Empfindlichkeit der Froschhaut. *Russk. fis. Zh.* **13**, 317–23. (Cited from *Ber. ges. Physiol.* **58**, 399, 1931.)

SPERRY, R. W. (1943). Effect of 180 degree rotation of the retinal field on visuomotor coordination. *J. Exp. Zool.* **92**, 263–79.

STÅHL, F. (1938a). Preliminary report on the colour changes and the incretory organs in the heads of some crustaceans. *Arch. Zool., Stockholm*, **30** B, no. 8.

STÅHL, F. (1938b). Über das Vorkommen von inkretorischen Organen und Farbwechselhormonen im Kopf einiger Crustaceen. *Acta Univ. lund.* N.F. Avd. 2, **34**, no. 12 (20 pp.).

STECHE, O. (1911). Die Färbung von *Dixippus morosus*. *Zool. Anz.* **37**, 60–1.

STEGGERDA, F. R. and SODERWALL, A. L. (1939). Relationship of the pars tuberalis to melanophore response in Amphibia (*Rana pipiens*). *J. Cell. Comp. Physiol.* **13**, 31–7.

STEHLE, R. L. (1934). Die Melanophoren-erweiternde Wirkung des Hypophysenextrakts. *Arch. exp. Path. Pharmak.* **175**, 466–70.

STEHLE, R. L. (1936). A method for obtaining a preparation of the melanophoric hormone of the pituitary gland. *J. Pharmacol.* **57**, 1–5.

STEHLE, R. L. and FRASER, A. M. (1935). The purification of the pressor and oxytocic hormones of the pituitary gland and some observations on the chemistry of the products. *J. Pharmacol.* **55**, 136–51.

STEPHENSON, E. M. (1932). Colour changes in crustacea. *Nature, Lond.*, **130**, 931.

STEPHENSON, E. M. (1934). Control of chromatophores in *Leander serratus*. *Nature, Lond.*, **133**, 912.

STOCKARD, C. R. (1915a). Differentiation of wandering mesenchymal cells in the living yolk-sac. *Science*, **42**, 537–41.

STOCKARD, C. R. (1915b). A study of wandering mesenchymal cells on the living yolk-sac and their developmental products: chromatophores, vascular endothelium and blood cells. *Amer. J. Anat.* **18**, 525–94.

STOPPANI, A. O. M. (1941a). *Estudios fisiologicos y farmacologicos sobre los melanoforos de los batracios.* 157 pp. Buenos Aires.

STOPPANI, A. O. M. (1941b). La regulación endocrina del color del *Bufo arenarum* Hensel. *Rev. Soc. argent. Biol.* **17**, 416–26.

STOPPANI, A. O. M. (1941c). La regulación nerviosa del color del *Bufo arenarum* Hensel. *Rev. Soc. argent. Biol.* **17**, 484–90.

STOPPANI, A. O. M. (1942a). Glándulas endocrinas y farmacologiá del color de los batracios. *Rev. Soc. argent. Biol.* **18**, 215–24.

STOPPANI, A. O. M. (1942b). Neuroendocrine mechanism of color change in *Bufo arenarum* Hensel. *Endocrinology*, **30**, 782–6.

STOPPANI, A. O. M. (1942c). Pharmacology of colour regulation in Amphibia and the importance of endocrine glands. *J. Pharmacol.* **76**, 118–25.

STRECKER, J. K. (1928). Field observations on the color changes of *Anolis carolinensis* Voight. *Contr. Baylor Univ. Mus.* **13**, 9 pp.

STSCHEGOLEW, G. G. (1927). Die Änderung der Färbung unter dem Einfluss des Lichtes bei *Protoclepsis tessellata* Braun 1805. *Rev. Zool. Russe*, **7**, no. 3, pp. 149–66.

STUTINSKY, F. (1934). Expansion des érythrophores chez *Phoxinus laevis*, par des produits non hypophysaires. *C.R. Soc. Biol., Paris*, **115**, 241–3.

STUTINSKY, F. (1935). Sur la physiologie des érythrophores du vairon action des extraits hypophysaires. *Bull. Soc. zool. Fr.* **60**, 173–87.

STUTINSKY, F. (1936). Effets de l'éclairement continu sur la structure de la glande pituitaire de la grenouille. *C.R. Soc. Biol., Paris*, **123**, 421–3.

STUTINSKY, F. (1937). Modifications histologiques de l'hypophyse de la grenouille après lésion infundibulaire. *C.R. Ass. Anat.* **32**, 396–406.

STUTINSKY, F. (1938). Neurocrine hypophysaire et 'réflexe photopituitaire' chez la grenouille. *C.R. Soc. Biol., Paris*, **127**, 409–11.

SULZBERGER, M. B. (1936). Zur Frage des Pigmenthormons und des antidiuretischen Prinzips der Hypophyse. *Klin. Wschr.* **15**, 489.

SUMNER, F. B. (1910). Adaptive color changes among fishes. *Bull. Zool. Soc. New York*, no. 42, pp. 699–701.

SUMNER, F. B. (1911). The adjustment of flatfishes to various backgrounds. A study of adaptive color change. *J. Exp. Zool.* **10**, 409–505.

SUMNER, F. B. (1933a). The differing effects of different parts of the visual field upon the chromatophore responses of fishes. *Biol. Bull. Woods Hole*, **65**, 266–82.

SUMNER, F. B. (1933b). Why do we persist in talking about the 'expansion' and 'contraction' of chromatophores? *Science*, **78**, 283–4.

SUMNER, F. B. (1934a). What are 'expansion' and 'contraction'? *Science*, **79**, 11.

SUMNER, F. B. (1934b). Does 'protective coloration' protect? Results of some experiments with fishes and birds. *Proc. Nat. Acad. Sci., Wash.*, **20**, 559–64.

SUMNER, F. B. (1934c). Studies of the mechanism of color changes in fishes. *Johnstone Mem. Vol.* pp. 62–80.

SUMNER, F. B. (1935a). Evidence for the protective value of changeable coloration in fishes. *Amer. Nat.* **69**, 245–66.

SUMNER, F. B. (1935b). Studies of protective color change. III. Experiments with fishes both as predators and prey. *Proc. Nat. Acad. Sci., Wash.*, **21**, 345–53.

SUMNER, F. B. (1937). Changeable coloration, its mechanism and biological value with special reference to fishes. *Sci. Month.* **45**, 60–4.

SUMNER, F. B. (1939a). Quantitative effects of visual stimuli upon pigmentation. *Amer. Nat.* **73**, 219–34.

SUMNER, F. B. (1939b). Human psychology and some things that fishes do. *Sci. Month.* **49**, 245–55.

SUMNER, F. B. (1940a). Quantitative changes in pigmentation, resulting from visual stimuli in fishes and Amphibia. *Biol. Rev.* **15**, 351–78.

SUMNER, F. B. (1940b). Further experiments on the relations between optic stimuli and increase or decrease of pigment in fishes. *J. Exp. Zool.* **83**, 327–43.

SUMNER, F. B. (1943). A further report upon the effects of the visual environment on the melanin content of fishes. *Biol. Bull. Woods Hole*, **84**, 195–205.

SUMNER, F. B. and DOUDOROFF, P. (1937). Some quantitative relations between visual stimuli and the production or destruction of melanin in fishes. *Proc. Nat. Acad. Sci., Wash.,* **23**, 211–19.

SUMNER, F. B. and DOUDOROFF, P. (1938a). Some effects of light intensity and shade of background upon the melanin content of *Gambusia. Proc. Nat. Acad. Sci., Wash.,* **24**, 456–63.

SUMNER, F. B. and DOUDOROFF, P. (1938b). The effects of light and dark backgrounds upon the incidence of a seemingly infectious disease in fish. *Proc. Nat. Acad. Sci., Wash.,* **24**, 463–6.

SUMNER, F. B. and DOUDOROFF, P. (1943). An improved method of assaying melanin in fishes. *Biol. Bull. Woods Hole,* **84**, 187–94.

SUMNER, F. B. and FOX, D. L. (1933). A study of variations in the amount of yellow pigment (xanthophyll) in certain fishes, and of the possible effects upon this of colored backgrounds. *J. Exp. Zool.* **66**, 263–301.

SUMNER, F. B. and FOX, D. L. (1935a). Studies of carotenoid pigment in fishes. II. Investigations of the effects of colored backgrounds and of ingested carotenoids on the xanthophyll content of *Girella nigricans. J. Exp. Zool.* **71**, 101–23.

SUMNER, F. B. and FOX, D. L. (1935b). Studies of carotenoid pigments in fishes. III. The effects of ingested carotenoids upon the xanthophyll content of *Fundulus parvipinnis. Proc. Nat. Acad. Sci., Wash.,* **21**, 330–40.

SUMNER, F. B. and KEYS, A. B. (1929). The effects of differences in the apparent source of illumination upon the shade assumed by a flatfish on a given background. *Physiol. Zoöl.* **2**, 495–504.

SUMNER, F. B. and WELLS, N. A. (1933). The effects of optic stimuli upon the formation and destruction of melanin pigment in fishes. *J. Exp. Zool.* **64**, 377–404.

SVERDLICK, J. (1942). Influencia de la hipófisie y de la suprarenal sobre el pigmento retiniano del *Bufo arenarum* (Hensel). *Rev. Soc. argent. Biol.* **18**, 207–14.

SWINGLE, W. W. (1921a). The relation of the pars intermedia of the hypophysis to pigmentation changes in anuran larvae. *J. Exp. Zool.* **34**, 119–41.

SWINGLE, W. W. (1921b). Homoplastic and heteroplastic endocrine transplants. *Anat. Rec.* **20**, 195–6.

TAIT, J. (1910). Colour change in the isopod, *Ligia oceanica. J. Physiol.* **40**, xl-xli.

TAIT, J. (1917). Body colour as affected by blood colour in amphipods and isopods, with some remarks on a bacterial infection of *Gammarus. Proc. Roy. Phys. Soc. Edinb.* **20**, 159–63.

TAKI, I. (1938). On the abnormal arrangement of scales and colour bands in a sole (*Zebrias*), with special reference to its adverse scales. *J. Sci. Hiroshima Univ.* Ser. B, Div. 1, **6**, 23–36.

TASKER, R. (1933). Origin of melanophores in *Platypoecilus. Anat. Rec.* **57**, 4 Suppl. p. 87.

TAYLOR, H. D. (1919). The tropistic action of blood vessels on the migration of chromatophores. *J. Exp. Med.* **29**, 133–8.

TEAGUE, R. S. and NOOJIN, R. O. (1938). Action of various drugs upon melanophores of *Rana pipiens. J. Pharmacol.* **63**, 36.

THÖRNER, W. (1929). Beobachtungen über peripheren Kreislauf und Melanophoren unter dem Einfluss von Degeneration und Kationenwirkung in Durchspülungsversuchen am Frosch. *Pflüg. Arch. ges. Physiol.* **222**, 52–70.

THUMANN, M. E. (1931). Die embryonale Entwicklung des Melanophorensystems bei *Brachydanio rerio* (Hamilton-Buchanan). *Z. mikr.-anat. Forsch.* **25**, 50–96.

TITSCHACK, E. (1922). Die sekundären Geschlechtsmerkmale von *Gasterosteus aculeatus* L. *Zool. Jb.* (Zool.), **39**, 83–148.

TOKURA, Y. (1933). Histological studies on the dermal functions with special reference to the changes of the colour of *Gekko japonicus* (Dumeril et Bibron). *J. Sci. Hiroshima Univ.* B, Div. 1, 2, 105–27.

TOMITA, G. (1936). Melanophore reactions to light during the early stages of the paradise fish, *Macropodus opercularis*. *J. Shanghai Sci. Inst.* Sec. IV, 2, 237–64.

TOMITA, G. (1938a). The physiology of color changes in fishes. I. The use of the angelfish as a test material. *J. Shanghai Sci. Inst.* Sec. IV, 4, 1–8.

TOMITA, G. (1938b). The physiology of color changes in fishes. II. The antidromic responses in the melanophore system in the angelfish. *J. Shanghai Sci. Inst.* Sec. IV, 4, 9–16.

TOMITA, G. (1940). The physiology of color changes in fishes. III. The reactions of melanophores to denervation in the angelfish, with special references to the melanophore innervation and to the antagonism of neurohumors. *J. Shanghai Sci. Inst.* Sec. IV, 5, 151–78.

TORRACA, L. (1914). L'azione dei raggi ultravioletti sulla pigmentazione della cute del tritone. *Int. Mschr. Anat. Physiol.* 30, 297–325.

TOUMANOFF, K. (1926). L'action combinée de l'obscurité et de la température sur la mélanogénèse chez *Dixippus morosus*. *C.R. Soc. Biol., Paris*, 94, 565–6.

TOUMANOFF, K. (1928). Le rapport entre la pigmentation et l'alimentation chez *Dixippus morosus* Br. et Redt. *C.R. Soc. Biol., Paris*, 98, 198–200.

TOYAMA, K. (1916). Itinino Mendel-seisitu ni tuite. *Nippon Ikusyugakkukwai Hôkoku*, 1, 1–9. (Cited from Goodrich, 1929, p. 99.)

TOZAWA, T. (1924). Notes on experiments on the color change of the goldfish. *Zool. Mag., Tokyo*, 36, 373–6 (Japanese).

TOZAWA, T. (1929). Experiments on the development of the nuptial coloration and pearl organs of the Japanese bitterling. *Fol. anat. japon.* 7, 407–17.

TRENDELENBURG, P. (1926). Weitere Versuche über den Gehalt des Liquor cerebrospinalis an wirksamen Substanzen des Hypophysenhinterlappens. *Arch. exp. Path. Pharmak.* 114, 255–9.

TROJAN, E. (1910). Ein Beitrag zur Histologie von *Phyllirhoe bucephala* Peron & Lesueur mit besonderer Berücksichtigung des Leuchtvermögens des Tieres. *Arch. mikr. Anat.* 75, 473–518.

TSUKAMOTO, R. (1925). A contribution to the study of the pharmacodynamical significance of the pigment cells of frogs. *J. Orient. Med.* 3, 60–1.

TUGE, H. (1937). The reactions of the melanophores of embryonic and larval salmon *Oncorhynchus keta*. *Sci. Rep. Imp. Univ. Sendai*, 4 ser. Biol. 12, 19–44.

TUSQUES, J. (1939). L'innervation des chromatophores. *C.R. Soc. Biol., Paris*, 130, 56–8.

TWITTY, V. C. (1935). Correlated genetic and embryological experiments on *Triturus*. II. The embryological basis of species differences in pigment pattern. *Anat. Rec.* 64, Suppl. 1, pp. 37–8.

TWITTY, V. C. (1936). Correlated genetic and embryological experiments on *Triturus*. I and II. *J. Exp. Zool.* 74, 239–302.

TWITTY, V. C. (1942). The role of genetic differentials in the embryonic development of Amphibia. *Biol. Symp.* 6, 291–310.

TWITTY, V. C. and BODENSTEIN, D. (1939). Correlated genetic and embryological experiments on *Triturus*. III. Further transplantation experiments on pigment development. IV. The study of pigment cell behavior *in vitro*. *J. Exp. Zool.* 81, 357–98.

UHLENHUTH, E. (1911). Zur Untersuchung des Farbensinnes. *Biol. Zbl.* 31, 767–71.

UYENO, K. (1922). Observations on the melanophores of the frog. *J. Physiol.* 56, 348–52.

VAN HERK, A. W. H. (1929). The segmental skin innervation of the flounder (*Pleuronectes flesus*). *Arch. néerl. Physiol.* pp. 470–500.

VAN HEUSEN, A. P. (1917). The skin of the catfish (*Amiurus nebulosus*) as a receptive organ for light. *Amer. J. Physiol.* **44**, 212–14.

VAN OORDT, G. J. (1923). Secondary sexual characters and testis of the ten-spined stickelback (*Gasterosteus pungitius* L.). *Proc. K. Akad. Wet. Amst.* **26**, 309–14.

VAN OORDT, G. J. (1924a). Die Veränderungen des Hodens während des Auftretens der secondären Geschlechtsmerkmale bei Fischen. I. *Gasterosteus pungitius* L. *Arch. mikr. Anat.* **102**, 379–405.

VAN OORDT, G. J. (1924b). The significance of the interstitium testis in fishes. *Proc. K. Akad. Wet. Amst.* **27**, 161–4.

VEIL, C. (1936a). Sur le mécanisme du changement de couleur chez les poissons. *J. Physiol. Path. gén.* **34**, 824–39.

VEIL, C. (1936b). Les nerfs pigmento-moteurs agissent par sécrétion d'un médiateur chimique de nature adrénalinique. *C.R. Soc. Biol., Paris,* **122**, 654–6.

VEIL, C. (1937). Hypophysectomie et changement de couleur chez le poisson chat. *C.R. Soc. Biol., Paris,* **124**, 111–13.

VEIL, C. (1938a). Evaluation de la quantité d'intermédine contenue dans l'organisme du poisson-chat. *C.R. Soc. Biol., Paris,* **127**, 42.

VEIL, C. (1938b). Action simultanée de l'adrénaline et de l'intermédine sur les mélanophores de la carpe. *C.R. Soc. Biol., Paris,* **127**, 44–6.

VEIL, C. et BEAUVALLET, M. (1930). Les mélanophores de la carpe et leurs chronaxies. *C.R. Soc. Biol., Paris,* **104**, 980–1.

VEIL, C. et COMANDON ET DE FONBRUNE (1933). Contractions rhythmiques des cellules pigmentaires sous l'action de divers poisons. (Presentation d'un film.) *Arch. Sci. Biol. Bologna,* **18**, 346–7.

VEIL, C. et MAY, R. M. (1937). Hypophysectomie et changement de couleur chez la torpille (*Torpedo marmorata*). *C.R. Soc. Biol., Paris,* **124**, 917–20.

VERNE, J. (1919). Formation expérimentale de mélanine chez les crustacés. *C.R. Soc. Biol., Paris,* **82**, 1319–21.

VERNE, J. (1923). Essai histochimique sur les pigments tégumentaires des crustacés décapodes. *Arch. Morph. gén. exp.* **16**, 168 pp.

VERNE, J. (1926). *Les pigments dans l'organisme animal.* 603 pp. Paris.

VERNE, J. et VILTER, V. (1935). Réactions pharmacodynamiques des mélanocytes de l'écaille isolée de *Carassius*. *C.R. Soc. Biol., Paris,* **119**, 1312–14.

VIALLI, M. (1927). Ricerche sulla fisiologia dei cromatofori dei pesci. *Biochim. Terap. sper.* **14**, 225–43.

VILTER, V. (1930). Action du rayonnement solaire d'altitude sur la mélanogenèse des betraciens. *C.R. Soc. Biol., Paris,* **103**, 593–7.

VILTER, V. (1931a). Modifications du système mélanique chez les axolotls soumis à l'action de fonds blancs ou noirs. *C.R. Soc. Biol., Paris,* **108**, 774–8.

VILTER, V. (1931b). Mécanisme de la mélanisation épidermique chez l'axolotl vivant sur fond noir. *C.R. Soc. Biol., Paris,* **108**, 836–9.

VILTER, V. (1931c). Origine des cellules mélanisées dans l'épithélium de l'axolotl soumis à l'action du fond noir. *C.R. Soc. Biol., Paris,* **108**, 941–3.

VILTER, V. (1932a). Migration des mélanophores dermiques dans l'épiderme chez l'axolotl. *C.R. Soc. Biol., Paris,* **110**, 938–40.

VILTER, V. (1932b). Les rapports entre les mélanophores et les terminations nerveuses 'pigmentomotrices'. *C.R. Soc. Biol., Paris,* **110**, 1286–8.

VILTER, V. (1933a). Interprétation des réactions chromatiques des mélanophores d'après leur localisation. *C.R. Soc. Biol., Paris,* **112**, 280–3.

VILTER, V. (1933b). La nature sympathique de contrôle neurohumoral de la pigmentation mélanique chez l'axolotl. *C.R. Soc. Biol., Paris,* **112**, 1207–9.

VILTER, V. (1933c). Les rapports entre le contrôle hormonal et neuro-humoral de la pigmentation mélanique chez l'axolotl. *C.R. Soc. Biol., Paris,* **112**, 1655–6.

24

VILTER, V. (1935). Le mélanoblaste dentritique des vertébrés et sa signification fonctionnelle. *Bull Soc. franç. Derm. Syph.* **42**, 1119–65.

VILTER, V. (1936). Déterminisme nerveux du dessin mélanique chez l'axolotl. *C.R. Soc. Biol., Paris,* **123**, 1137–8.

VILTER, V. (1937*a*). Les rapports entre les localisations rétiniennes et la polarisation dorsoventrale de la livrée mélanique chez l'axolotl. *C.R. Soc. Biol., Paris,* **124**, 47–8.

VILTER, V. (1937*b*). Réglage sympathicohypophysaire de la pigmentation mélanique chez les sélaciens. *C.R. Soc. Biol., Paris,* **126**, 794–5.

VILTER, V. (1937*c*). Rapports entre les champs rétiniens et l'activité sympathique locale de la peau de l'axolotl. *C.R. Ass. Anat.* **32**, 448–61.

VILTER, V. (1937*d*). Recherches histologiques et physiologiques sur la fonction pigmentaire des sélaciens. *Bull. Soc. sci. Arcachon,* **34**, 65–136.

VILTER, V. (1938*a*). Déterminisme mélano-constricteur de bandes d'assombrissement consécutives aux sections nerveuses dans la nageoire dorsale du *Gobius*. *C.R. Soc. Biol., Paris,* **129**, 1166–8.

VILTER, V. (1938*b*). Recherches sur le déterminisme physiologique de dessin mélanique de l'axolotl (*Amblystoma mexicanum*). *Arch. Anat., Strasbourg,* **26**, 1–252.

VILTER, V. (1939*a*). Configuration des dermatomes pigmento-moteurs chez les téléostéens et modalités de leur recouvrement reciproque. *C.R. Soc. Biol., Paris,* **130**, 388–90.

VILTER, V. (1939*b*). Evolution des bandes sombres provoquées par la section de nerfs pigmento-moteurs chez les téléostéens. Intervention de la circulation en tant que vecteur des hormones pigmento-motrices. *C.R. Soc. Biol., Paris,* **130**, 391–3.

VILTER, V. (1940). Inversion de la polarité pigmentaire des téguments chez *Synodontis batensoda* Rup. Poisson nageant sur le dos. *Bull. Acad. Sci. Lett. Montpellier,* pp. 73–7.

VILTER, V. (1941*a*). Polarisation dorso-ventrale de la livrée pigmentaire, sa physiologie et ses origines. *Bull. Mus. Hist. nat. Marseille,* **1**, 157–87, 259–71.

VILTER, V. (1941*b*). La livrée pigmentaire de l'anguille et sa régulation physiologique. I. Organisation cellulaire du système xanthomélanocytique. *Bull. Histol. Tech. micr.* **18**, 8–24.

VILTER, V. (1941*c*). La livrée pigmentaire de l'anguille et sa régulation physiologique. II. Physiologie de la livrée xantho-mélanique dans ses rapports avec la morphogenèse pigmentaire. *Bull. Histol. Tech. micr.* **18**, 145–66.

VOGEL, H. H., Jr. (1940). Autoplastic and homoplastic transplantation of skin in adult *Rana pipiens* Schreber. *J. Exp. Zool.* **85**, 437–73.

VOIGT, E. (1934). Die Fische aus der mitteleozänen Braunkohle des Geiseltales, mit besonderer Berücksichtigung der erhaltenen Weichteile. *Nova Acta Leop. Carol.* N.F. **2**, 21–146.

VOIGT, E. (1935). Die Erhaltung von Epithelzellen mit Zellkernen, von Chromatophoren und Corium in fossiler Froschhaut aus der mitteleozänen Braunkohle des Geiseltales. *Nova Acta Leop. Carol.* N.F. **3**, 339–60.

VUNDER, P. A. (1931). Trans. *Dynamics of Development.* (Cited from Etkin, W. and Rosenberg, L. 1938.)

WAGNER, K. (1911). Beiträge zur Entstehung des jugendlichen Farbkleides der Forelle (*Salmo fario*). *Int. Rev. Hydrobiol.* Suppl. 2, Ser. (zu. Bd. **4**), 32 pp.

WALD, G. and DU BUY, H. G. (1936). Pigments of the oat coleoptile. *Science,* **84**, 247.

WALLS, G. L. (1942). The vertebrate eye and its adaptive radiation. *Bull. Cranbrook Inst. Sci.* **19**, 785 pp.

WARING, H. (1936a). Colour changes in the dogfish (*Scyllium canicula*). *Trans. Lpool Biol. Soc.* **49**, 17–64.

WARING, H. (1936b). A preliminary study of the melanophore-expanding potency of the pituitary gland in the frog and dogfish. *Trans. Lpool Biol. Soc.* **49**, 65–90.

WARING, H. (1936c). Colour in the dogfish *Scyllium canicula*. *Nature, Lond.*, **138**, 1100.

WARING, H. (1938). Chromatic behaviour of elasmobranchs. *Proc. Roy. Soc.* B, **125**, 264–82.

WARING, H. (1940). The chromatic behaviour of the eel (*Anguilla vulgaris* L.). *Proc. Roy. Soc.* B, **128**, 343–53.

WARING, H. (1942). The co-ordination of vertebrate melanophore responses. *Biol. Rev.* **17**, 120–50.

WARING, H. and LANDGREBE, F. W. (1941). On chromatic effector speed in *Xenopus* and *Anguilla* and the level of melanophore expanding hormone in eel blood. *J. Exp. Biol.* **18**, 80–97.

WARING, H., LANDGREBE, F. W. and BRUCE, J. R. (1942). Chromatic behaviour of *Scyllium canicula*. *J. Exp. Biol.* **18**, 306–16.

WARREN, A. E. (1932). Xanthophores in *Fundulus*, with special consideration of their 'expanded' and 'contracted' phases. *Proc. Nat. Acad. Sci., Wash.*, **18**, 633–9.

WEBER, R. (1923). Die Chromatophoren von *Limax agrestis* L. *Zool. Jb.* (Abt. Physiol.), **40**, 241–92.

WEESE, A. O. (1917). An experimental study of the reactions of the horned lizard, *Phrynosoma modestum* Gir., a reptile of the semi-desert. *Biol. Bull. Woods Hole*, **32**, 98–116.

WEIDENREICH, F. (1912). Die Lokalisation des Pigmentes und ihre Bedeutung in Ontogenie und Phylogenie der Wirbeltiere. *Z. Morph. Anthr.* **2**, 59–140.

WEIDENREICH, F. (1927). Es gibt Rochen, die ihre Farbe auch auf der Bauchseite zu wechseln vermögen. *Natur u. Mus.* **57**, 46–8.

WEISSBERGER, A. und BACH, H. (1932). Ueber das Pigment der Goldfischhaut. *Naturwissenschaften*, **20**, 350.

WELLS, G. P. (1932). Colour response in a leech. *Nature, Lond.*, **129**, 686–7.

WELSH, J. H. (1930). Diurnal rhythm of the distal pigment cells in the eyes of certain crustaceans. *Proc. Nat. Acad. Sci., Wash.*, **16**, 386–95.

WELSH, J. H. (1935). Further evidence of a diurnal rhythm in the movement of pigment cells in eyes of crustaceans. *Biol. Bull. Woods Hole*, **68**, 247–52.

WELSH, J. H. (1936). Diurnal movements of the eye pigments of *Anchistioides*. *Biol. Bull. Woods Hole*, **70**, 217–27.

WELSH, J. H. (1937). The eye-stalk hormone and rate of heart beat in crustaceans. *Proc. Nat. Acad. Sci., Wash.*, **23**, 458–60.

WELSH, J. H. (1939). The action of eye-stalk extracts on retinal pigment migration in the crayfish, *Cambarus bartoni*. *Biol. Bull. Woods Hole*, **77**, 119–25.

WELSH, J. H. (1941). The sinus gland and 24-hour cycles of retinal migration in the crayfish. *J. Exp. Zool.* **86**, 35–49.

WELSH, J. H. and CHACE, F. A., Jr. (1937). Eyes of deep sea crustaceans. I. Acanthephyridae. *Biol. Bull. Woods Hole*, **72**, 57–74.

WELSH, J. H. and OSBORN, C. M. (1937). Diurnal changes in the retina of the catfish, *Ameiurus nebulosus*. *J. Comp. Neurol.* **66**, 349–59.

VON DER WENSE, T. F. (1938). Wirkungen und Vorkommen von Hormonen bei wirbellosen Tieren. *Zwangl. Abh. Inn. Sekretion*, **4**, 80 pp.

WERNER, F. (1930). Ueber das Vorkommen von Unter- und Überpigmentierung bei niederen Wirbeltieren. *Zool. Jb.* (Abt. System.), **59**, 647–62.

WHETHAM, E. O. (1933). Factors modifying egg production with special reference to seasonal changes. *J. Agric. Sci.* **23**, 383–418.

WIGGLESWORTH, V. B. (1939). *The Principles of Insect Physiology*. 434 pp. New York.

WILLIAMS, C. M., BARNESS, L. A. and SAWYER, W. H. (1943). The utilization of glycogen by flies during flight and some aspects of the physiological ageing of *Drosophila*. *Biol. Bull. Woods Hole*, **84**, 263–72.

WILSON, F. H. (1939). Color changes in blindfolded American chameleons, *Anolis carolinensis* (Cuvier). *Anat. Rec.* **75**, Suppl. p. 62.

WILSON, F. H. (1940). Color changes in blindfolded anoles. *Copeia*, pp. 151–3.

WINGE, Ö. (1922a). A peculiar mode of inheritance and its cytological explanation. *J. Genet.* **12**, 137–44.

WINGE, Ö. (1922b). One-sided masculine and sex-linked inheritance in *Lebistes reticulatus*. *J. Genet.* **12**, 145–62.

WINGE, Ö. (1923). Crossing-over between the *X*- and the *Y*-chromosome in *Lebistes*. *J. Genet.* **13**, 201–17.

WINGE, Ö. (1927). The location of eighteen genes in *Lebistes reticulatus*. *J. Genet.* **18**, 1–43.

WINGE, Ö. (1930). On the occurrence of *XX* males in *Lebistes*, with some remarks on Aide's so-called 'non-disjunctional' males in *Aplocheilus*. *J. Genet.* **23**, 69–76.

WINKLER, F. (1910a). Beobachtungen über die Bewegungen der Pigmentzellen. *Arch. Derm. Syph., Wien*, **100**.

WINKLER, F. (1910b). Studien über Pigmentbildung. I. Die Bildung der verzweigten Pigmentzellen in Regenerate des Amphibienschwanzes. II. Transplantationsversuche an pigmentierter Haut. *Arch. EntwMech. Org.* **29**, 616–31.

WORONZOWA, M. A. (1928a). On the degree of the specificity of pituitary hormone in the skin pigment reaction of axolotls. *Trans. Lab. Exp. Biol. Zoopark Moscow*, **4**, 105.

WORONZOWA, M. A. (1928b). Morphogenetic analysis of the colour in the white axolotl. *Trans. Lab. Exp. Biol. Zoopark, Moscow*, **4**, 124.

WORONZOWA, M. A. (1929). Morphogenetische Analyse der Färbung bei weissen Axolotln. *Roux Arch. Entw. Mech. Organ.* **115**, 93–109.

WORONZOWA, M. A. (1932). Analyse der weissen Fleckung bei Amblystomen. *Biol. Zbl.* **52**, 676–84.

WUNDER, W. (1930). Experimentelle Untersuchungen am dreistachligen Stichling (*Gasterosteus aculeatus* L.) während der Laichzeit. *Z. Morph. Ökol. Tiere*, **16**, 453–98.

WUNDER, W. (1931). Experimentelle Erzeugung des Hochzeitskleides beim Bitterling (*Rhodeus amarus*) durch Einspritzung von Hormonen. *Z. vergl. Physiol.* **13**, 696–708.

WUNDER, W. (1934). Beeinflussung der sekundären Geschlechtsmerkmale des Bitterlings (*Rhodeus amarus*) durch Hormone und andere Reize. *Med. Klin.* pp. 1–7.

WYKES, U. (1936). Observations on pigmentary coordination in elasmobranchs. *J. Exp. Biol.* **13**, 460–6.

WYKES, U. (1937). The photic control of pigmentary responses in teleost fishes. *J. Exp. Biol.* **14**, 79–86.

WYKES, U. (1938). The control of photo-pigmentary responses in eyeless catfish. *J. Exp. Biol.* **15**, 363–70.

WYMAN, L. C. (1922). The effect of ether upon the migration of the scale pigment and the retinal pigment in the fish, *Fundulus heteroclitus*. *Proc. Nat. Acad. Sci., Wash.*, **8**, 128–30.

WYMAN, L. C. (1924a). Blood and nerve as controlling agents in the movements of melanophores. *J. Exp. Zool.* **39**, 73–132.

WYMAN, L. C. (1924b). The reactions of the melanophores of embryonic and larval *Fundulus* to certain chemical substances. *J. Exp. Zool.* **40**, 161–80.

YAMAMOTO, K. (1931). On the physiology of the peritoneal melanophore of the fish. *Mem. Coll. Sci. Kyoto*, B, **7**, 189–203.

YAMAMOTO, K. (1937). On the physiology of the peritoneal melanophore of the frog tadpole. *Mem. Coll. Sci. Kyoto*, B, **12**, 175–86.

YAMAMOTO, T. (1933). Pulsations of melanophores in the isolated scales of *Oryzias latipes* caused by the increase of the ion quotient CNa/CCa. *J. Fac. Sci. Univ. Tokyo*, IV, **3**, 119–28.

YOUNG, H. M. (1934). Rapid pigment appearance in Ohio red bellied dace as a test for intermedin. *Amer. J. Clin. Path.* **4**, 485–91.

YOUNG, J. Z. (1929). Fenomeni istologici consecutivi alla sezione dei nervi nei cefalopodi. *Boll. Soc. Biol. Sper.* **4**, 741–4.

YOUNG, J. Z. (1931). On the autonomic nervous system of the teleostean fish *Uranoscopus scaber*. *Quart. J. Micr. Sci.* **74**, 491–535.

YOUNG, J. Z. (1933). The autonomic nervous system of selachians. *Quart. J. Micr. Sci.* **75**, 571–624.

YOUNG, J. Z. (1935). The photoreceptors of lampreys. II. The functions of the pineal complex. *J. Exp. Biol.* **12**, 254–70.

YOUNG, J. Z. and BELLERBY, C. W. (1935). The response of the lamprey to injection of anterior lobe pituitary extract. *J. Exp. Biol.* **12**, 246–53.

ZACHARIAS, O. (1913). Zu dem Umfärbungsphänomen der Stabheuschrecke *Dixippus morosus*. *Biol. Zbl.* **33**, 104–5.

ZAHL, P. A. and DAVIS, D. D. (1932). Effects of gonadectomy on the secondary sexual characters in the ganoid fish, *Amia calva* Linnaeus. *J. Exp. Zool.* **63**, 291–307.

ZIESKI, R. (1932). Einfluss der Entfernung von Hypophyse oder Augen auf den Farbwechsel des Laubfrosches (*Hyla arborea* L.). *Z. vergl. Physiol.* **17**, 606–43.

ZONDEK, B. (1935*a*). Chromatophorotropic principle of the pars intermedia of the pituitary. *J. Amer. Med. Ass.* **104**, 637–8.

ZONDEK, B. (1935*b*). Chromatophorotropic principle of the pars intermedia of the pituitary. *Glandular Physiology and Therapy* (Amer. Med. Assoc.), pp. 133–8.

ZONDEK, B. und KROHN, H. (1932*a*). Hormon des Zwischenlappens der Hypophyse (Intermedin). I. Die Rotfärbung der Elritze als Testobjekt. *Klin. Wschr.* **11**, 405–8.

ZONDEK, B. und KROHN, H. (1932*b*). Hormon des Zwischenlappens der Hypophyse (Intermedin). II. Intermedin im Organismus (Hypophyse, Gehirn). *Klin. Wschr.* **11**, 849–53.

ZONDEK, B. und KROHN, H. (1932*c*). Hormon des Zwischenlappens der Hypophyse (Intermedin). III. Zur Chemie, Darstellung und Biologie des Intermedins. *Klin. Wschr.* **11**, 1293–8.

ZONDEK, B. und KROHN, H. (1932*d*). Ein Hormon der Hypophyse. Zwischenlappenhormon (Intermedin). *Naturwissenschaften*, **20**, 134–6.

ZOOND, A. and BOKENHAM, N. A. H. (1935). Studies in reptilian colour response. II. The role of retinal and dermal photoreceptors in the pigmentary activity of the chameleon. *J. Exp. Biol.* **12**, 39–43.

ZOOND, A. and EYRE, J. (1934). Studies in reptilian colour response. I. The bionomics and physiology of the pigmentary activity of the chameleon. *Philos. Trans.* B, **223**, 27–55.

INDEX

Milton Keynes UK
Ingram Content Group UK Ltd.
UKHW041521181024
449640UK00009B/118